COMSOL® 5

FOR ENGINEERS

COMSOL® 5
FOR ENGINEERS

Mehrzad Tabatabaian, PhD, PEng

MERCURY LEARNING AND INFORMATION
Dulles, Virginia
Boston, Massachusetts
New Delhi

Publisher: David Pallai
MERCURY LEARNING AND INFORMATION
22841 Quicksilver Drive
Dulles, VA 20166
info@merclearning.com
www.merclearning.com
(800) 232-0223

M. Tabatabaian. *COMSOL®5 for Engineers.*
ISBN: 978-1-942270-42-3

The publisher recognizes and respects all marks used by companies, manufacturers, and developers as a means to distinguish their products. All brand names and product names mentioned in this book are trademarks or service marks of their respective companies. Any omission or misuse (of any kind) of service marks or trademarks, etc. is not an attempt to infringe on the property of others.

Library of Congress Control Number: 2015944869

151617321 This book is printed on acid-free paper.

Our titles are available for adoption, license, or bulk purchase by institutions, corporations, etc. For additional information, please contact the Customer Service Dept. at 800-232-0223 (toll free).

All of our titles are available in digital format at authorcloudware.com and other digital vendors. Companion files (figures and code listings) for this title are also available by contacting info@merclearning.com.

The sole obligation of MERCURY LEARNING AND INFORMATION to the purchaser is to replace the disc, based on defective materials or faulty workmanship, but not based on the operation or functionality of the product.

To my family
for their support and encouragement.

CONTENTS

PREFACE

This is an updated edition of the previously published *COMSOL for Engineers* [1]. Since the publication of the first edition, COMSOL has published its version 5, which includes a revolutionary tool, i.e. *Application Builder*. This tool enables users to build apps based on COMSOL models that can be run in any operation system (Windows, Mac, mobile/iOS), specifically on the COMSOL Server. Apps could be used for educational and practical design purposes, and represent the essence of a model in application. Therefore, we were motivated to publish this new edition (*COMSOL 5 for Engineers*) to bring the *Application Builder* introduction and implementation, as well as other new features of COMSOL 5, to the readers' use.

In this edition COMSOL 5 is used for modeling. In addition to the model examples, the COMSOL *Application Builder* is used for building apps. We rearranged the chapters from the previous edition in order to make it easier for readers to locate models according to the physics involved. New features in this edition include:

- Expanded Finite Element Method (FEM) theory and more examples.

- COMSOL 5 new features overview, graphical user interface (GUI), and how to build COMSOL app for a model.

- Built apps for selected model examples, with parameterization of these models.

- New and modified solved model examples, in addition to the models given in the previous edition.

- More models and apps, available from the companion disc.

Our objective is to provide a collection of examples and modeling guidelines through which readers can build their own models.

We took a *flexible-level* approach for presenting the materials along with using practical examples. The mathematical fundamentals, engineering principles, and design criteria are presented as integral parts of examples. We have added references that contain more in-depth physics, technical information, and data; these are referred to throughout the book and used in the examples. This approach allows readers to learn the materials at their desired level of complexity.

COMSOL 5 for Engineers could be used as a textbook complementing another text that provides background training in engineering computations and methods, such as FEM. Examples provided in this book could be considered as "lessons" for which background physics could be explained in more detail. Exercises or their variations, could be used for homework assignments.

We start each chapter with an overview, background physics, and mathematical models to set the foundation. We then present the relevant modeling techniques and materials through several examples. The examples are put into groups with relevance to the required physics concepts. Several exercise questions follow and are relevant to each example. We use the COMSOL 5 software tool for solving the examples. Where suitable, we also compare the modeling results with existing analytical, experimental, or other relevant models. Detailed steps are provided to build the relevant model for each example, but it is recommended that readers, especially students, go through all of the models to master applications of COMSOL. The purpose of using COMSOL software is to introduce this tool to engineering students, engineers, and researchers.

This book is composed of the following chapters:

Chapter 1: Introduction to Finite Element Method

In this chapter, we provide a summary of FEM and its main merits. Ritz method, along with an example, is given for covering some background for FEM. This chapter is intended to help the reader understand some technical features of COMSOL. It also provides a common level of understanding of this popular and powerful engineering computational method for readers with different educational backgrounds.

Chapter 2: COMSOL 5 and Application Builder

In this chapter, we introduce the main features and structure of COMSOL 5, including modules available and the new *Application Builder* tool. Main references for further readings for interested readers are also given. Additional details for using this modeling tool are provided when it is used for solving an example and for future examples later in the book.

Chapters 3: Models Examples for Flexible Structures, Parts, and Assembly

Chapters 4: Models Examples for Internal Fluid Flows: Steady and Transient

Chapter 5: Model Examples for Heat Transfer in Media: Steady and Transient

Chapters 6: Models Examples for Electrical Circuits and Generator

Chapters 7: Models Examples for Complex-Multiphysics Systems

In these chapters, we use COMSOL 5 to solve examples that represent "practical" engineering problems involving fluids, solids, and electrical networks. Several examples and step-by-step instructions to build the models in COMSOL and an interpretation of results are presented. The *Application Builder* tool is used for building apps for select models. Readers will find it useful to understand the preceding chapters before attempting the content in following chapters.

The models available from the companion disc could be used / uploaded with the newer software version.

Mehrzad Tabatabaian, PhD, PEng
Vancouver, BC
July 2015

ABOUT THE AUTHOR

D r. Mehrzad Tabatabaian is a faculty member with several years of teaching experience in the Mechanical Engineering Department, School of Energy at the British Columbia Institute of Technology (BCIT). In addition to teaching courses in the mechanical engineering curriculum, including thermodynamics, energy systems management modeling, and strength of materials, he also does research on renewable energy systems. Dr. Tabatabaian is the School of Energy Research Committee Chair and actively involved in their energy-initiative activities. He has authored several textbooks and published papers in various scientific journals and conferences. He holds several patents in the energy field. Dr. Tabatabaian's recent focus is on wind and solar power which has resulted in registered patents.

Recently, he volunteered to aid in establishing a new division at APEGBC, Division for Energy Efficiency and Renewable Energy (DEERE). He offers several PD seminars for the APEGBC members on the subjects of wind power, solar power, renewable energy, and the finite element modeling method.

Mehrzad Tabatabaian received his BEng from Sharif University of Technology and advanced degrees from McGill University (MEng and PhD). He has been an active academic, professor, and engineer in leading alternative energy, oil, and gas industries. He also has a Leadership Certificate from the University of Alberta, holds an APEGBC P.Eng license, and is an active member of ASME.

CHAPTER 1

INTRODUCTION TO FINITE ELEMENT METHOD

OVERVIEW AND INTRODUCTION

F EM is the dominant computational method in engineering and applied science fields. Other methods, including finite-volume, finite-difference, boundary element, and collocation, are also used in practice. To provide general readers with a background for applications of FEM, either directly or with application of a software tool, we provide an introduction to the FEM foundation and principles in this chapter. We also refer readers who are interested in further reading on this subject to a selection of available textbooks and references.

Since the mid-20th century a new definition of modeling has gradually emerged [2], [3]. This definition is a direct consequence of the development of advanced computational methods as well as huge advances in digital computers in terms of their CPUs and graphics processing power. As a result, computer modeling is synonymous with "modeling." The combination of advanced computational methods, applied mathematics, and powerful computers has created a valuable tool for engineers and applied scientists to model their designs/products before manufacturing them. The state-of-the-art modeling technologies that we currently enjoy using our laptops and mobile computers equipped with powerful software packages are the result of vast progress and advances in applied mathematics, computer science, engineering methods of analysis, and, of course, capital and business investment in these fields. For example, not long ago, just analyzing a tapered beam would take quite an amount of time and resources, whereas now an engineer can perform a similar—and accurate—analysis in few minutes!

The mathematical models or governing equations relevant to the physical phenomena involved in a given problem are the foundation of modeling. The mathematical model may be a system of algebraic equations, ordinary differential equations (ODEs), partial differential equations (PDEs), or a more complex form of differential-integral equations in the form of a functional. Some of these equations are listed in Table 1.1 as examples where summation convention of indices is applicable. Among these models the ODEs and PDEs require application of suitable computational methods, for example, FEM, to find approximate solutions for a set of given boundary conditions for a given domain or geometry. It should be noted here that for some simple cases (e.g. simple geometries) exact solutions of the ODE/PDEs may exist, serving as valuable resources that can be used to validate the corresponding modeling results. However, in practice and especially for engineering applications, we use methods like **FEM** to find solutions.

$\sigma_{ij,j} + \kappa_i = 0$	Equilibrium equation (elliptic)
$\rho\left(\dfrac{\partial u_i}{\partial t} + u_j u_{i,j}\right) = p_{,i} + \mu u_{i,jj}$	Navier-Stokes equation (viscous fluid motion)
$\dfrac{\partial T}{\partial t} = \alpha T_{,jj}$	Diffusion-Heat equation (parabolic)
$\dfrac{\partial^2 \psi}{\partial t^2} = c\psi_{,jj}$	Wave equation (hyperbolic)
$F = k\,\Delta$	Hooke's law (algebraic)
$V = RI$	Ohm's law (algebraic)
$\nabla.E = 4\pi\rho$ $\nabla.B = 0$ $-\nabla \times E = \dfrac{1}{c}\dfrac{\partial B}{\partial t}$ $-\nabla \times B = \dfrac{1}{c}\dfrac{\partial E}{\partial t} + \dfrac{4\pi}{c}J$	Maxwell's equations (electromagnetics)

TABLE 1.1 Examples of Mathematical Models for Different Phenomena

The FEM method becomes more valuable when a problem has complex geometry and/or complex boundary conditions.

FEM BACKGROUND

FEM computation procedure starts with dividing the geometry of the problem at hand into several subdivisions. This step in the process is called *meshing* or discretization. As shown in Figure 1.1, the subdivisions or *elements* are simple geometrical shapes, such as triangle, quadrilateral, tetrahedron, or hexagon. Then the variation of the quantity in question that is governed by the relevant PDE/ODE (e.g. displacement, temperature, fluid pressure) is approximated using simple functions (usually polynomial functions) with the values of the dependent variables at the *nodes* or vertices of elements.

Using the calculus of variations to minimize a variational principle [4] or a weighted-residual method, the governing PDEs are transformed into algebraic equations for each element. The equations obtained for each element are then collected to form the *global* system of algebraic equations that can be solved using available standard and advanced solvers. The solutions are the nodal values of the quantity in question (i.e. displacement, temperature, pressure). These nodal values can be used to calculate a quantity's values at any point inside each element and therefore over the whole geometrical domain of the problem.

Several advances have been made in the field of computer-aided drafting (CAD), automatic meshing of complex geometries, robust solvers, elements formulations, post-processing, and overall "reliable" computational software packages. COMSOL is one of these packages that includes multiphysics modeling facilities. Recently, commercial CAD packages (e.g. Autodesk®, SolidWorks®) are equipped with some modeling facilities, as well.

Finding an approximate numerical solution for variations of a quantity in a continuum is a major task in engineering design of products and processes,

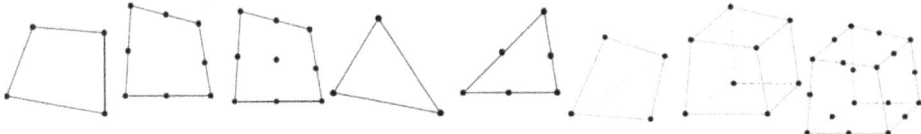

FIGURE 1.1 Examples of mesh *elements* and *nodes* (2D and 3D).

for example: displacements of a plate under loading, velocity of a fluid flowing in a pipe, temperature variation across a slab. The governing equations for each case are well known, as some examples are given in Table 1.1. However, finding the exact solutions for these equations is not always possible and specifically in complex geometry of practical problems. This challenge requires approximation of the solution, but in a way that the solution is reasonable and has small deviations from the exact solution, or at least with acceptable engineering "error." At the same time the numerical procedure should be reliable such that from problem to problem we can apply the same procedure for finding the numerical approximate solutions. Several methods have been developed and proposed, to achieve this goal. One of these methods is the Ritz method [4] and [5] (also referred to as the Rayleigh-Ritz method), which is a numerical tool developed and could be considered as the first generation of attempts that led to modern FEM. In the next section we explain the Ritz method, mainly through a numerical example.

THE RITZ METHOD

This method could be very useful in terms of understanding the foundation of FEM, both technically and historically [6]. This method starts with a variational function, or functional, that should be minimized (in general extremized) to recover the governing equations for a given phenomenon, for example, for a structure the minimization of total potential energy could be used [4], knowing that for a stable system the total potential energy is a stationary as well as a minimum. In this method a trial function, which involves some unknown coefficients, is selected and proposed. This function should satisfy the essential or kinematic boundary conditions of the problem at hand. Then by applying the trail function to the relevant functional integral and minimizing it, with reference to the unknown coefficients, we get a system of equations for the unknowns involved in the trail function. This method is applicable to simple geometry, but when more complex geometries are involved its applications are more challenging, although in principle possible. We use an example to demonstrate the Ritz method application and its relation to FEM.

Example 1.1: Ritz method application-displacement of a cantilever beam

We consider a cantilever beam with length L that is loaded at the free end by a concentrated load P, as shown in Figure 1.2. We assume a bending

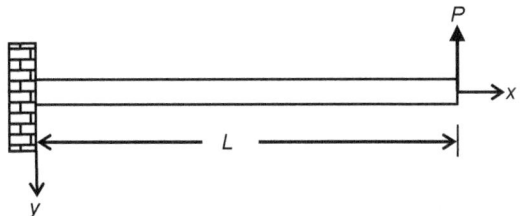

FIGURE 1.2 A cantilever beam with point load at the free end.

rigidity of EI, where E is the modulus of elasticity and I is the cross-sectional moment of inertia. We would like to calculate the deflection and rotation of the beam at its free end using the Ritz method.

The first step is to form the total potential energy functional. If the displacement and rotation at the free end are $y(L) = \Delta$ and $y'(L) = \theta$, respectively, then the total potential energy of the beam is given by $\Pi_{total} = -Py + \int_0^L \frac{EI}{2}(y'')^2 dx$. Now we introduce a trial function, like $\tilde{y}(x) = C_1 x^2$. This function satisfies the kinematic (or rigid) boundary condition of the beam, i.e. $\tilde{y}(0) = \tilde{y}'(0) = 0$. In other words, any trial function that is selected should have zero values for displacement and slope at the base of the beam. For any given value of C_1 we can calculate the total potential energy. However, we are interested in the specific value of C_1 that minimizes Π. This can be achieved by applying $\frac{\partial \Pi}{\partial C_1} = 0$, as follows;

$$\Pi = -P\tilde{y}(L) + \int_0^L \frac{EI}{2}(\tilde{y}'')^2 dx = -PC_1 L^2 + 2EIC_1^2 L$$

$$\frac{\partial \Pi}{\partial C_1} = -PL^2 + 4EIC_1 L = 0 \Rightarrow C_1 = \frac{PL}{4EI}$$

Therefore, the trial function reads $\tilde{y}(x) = \frac{PL}{4EI}x^2$ and displacement and slope

at the free end of the cantilever are $\tilde{y}(L) = \frac{PL^3}{4EI}$ and $\tilde{y}'(L) = \frac{PL^2}{2EI}$, respec-

tively. The exact value of displacement at the free end is $y_{exact}(L) = \frac{PL^3}{3EI}$;

therefore, the trial function has an error of 25%, for the displacement.

The trial function, however, gives the exact value of the slope at the free end. In order to improve the results, we add another higher-order term to the trial function, i.e. $\tilde{y}(x) = C_1 x^2 + C_2 x^3$. When using this trial function and repeating the calculations, similar to what is explained above and using $\dfrac{\partial \Pi}{\partial C_1} = 0$ and $\dfrac{\partial \Pi}{\partial C_2} = 0$, we can calculate the values of $C_1 = \dfrac{PL}{2EI}$ and $C_2 = \dfrac{-P}{6EI}$. Therefore, the trial function reads $\tilde{y}(x) = \dfrac{P}{6EI} x^2 (3L - x)$. Using this function displacement and slope at the free end of the cantilever are $\tilde{y}(L) = \dfrac{PL^3}{3EI}$ and $\tilde{y}'(L) = \dfrac{PL^2}{2EI}$, respectively. Both of these values are equal to the exact solution.

This example demonstrates the application of the Ritz method, as a procedure for calculating displacement and slope of a cantilever beam. There are two requirements for application of the Ritz method:

1. A functional integral should be available and calculated based on the physics of the problem at hand. For this example it was "easy" to find the total potential function. For more complex problems the relevant functional may not be easily available, or it may not even be possible to calculate it. In the next section we will discuss and explain how to relax this requirement, using an approach with FEM.

2. Trial function is given, in general form, as $\tilde{y}(x) = \sum_{i=1}^{N} C_i f_i(x)$. As the value of N increases, the solution error decreases. We also need to have $\dfrac{\partial \Pi}{\partial C_i} = 0$ for $(i = 1, ..., N)$ in order to calculate the so-called Ritz coefficients, C_i. It should be noted that in the Ritz method we consider the entire domain as one "element," and we increase the number of Ritz coefficients to find a solution. In other words, we decrease the hidden constraints in the trial function such that it is "flexible" enough to recover an approximate solution, with zero or minimal error, within the domain of the problem. In FEMs, we do the opposite; i.e. we break down the whole domain into small elements and consider simple polynomials, as trial functions, to calculate the final solution. This "simple" shift in the numerical procedure has huge benefits and actually is more practically useful, as we discuss in the next section. This useful hint was suggested by Courant [7].

FEM FORMULATION

In general, there are two types of approaches to implement the FEM procedure described above:

1. Minimization of a functional-variational principle approach: In this approach, we use an integral expression instead of a PDE, and by using the calculus of variations we minimize this integral expression or the functional. It can be shown [8] that this approach yields the same corresponding PDE or the governing equations (Euler-Lagrange equations) for a given problem. For example, the principle of minimum potential energy is a subset of this approach. This approach is used mostly in structural mechanics where we have the functional or can simply calculate it (e.g. using principle of virtual work). The matrix approach to finite element formulation can also be considered a subset of this approach.

2. Weighted residual approach:For many engineering problems either it is difficult or impossible to find an integral expression or functional that can be minimized and result in the governing or equilibrium equations. An example is the Navier-Stokes equations that govern viscous fluid flow. For these types of problems we use a weighted residual method and start the formulation directly from the corresponding PDEs. This method is mostly used for fluid flow, heat transfer, and nonlinear types of PDEs. See references to classical texts for further readings on FEM, like [9], [10].

Both of these methods are applicable for finite element formulation. The flowchart in Figure 1.3 depicts the process. We will use both of these methods (matrix and weighted residual) in the following examples to demonstrate the step-by-step FEM procedure accordingly. These simple examples are intended to serve the objectives of this chapter. More complex examples will be presented in future chapters.

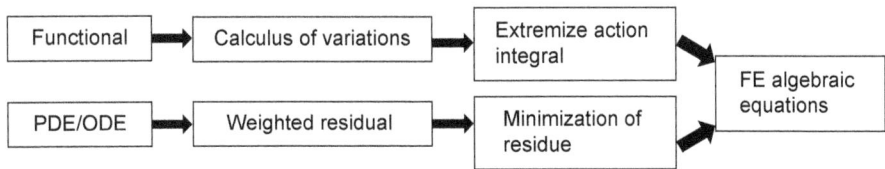

FIGURE 1.3 General process for finite element formulation.

MATRIX APPROACH

As mentioned above, the matrix approach can be considered a subset of the variational approach. For structures having members that can be considered as elements, such as trusses, the method is very straightforward. For analysis of a plate, which does not have obvious members or elements, we use the overarching variational approach. This method has commonly and traditionally been used for large structures such as aircrafts and tall buildings. This approach has its roots in the 1950s when pioneering engineers such as Turner et al. [11] and Clough [12] developed and applied it to the analysis of aircraft structures.

Example 1.2: Analysis of a 2D truss

For the 2D structure shown in Figure 1.4, calculate the displacement of the nodes/joints. Each node has 2 degrees of freedom (d.o.f.), which are displacements in x and y directions. All members of the truss have the same cross-section area A and modulus of elasticity E. Member 2–3 has a length of L, and vertical load P is applied at node 1. After finding the symbolic solution, numerical values can be assigned to the variables L, E, A, P for practical applications.

Solution: This example can be solved using methods discussed in subjects like Statics and Strength of Materials. However, we would like to use this example for demonstrating the application of FEM using the matrix approach. The following steps explain this procedure:

First, we assign node numbers (1, 2, 3) and element numbers (e1, e2, e3) to the nodes and members of this truss, as shown in Figure 1.4. The order is arbitrary; it will affect the resultant global matrix in general but

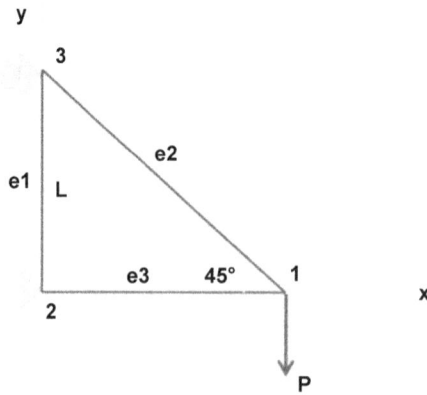

FIGURE 1.4 A 2D structure with truss elements.

not the final solution. Since each node has 2 d.o.f., each element/member has 4 d.o.f. and hence a 4 x 4 stiffness matrix. For reference, by definition the elements of stiffness matrix are forces per unit displacement for each node, and the direction of the force and displacement will determine the location of each element in the matrix. A general member i–j with orientation angle α is shown in Figure 1.5. The relationship for forces $\boldsymbol{F} = (F_x, F_y)$ applied at each node (i or j) and the corresponding displacement $\boldsymbol{u} = (u, v)$ can be written in a matrix format using the equilibrium equation $\{\boldsymbol{F}\} = [\boldsymbol{K}]\{\boldsymbol{u}\}$, where $[\boldsymbol{K}] = K_{ij}$ is the stiffness matrix of the truss members. These quantities are defined in the global system of coordinates (x, y).

For example, K_{11} is the force at node i = 1 in the x-direction due to a unit displacement at node i = 1 in the x-direction. Similarly, K_{12} is the force at node i = 1 in the x-direction due to a unit displacement at node 1 in the y-direction. Since $[\boldsymbol{K}]$ is a 4 x 4 matrix and truss elements experience only compression or tension load, then we need to transform these loads into the global system of coordinates [8] to relate them to $\{\boldsymbol{F}\}$ and $\{\mathbf{u}\}$, as shown below:

$$\frac{AE}{L}\overbrace{\begin{bmatrix} c^2 & cs & -c^2 & -cs \\ cs & s^2 & -cs & -s^2 \\ -c^2 & -cs & c^2 & cs \\ -cs & -s^2 & cs & s^2 \end{bmatrix}}^{K_{ij}}\begin{Bmatrix} u_i \\ v_i \\ u_j \\ v_j \end{Bmatrix} = \begin{Bmatrix} F_{ix} \\ F_{iy} \\ F_{jx} \\ F_{jy} \end{Bmatrix}$$

Where $c = Cos(\alpha) = \dfrac{x_j - x_i}{L}, s = Sin(\alpha) = \dfrac{y_j - y_i}{L}, L = \sqrt{x_j - x_i)^2 + (y_j - y_i)^2}$,

and u_i is displacements at node i in x-direction and v_i is displacements at node i in y-direction. Using this formulation for each element/member of the truss, we can calculate their corresponding stiffness matrices. Note that angle α is measured counterclockwise, or using the right-hand rotation rule for the system of coordinates shown in Figure 1.4.

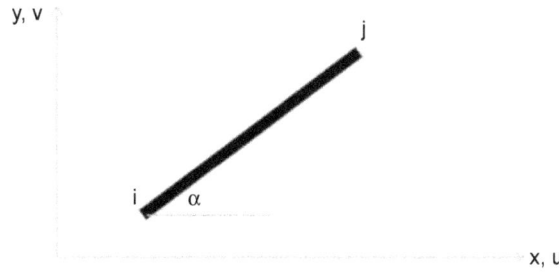

FIGURE 1.5 A general element with orientation with respect to x, y global coordinates.

▪ Member e1, node $i = 2$, node $j = 3$, $\alpha = 90°$, x2 = y2 = 0, x3 = 0, y3 = L; s = sin(90) = 1, c = cos(90) = 0; then;

$$k_{e1} = \frac{AE}{L} \begin{bmatrix} 0 & 0 & 0 & 0 \\ 0 & 1 & 0 & -1 \\ 0 & 0 & 0 & 0 \\ 0 & -1 & 0 & 1 \end{bmatrix}$$

▪ Member e2, node i = 1, node j = 3, $\alpha = (180 - 45) = 135°$, x1 = L, y1 = 0, x3 = 0, y3 = L ; s = sin(135) = $\sqrt{2}/2$, c = cos(135) = $-\sqrt{2}/2$; then

$$k_{e2} = \frac{AE}{L\sqrt{2}} \begin{bmatrix} 0.5 & -0.5 & -0.5 & 0.5 \\ -0.5 & 0.5 & 0.5 & -0.5 \\ -0.5 & 0.5 & 0.5 & -0.5 \\ 0.5 & -0.5 & -0.5 & 0.5 \end{bmatrix}$$

▪ Member e3, node i = 1, node j = 1, $\alpha = 180°$, x1 = L, y1 = 0, x2 = y2 = 0; s = sin(0) = 0, c = cos(180) = -1; then

$$k_{e3} = \frac{AE}{L} \begin{bmatrix} 1 & 0 & -1 & 0 \\ 0 & 0 & 0 & 0 \\ -1 & 0 & 1 & 0 \\ 0 & 0 & 0 & 0 \end{bmatrix}$$

Now we need to combine the members' stiffness matrices, which are written in the global system of coordinates, to get the *global stiffness matrix* of the structure. For this we need to consider the degrees of freedom at each node (i.e. 2) and the total number of nodes (i.e. 3). Hence the global stiffness matrix would be 6 × 6 since (2 × 3 = 6). We write a 6 × 6 matrix and fill its elements as follows:

▪ Element 1 has the nodes 2 and 3, so its stiffness matrix will occupy rows and columns (2i – 1 and 2i); 2 × 2 – 1 = 3̲ and 2 × 2 = 4̲ and 2 × 3 – 1 = 5̲ and 2 × 3 = 6̲

In row 3, columns 3, 4, 5, and 6 put (0, 0, 0, 0)

In row 4, columns 3, 4, 5, and 6 put (0, 1, 0, –1)

In row 5, columns 3, 4, 5, and 6 put (0, 0, 0, 0)

In row 6, columns 3, 4, 5, and 6 put (0, –1, 0, 1)

▪ Element 2 has the nodes 1 and 3, so its stiffness matrix will occupy rows and columns (2i − 1 and 2i); $2 \times 1 - 1 = \underline{1}$ and $2 \times 1 = \underline{2}$ and $2 \times 3 - 1 = \underline{5}$ and $2 \times 3 = \underline{6}$

In row 1, columns 1, 2, 5, and 6 put $(1, -1, -1, 1) / 2\sqrt{2}$

In row 2, columns 1, 2, 5, and 6 put $(-1, 1, 1, -1) / 2\sqrt{2}$

In row 5, columns 1, 2, 5, and 6 put $(-1, 1, 1, -1) / 2\sqrt{2}$

In row 6, columns 1, 2, 5, and 6 put $(1, -1, -1, 1) / 2\sqrt{2}$

▪ Element 3 has the nodes 1 and 2, so its stiffness matrix will occupy rows and columns (2i − 1 and 2i); $2 \times 1 - 1 = 1$ and $2 \times 1 = 2$ and $2 \times 2 - 1 = 3$ and $2 \times 2 = 4$

In row 1, columns 1, 2, 3, and 4 put $(1, 0, -1, 0)$

In row 2, columns 1, 2, 3, and 4 put $(0, 0, 0, 0)$

In row 3, columns 1, 2, 3, and 4 put $(-1, 0, 1, 0)$

In row 4, columns 1, 2, 3, and 4 put $(0, 0, 0, 0)$

The resulting global stiffness matrix is:

$$K_{global} = \frac{AE}{L} \begin{bmatrix} 1/2\sqrt{2}+1 & -1/2\sqrt{2} & -1 & 0 & -1/2\sqrt{2} & 1/2\sqrt{2} \\ -1/2\sqrt{2} & 1/2\sqrt{2} & 0 & 0 & 1/2\sqrt{2} & -1/2\sqrt{2} \\ -1 & 0 & 1 & 0 & 0 & 0 \\ 0 & 0 & 0 & 1 & 0 & -1 \\ -1/2\sqrt{2} & 1/2\sqrt{2} & 0 & 0 & 1/2\sqrt{2} & -1/2\sqrt{2} \\ 1/2\sqrt{2} & -1/2\sqrt{2} & 0 & -1 & -1/2\sqrt{2} & 1/2\sqrt{2} \end{bmatrix}$$

Now we apply the boundary conditions and known forces applied on the truss. These are $u_2 = v_2 = u_3 = 0$. The final system of algebraic equations is:

$$\frac{AE}{L} \begin{bmatrix} 1/2\sqrt{2}+1 & -1/2\sqrt{2} & -1 & 0 & -1/2\sqrt{2} & 1/2\sqrt{2} \\ -1/2\sqrt{2} & 1/2\sqrt{2} & 0 & 0 & 1/2\sqrt{2} & -1/2\sqrt{2} \\ -1 & 0 & 1 & 0 & 0 & 0 \\ 0 & 0 & 0 & 1 & 0 & -1 \\ -1/2\sqrt{2} & 1/2\sqrt{2} & 0 & 0 & 1/2\sqrt{2} & -1/2\sqrt{2} \\ 1/2\sqrt{2} & -1/2\sqrt{2} & 0 & -1 & -1/2\sqrt{2} & 1/2\sqrt{2} \end{bmatrix} \begin{Bmatrix} u_1 \\ v_1 \\ 0 \\ 0 \\ 0 \\ v_3 \end{Bmatrix} = \begin{Bmatrix} 0 \\ -p \\ R_{2x} \\ R_{2y} \\ R_{3x} \\ 0 \end{Bmatrix}$$

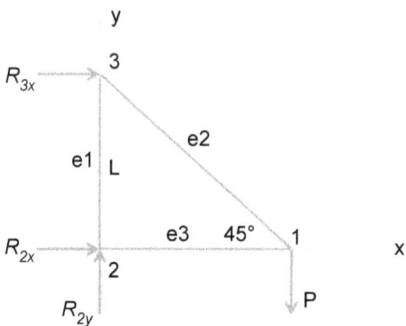

FIGURE 1.6 Free-body diagram, reaction forces, applied load.

Rs are reaction forces at the supports, as shown in Figure 1.6. We have, as a result, six unknowns $(u_1, v_1, v_3, R_{2x}, R_{2y}, R_{3x})$ and six equations that can be solved for the unknowns.

The solution of Example 1.2 is complete here. We should also note that the stiffness matrix method or displacement method (as compared with flexibility matrix method or force method) is very suitable for computer programing. However, we need to obtain the reduced form of the global stiffness matrix to avoid the singular matrix, which means to avoid rigid-body motion. This task is performed by implementing the boundary conditions.

General Procedure for Global Matrix Assembly

When the number of elements for a structure is large, combining the stiffness matrices of the elements becomes a difficult task sometimes prone to error. To help with this task and get the right global stiffness matrix, we present a formula that can be used for any number of nodes and d.o.f. in a structure.

Let's assume that we have n number of d.o.f. (n can take a value from 1 to 6) for each node. Then for a triangular element, which has 3 nodes, we have a $3n$ x $3n$ stiffness matrix. The values of stiffness matrix components of this triangular element should find their right places in the global stiffness matrix as follows (say, for node number j):

$$nj - (n - k)$$

where k is an integer; $k = 1, 2, \ldots, n$. Similarly for an element with 4 nodes, such as a quadrilateral, we have a $4n$ x $4n$ stiffness matrix, for which the above equations can be used. We demonstrate the application of this useful formula in the next example.

Example 1.3: Global matrix for triangular elements

The elements of the stiffness matrix for the triangular shape element are given as $[k] = k_{ij}$, as shown in Figure 1.6. This element has 3 nodes

(1, 3, 5). Calculate the location of each element of [k] in the global stiffness matrix when the total number of nodes is 5 (i.e. the whole structure has 5 nodes) and the d.o.f. for each node is 3.

Solution: Since each triangular element has 3 nodes, then its element stiffness matrix is 9×9:

$$[k] = \begin{bmatrix} k_{11} & k_{12} & \cdots & k_{19} \\ k_{21} & k_{22} & \cdots & k_{29} \\ \vdots & \vdots & \vdots & \vdots \\ k_{91} & k_{92} & \cdots & k_{99} \end{bmatrix}; k_{ij} = k_{ji}$$

Here we assume that we have the numerical values of k_{ij}. Using the formula $\boldsymbol{nj - (n - k)}$, for $(n = 3, k = 1, 2, 3)$ we have:

for j = 1; $\{3 \times 1 - (3 - 1) = \underline{\mathbf{1}}, 3 \times 1 - (3 - 2) = \underline{\mathbf{2}}, 3 \times 1 - (3 - 3) = \underline{\mathbf{3}}\}$

for j = 3; $\{3 \times 3 - (3 - 1) = \underline{\mathbf{7}}, 3 \times 3 - (3 - 2) = \underline{\mathbf{8}}, 3 \times 3 - (3 - 3) = \underline{\mathbf{9}}\}$

for j = 5; $\{3 \times 5 - (3 - 1) = \underline{\mathbf{13}}, 3 \times 5 - (3 - 2) = \underline{\mathbf{14}}, 3 \times 5 - (3 - 3) = \underline{\mathbf{15}}\}$

Since the global stiffness matrix is 15×15 ($3 \times 5 = 15$), the above elements would take the following places in the global stiffness matrix: {1, 2, 3, 7, 8, 9, 13, 14, 15}. The procedure to place these elements in the global stiffness matrix is as follows: We place the elements of row 1 from k_{ij} (k_{11}, $k_{12}, \ldots k_{19}$) in row 1 and columns (1, 2, 3, 7, 8, 9, 13, 14, 15) of the global matrix. Similarly, elements of row 2 of k_{ij} are placed in row 2 and columns (1, 2, 3, 7, 8, 9, 13, 14, 15) of the global matrix. We continue in the same manner for rows 3, 7, 8, 9, 13, 14, and 15. The resulting global stiffness matrix $\mathbf{K}$, with elements of k_{ij}, is shown in Figure 1.7.

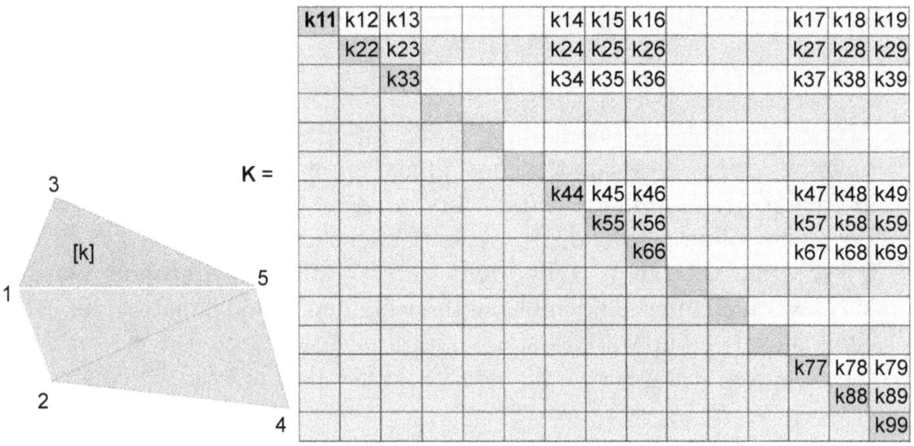

FIGURE 1.7 A triangular shape element and global stiffness matrix.

WEIGHTED RESIDUAL APPROACH

As mentioned in previous section, for any PDE, in principle, there exists an integral equation that can be minimized using variational principle, and the resulting Euler-Lagrange equation will be the same as the original PDE. Minimization of the corresponding integral is, usually but not always, easier than solving the original PDE. However, finding this integral equation is not always easy or for some equations not yet obtained (e.g. Navier-Stokes). It can be shown [13] that the method of weighted residual is an equivalent approach that can be applied to any PDE and is also more suitable for computer programming. There are several weighted residual methods; among them we will discuss the Galerkin method, which is the most commonly used FEM in engineering.

GALERKIN METHOD

Assume that a PDE is represented by $L(\Phi) = 0$, where Φ is a dependent variable and in general can be a function of space and time, and L is the differential operator (e.g., ∇^2, $\nabla\cdot$, etc.). Now, if we take the approximate solution of the function Φ to be $\tilde{\Phi}$, then $L(\tilde{\Phi}) \neq 0$, obviously. Let's assume the error to be R; then we have:

$$L(\tilde{\Phi}) = R$$

By using a weighted residual method we minimize this error over the entire domain in which the PDE applies, and therefore the approximated solution asymptotes to the exact one. The integral of R over the computational domain V using a weight function W can be written as:

$$\int_V RW \, dV = 0$$

For the Galerkin method, W is the same as *shape functions* or trial functions. We will explain the concept and applications of shape functions in Example 1.4, but interested readers can refer to finite element texts [8]. For finite-volume method, $W = 1$; for Collocation method, $W = \delta(x)$; and for spectral method, W = Fourier series. By dividing the whole domain into several elements and applying the weighted residual method for each element, we end up with a system of algebraic equations. This system then can be solved using known boundary and initial values for the dependent variable Φ. It should be mentioned here that the guessed function $\tilde{\Phi}$ should satisfy the initial and boundary conditions, as well. The Galerkin method should include the essential (or kinematic constraints) and will result in the nonessential (or natural dynamic) ones.

SHAPE FUNCTIONS

As mentioned previously and demonstrated in Example 1.4, obviously the choice of shape functions is an important step in the FE analysis. The shape function can be piecewise linear, quadratic, or cubic polynomials. More complex functions are also considered for different types of elements. We can use higher-order polynomials for the shape functions and/or increase the number of elements to reach an acceptable accuracy level and convergence [14].

In finite element software packages, this choice is referred to as the "element type" for modeling.

In this section, we go a bit deeper to explain the concept and application of shape functions.

As mentioned previously in FE, we use piecewise functions (usually polynomials) for approximating the variation of the dependent variables inside elements. Let's assume a linear function as $u = a_1 + a_2 x$, which defines a linear variation for u over the domain of an element with length L. Also let u_1 and u_2 be the values of u at nodes or end points of the element. Then we have

$$u\big|_{x=0} = u_1 = a_1$$

$$u\big|_{x=L} = u_2 = u_1 + a_2 L \Rightarrow a_2 = \frac{u_2 - u_1}{L}$$

$$\therefore \quad u = u_1(1 - \frac{x}{L}) + u_2 \frac{x}{L}$$

As it should, this function satisfies the boundary conditions. We define $N_1 = 1 - \frac{x}{L}, N_2 = \frac{x}{L}$ as shape function for function u. Then we can write the function u as

$$u = N_1 u_1 + N_2 u_2 \quad or$$

$$u = [N_1 \quad N_2]\begin{Bmatrix} u_1 \\ u_2 \end{Bmatrix}$$

Note that the shape function $N_1 = 1$ and $N_2 = 0$ at $x = 0$ and $N_2 = 1$ and $N_1 = 0$ at $x = L$. This is a very important property of shape functions. In general for multiple elements, we can write

$$u = \sum_{i=1}^{n} u_i N_i$$

where n is the total number of nodes of the elements. This approach can be expanded for 2D or 3D geometries [8].

CONVERGENCE AND STABILITY

The solution of an FE model should asymptote/approach to the corresponding exact solution, with "acceptable" accuracy. In other words, the results should converge toward the "exact" solution in a "stable" manner. The stability of the solution indicates that the error should not oscillate in a way that the answer becomes infinite. We want the answer of the FE model approaches exact solution in a monotonic or "damped" oscillation manner. There are two ways to achieve the convergence:

- **h-type** convergence, or by increasing the number of elements or the resolution of the mesh; and

- **p-type** convergence, or by increasing the element interpolation/shape function order, which employs higher order polynomials for shape functions.

Usually p-type approach provides a faster convergence, for a given problem, toward exact solution, but in practice h-type is used more often because computer time and power are readily available.

Example 1.4: Heat transfer in a slender steel bar

A slender steel bar with length L is given, as shown in Figure 1.8. At $x = 0$ the bar is subjected to heat flux q and at the other end is kept at a fixed

FIGURE 1.8 A slender steel bar with heat flux and temperature boundary conditions.

temperature T_L. All other surfaces are thermally insulated. Calculate the temperature distribution along the bar using FEM.

Solution: The temperature T satisfies the following differential equation and boundary conditions (steady state, 1D):

$$-k\frac{d^2T}{dx^2} = 0 \ , \ T = T_L \ @ \ x = L \ \text{and} \ -k\frac{dT}{dx} = q \ @ \ x = 0$$

Note that this problem has an analytical solution that can be easily found by integrating the PDE twice, or $= T = T_L + \frac{q}{k}(L-x)$. The exact solution, if available, is useful for checking the FE results. However, in this example we want to use FEM to find the solution. This is done by following the steps below:

- We divide the steel bar into two sections, or *elements* (e1 and e2), hence we will have 3 nodes, $i = 1, 2, 3$, as shown in Figure 1.9.

- Within each element, we approximate the temperature distribution by using a simple function, as: $\tilde{T}(x) = \sum_{i=1}^{3} N_i(x)T_i = N_1T_1 + N_2T_2 + N_3T_3$

N_i is the linear shape function that defines the temperature distribution within element e_i and T_i is the value of temperature at node i. The approximated function $\tilde{T}(x)$ should satisfy the boundary conditions, always. The shape function N_i should be equal to 1 at node i and zero at the other node of corresponding element.

Now we introduce the $\tilde{T}(x)$ into the governing equation, which results in a residue since the approximate function will not satisfy the PDE, hence $-k\frac{d^2\tilde{T}}{dx^2} = R \neq 0$. We cannot force the residue to vanish at any location on the steel bar, but we can have its integral (or its total net value) vanish.

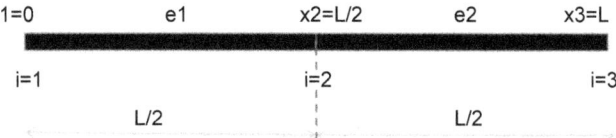

FIGURE 1.9 Elements and nodes numbers for the steel bar.

To do this we use the method of minimization of weighted residuals, as mentioned previously. Therefore we will have

$$\int W[-k\frac{d^2\tilde{T}}{dx^2}]dx = \int N(x)R\,dx = 0$$

We integrate the above equation, using the method of integration by parts

$$\int N[-k\frac{d^2\tilde{T}}{dx^2}]dx = \int [k\frac{d\tilde{T}}{dx}\frac{dN}{dx}dx] - kN\frac{d\tilde{T}}{dx}\Big|_0^L = 0$$

This is called *weak* formulation for FEM, since the second-order derivative is transformed into first-order one (hence a "weaker" constraint on the variation of the independent variable, in this case the temperature $\tilde{T}(x)$). Now we substitute for

$$\tilde{T}(x) = \sum_{i=1}^{n+1} N_i(x)T_i = N_1T_1 + N_2T_2 + N_3T_3$$

then

$$\int_0^{L/2} [k\frac{dN_i}{dx}(\sum_{j=1}^{3}\frac{dN_j}{dx}\tilde{T}_j)]dx + [N_i(-k\frac{d\tilde{T}}{dx})]_0^{L/2}$$

$$+ \int_{L/2}^{L} [k\frac{dN_i}{dx}(\sum_{j=1}^{3}\frac{dN_j}{dx}\tilde{T}_j)]dx + [N_i(-k\frac{d\tilde{T}}{dx})]_{L/2}^{L} = 0 \quad \textit{for } i = 1,2,3$$

Using the shape functions definition, we have as a result a system of 3 equations for 3 unknown values $\tilde{T}_1, \tilde{T}_2, \tilde{T}_3$. The final equations in matrix form are

$$\begin{bmatrix} 2k/L & -2k/L & 0 \\ -2k/L & 4k/L & -2k/L \\ 0 & -2k/L & 2k/L \end{bmatrix}\begin{bmatrix} \tilde{T}_1 \\ \tilde{T}_2 \\ \tilde{T}_3 \end{bmatrix} = \begin{bmatrix} q \\ 0 \\ -q_L \end{bmatrix}$$

But $\tilde{T}_3 = T_L$; therefore, the third equation can be solved separately. Finally, we have

$$\begin{bmatrix} 2k/L & -2k/L \\ -2k/L & 4k/L \end{bmatrix}\begin{bmatrix} \tilde{T}_1 \\ \tilde{T}_2 \end{bmatrix} = \begin{bmatrix} q \\ 0 \end{bmatrix} + \frac{2k}{L}T_L\begin{bmatrix} 0 \\ 1 \end{bmatrix}$$

Or $\tilde{T}_1 = qL/k + T_L$ and $\tilde{T}_2 = qL/2k + T_L$.

We see that these are actually equal to the values of exact solutions at the nodes $i = 1, 2, 3$ since the exact variation of temperature in the rod is a linear one, similar to our assumed shape functions. The solution of this example is complete here.

This example demonstrates an application of Galerkin method. Readers may want to review and pay attention to the process of FEM, weak formulation of FE, and weighted residual minimization among other concepts presented in this example. For more examples, readers may refer to reference [4].

EXERCISES

Problem 1.1: Solve Example 1.1 using the Ritz method, for a cantilever beam with a concentrated load exerted at its mid-span and pointed upward. Compare your results with exact solution.

Problem 1.2: Solve Example 1.2 for numerical values A = 10 cm², E = 200GPa, L = 1.5 m, P = 200 N.

Problem 1.3: Repeat Problem 1.2 for α = 53°.

Problem 1.4: Using the method described in Example 1.3, write down the global stiffness matrix for the three-element structure as shown below. [M] and [L] are elements' stiffness matrices.

Problem 1.5: Renumber the nodes of the plate structure shown in Figure 1.7 and repeat the assembly of global stiffness matrix. Compare your results with those of Problem 1.4.

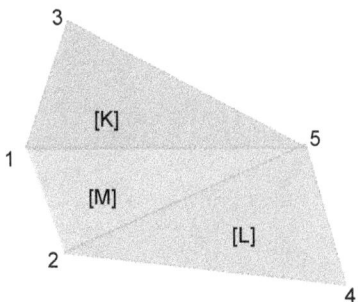

Problem 1.6: Solve Example 1.4 with a set of new boundary condition given as T = T₁ at x = 0, and T = T₂ at x = L.

CHAPTER 2

COMSOL 5 AND APPLICATION BUILDER

OVERVIEW

In this chapter, we introduce COMSOL 5 and its features as a software tool for modeling. The objective is to provide a "tour" of this software package and introduce its features, modules, and facilities, with emphasis on the Application Builder, to readers as well as provide guidelines for building models using COMSOL. To demonstrate COMSOL module applications, we will provide several modeling examples and their applications in detail in the next chapter. Because it would be exhaustive to include all features available in COMSOL in a single book, our main objective is to provide a collection of examples and modeling guidelines through which readers can build their own models.

COMSOL, which is a finite-element-based modeling tool, has a well-developed GUI and several modules for modeling common and advanced types of physics involved in engineering and applied science practices. Its history goes back to when this package was called FEMLAB and was written based on MATLAB™ whereas newer versions are stand-alone packages. The latest version is COMSOL 5 series. In version 5, the user interface was upgraded and more modules and stronger, smarter solvers were added. One of these additions is the revolutionary Application Builder tool. This tool enables users to build apps based on a model. We will discuss and provide an example for building an app in a further section. Meshing with COMSOL is almost seamless and "automatic," yet it gives users the choice of having custom-designed mesh both for structured and unstructured types. It has a rich materials property database and yet allows users to define their own database for their desired materials.

Another major feature of COMSOL is the ability to solve any PDE/ODE that users might have and that may not fit into classical governing equations (e.g. wave, heat, equilibrium). A so-called (0D) allows users to solve problems that do not have space as a relevant defined dimension, such as electric or thermal equivalent networks. Among new features are seamless GUI interactions with CAD packages, which allows COMSOL to be run directly through a CAD software package interface such as SolidWorks™ and some Autodesk™ products; new meshing facilities for large assemblies; creating geometry from imported meshes; more features and tools for built-in CAD modules; new turbulence models; and several new features for multiphysics modeling. The highlights of the new features are the Application Builder and COMSOL Server™ (a cloud-type facility). We will discuss and explain these two new facilities in further sections.

The author's experience with this package includes its ongoing improvement in features and especially the solvers, which makes it an efficient modeling tool for small to medium-size problems (in terms of geometrical size) and with multiphysics involved. In COMSOL, users can see the governing equations for the type of physics they solve right on the interface—a feature that is very helpful for assigning right values to the variables and boundary conditions as well as knowing what type of equations you are solving using the FEM. Building the geometry of a model is possible either by using CAD facilities available in COMSOL or using live communication modules such as LiveLink™. LiveLinks are available for major CAD packages (e.g. Inventor™, Revit™, SolidWorks™, AutoCAD™), as well as MATLAB™ and Excel™. The post-processing features allow users to study and see the modeling results and analyze those using color-coded surface graphs and data line graphs, among others. The Reports feature is very useful, enabling users to generate a file using modeling results in common word processing formats.

COMSOL has comprehensive and rich Help documentation as well as tutorials. COMSOL provides free workshops and webinars for new users, as well as more extensive training courses for a fee. More information about the features, models gallery, and tutorials is available on the COMSOL website (www.comsol.com). In the following sections, we will take a tour of COMSOL features and modules and introduce some of its features.

COMSOL 5 DESKTOP INTERFACE

After purchasing the product license you can install COMSOL on your machine, either PC or Mac. When launching COMSOL, the New window, similar to the one shown in Figure 2.1, will open. COMSOL 5 has a new button, Application Wizard, for Application Builder, but version 5.1 does not have this button. The default is for building a new model, either using the Model Wizard or a Blank Model. Users can also click on the File > Open, from the menu toolbar, and open an old file. It is recommended to start a new model using Model Wizard, since it will guide the users through the steps necessary for building a model.

After clicking on Model Wizard icon a new window, Select Space Dimension, will open as shown in Figure 2.2. Users can choose the physical dimension of the model by clicking on the relevant icon, which includes 0D, 1D, and 2D axisymmetric cases, as well.

After selecting a space dimension, the Select Physics window opens as shown in Figure 2.3. Users can choose the physics, or multiphysics, related to the problem at hand and assign it to the model. Available physics choices depend on the user's purchased license, including Modules.

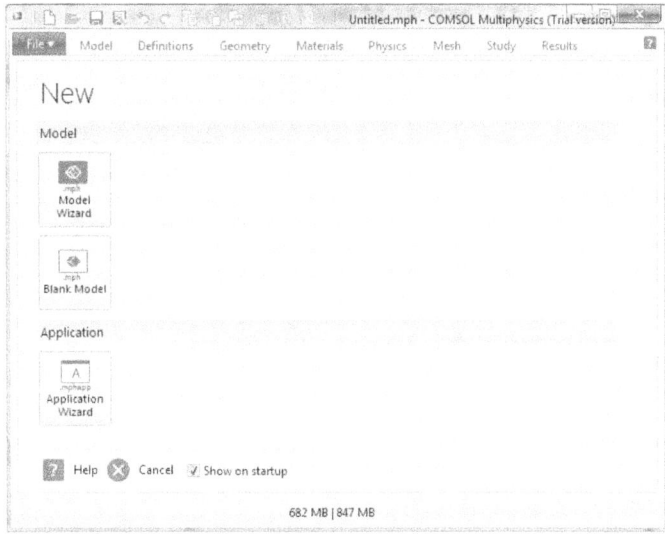

FIGURE 2.1 New window opens when launching COMSOL 5.

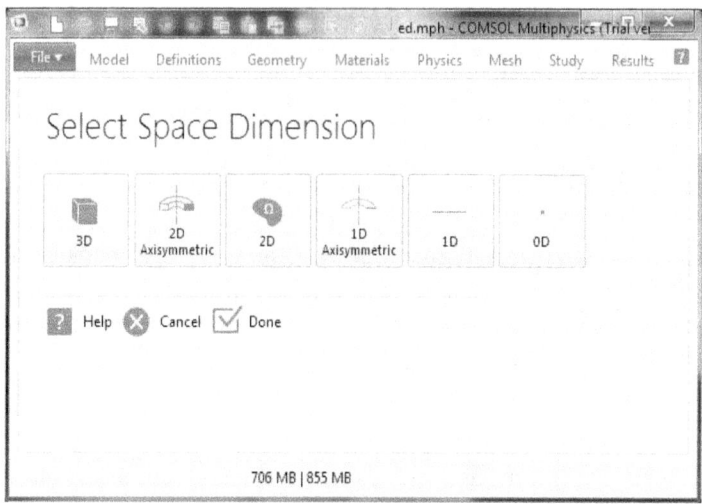

FIGURE 2.2 Select Space Dimension window in COMSOL 5.

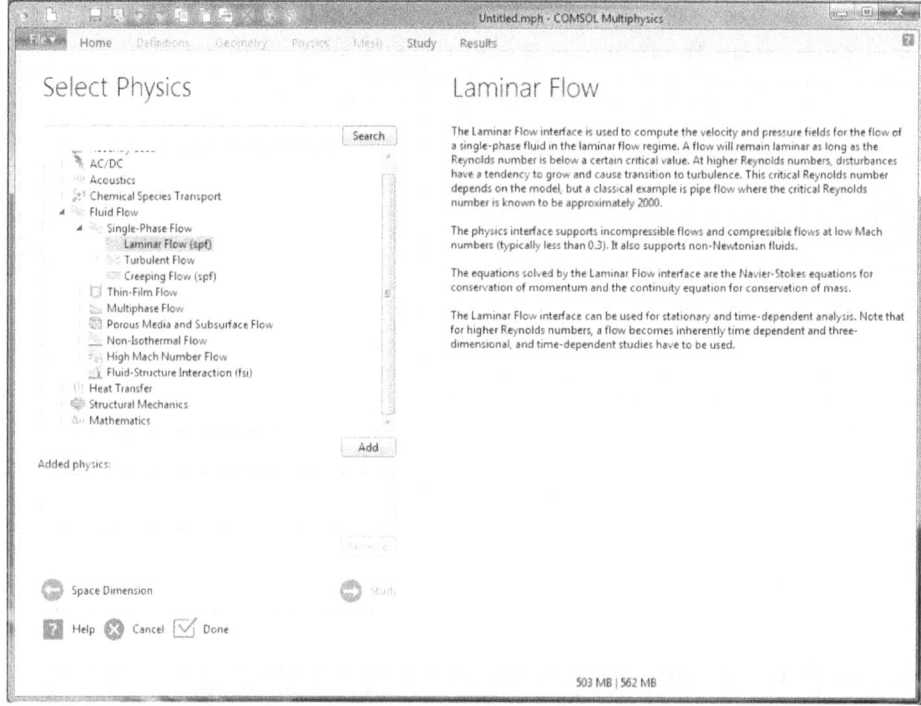

FIGURE 2.3 Select Physics window in COMSOL 5.

After selecting the desired physics, click on the Add button to add the physics to the model. For any selected choice/physics a brief explanation appears on the right side of the corresponding window.

Next we should select the type of solver for solving the model equations. In COMSOL, in general, solvers are selected by default based on the type of physics involved and are called Study. Yet users have the option to choose a different solver from the available list in COMSOL. This feature adds to the flexibility of COMSOL as a comprehensive modeling tool. After clicking on the Study button, located at the bottom-right corner of the Select Physics window, the Select Study window appears, as shown in Figure 2.4. At this stage, users have the option of going back to previous steps and modifying them if desired, accordingly. When options selected are finalized, click on the Done button, located at the bottom-right corner of the Select Study window, to proceed.

The main interface or COMSOL Desktop will show up. This interface includes a Quick Access Toolbar menu on the top, and a ribbon bar that changes according to the selected tab. The Quick Access Toolbar could

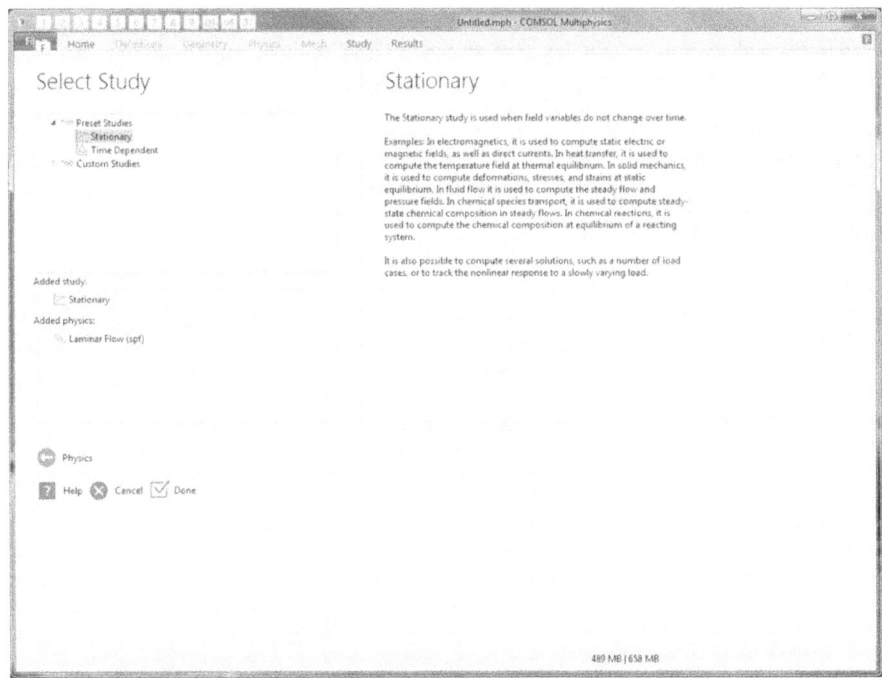

FIGURE 2.4 Select Study window in COMSOL 5.

be moved to be placed under the Ribbon bar, as well. The main toolbar items are listed according to the usual sequence used for building a model; Model, Definitions, Geometry, Materials, Physics, Mesh, Study, Results, as shown in Figure 2.5. The ribbon bar under the Model tab lists the modeling sequence actions required, as well. In COMSOL version 5.1, the Model and Application windows are integrated. Users are referred to the COMSOL manual for highlights of new features in version 5.1.

A useful item in the Ribbon, under the Model tab, is the Layout. Users can choose their Desktop Layout by clicking on this icon, and choose, for example, Reset Desktop. In addition, a COMSOL Desktop, as shown in Figure 2.6, has three main sections or sub-windows; Model Builder, Settings, and Graphics, which appear from left to right, respectively, in the default Desktop layout. The **Model Builder** or model tree window works as a registry for bookkeeping the model features, data, physics, mesh, study, results, etc., and can be used to quickly access model features or modify them if needed. The **Settings** window changes according to the item selected in the Model Builder. The third window is the **Graphics** window, which shows the geometry of the model, allows selection of domains and boundaries, and shows modeling results including graphs. Any of these three windows could be separated from the Desktop and moved around by making them Float. This feature is available by right-clicking on the corresponding window's title section and selecting the Float option. Additional Information windows, like Messages, Progress, Log, Table, and External Process, are also available under the More Windows tab from the Model toolbar tab.

A user starts building a model by operating in the COMSOL Desktop. The sequence of actions is listed and recorded in the Model Builder window, which becomes handy for editing and maneuvering around the model.

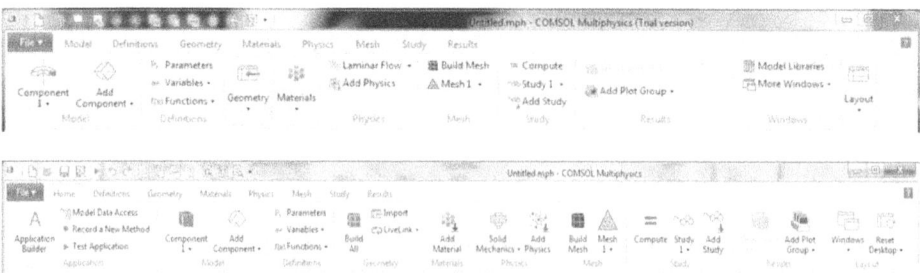

FIGURE 2.5 Toolbar and Ribbon for selected Model tab in COMSOL 5 Desktop.

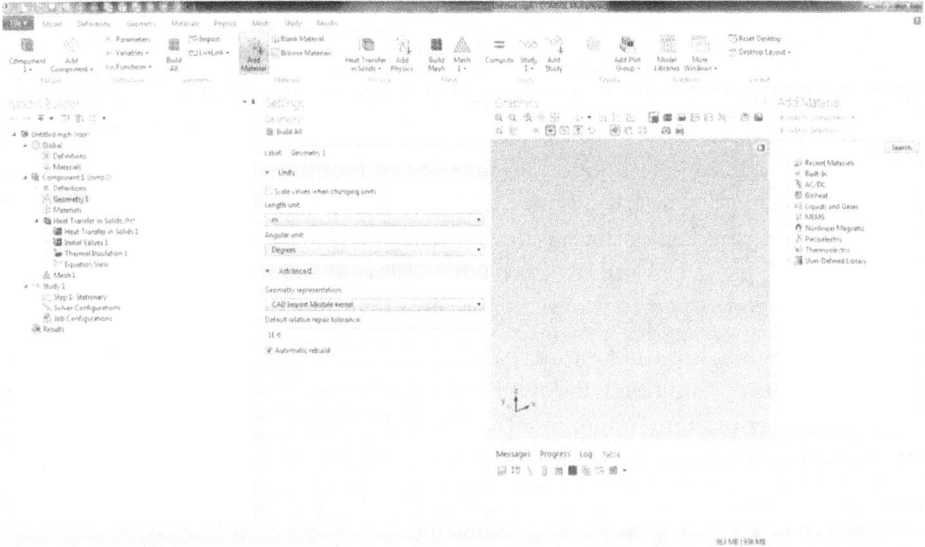

FIGURE 2.6 COMSOL 5 Desktop with Geometry data entry window.

The graphics window will show the results in different stages during construction from geometry to final modeling results.

Building model geometry can be done in two ways in COMSOL: by using the drawing tools to build relatively simple geometry inside COMSOL, or by importing geometry from a commercial CAD package.[1] The main module for seamlessly importing a geometry file is called *LiveLink*™ *Interfaces* and should be purchased as part of the user license. When a user has a solid model of a geometry open in a CAD package, through LiveLink it is possible to seamlessly import the file as well as communicate between the parameters of the CAD geometry in COMSOL and update it if needed. The LiveLink modules make the CAD package an integral part of COMSOL; the modules enable users to run COMSOL from the interface of a CAD package.

LiveLink modules are extremely useful and are recommended for users who have more complex model geometry and would like to use their favorite CAD package for building it. In addition, users can import a solid model

---

[1]Inventor, SolidWorks, Pro/Engineer, SpaceClaim, Creo Parametric. Please check www.comsol.com for updates.

in conventional format such as Parasolid, STEP, IGES, VRML, and STL. See the COMSOL manuals for CAD import modules.

Several Unit Systems of measurement are available in COMSOL. Users can choose their desired Units by clicking on the (*root*) node (usually accompanied by the model file name) at the top of Model Builder window and scrolling down in the Unit System section, in Settings Root window, to find the desired selection. When a Unit System (e.g. *SI*) is chosen at the *root* level, users can enter the data in different units, but COMSOL automatically converts it to the main selected Unit System.

By clicking on the Geometry tab from the toolbar or Geometry 1 node in the Model Builder window, the Settings Geometry window will open again. After building the geometry (or importing it), material properties can be added to the geometry by either entering the data directly or using the database available in COMSOL. This is done by clicking on the Materials tab in the toolbar window and choosing the desired material from the Add Material window, as shown in Figure 2.7. To add the material selected, simply click on the Add to Component icon. Materials are categorized for different applications, and users also can add their own material library to the list.

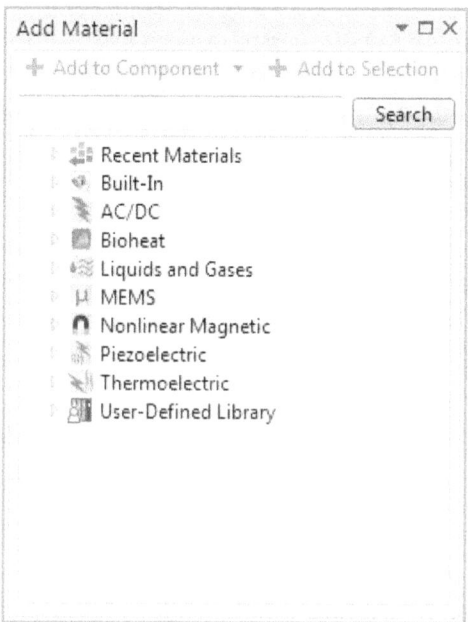

FIGURE 2.7 Add Material window in COMSOL 5.

Boundary and Initial conditions can be selected by clicking on Boundaries under the Physics tab, from the toolbar. Alternatively this tool is available by right-clicking on the selected/corresponding physics in the Model Builder window. When selected, the conditions can be entered through the Settings window and implemented to the boundaries by clicking the corresponding edges or surfaces of the model geometry in the Graphics window. A typical temperature boundary selection is shown in Figure 2.8.

Meshing, a process for dividing the geometry into finite elements, is a major step in modeling. The default meshing is automatically selected based on the physics involved. For example, for a fluid flow domain COMSOL automatically assigns boundary-layer-type elements to the solid boundaries. Users can accept or modify the settings of the default mesh, as well. Meshing facility in COMSOL is generally done using *free* meshing or *mapped* mesh. Free meshing is unstructured mesh, and mapped meshing is similar to structured mesh, as two examples are shown in Figure 2.9. Users have options to modify the resolution of the mesh for different regions of the geometry. Mixed or hybrid meshing is also possible by combining structured and unstructured mesh types. Users can also define the resolution or number of meshes/elements in different regions of the geometry.

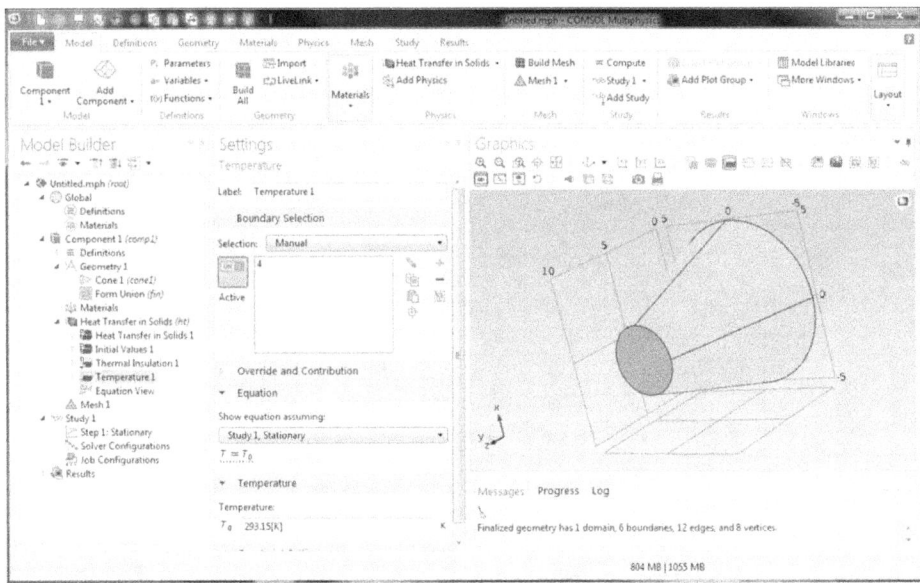

FIGURE 2.8 Boundary condition selection and implementation in COMSOL 5.

COMSOL has adaptive-mesh facility, a feature that customizes the mesh resolution during the solution process for complex geometry and/or physics. Clicking on the Mesh tab in the toolbar, or Mesh 1 node in the Model Builder, opens the Settings Mesh window, as shown in Figure 2.10. Users

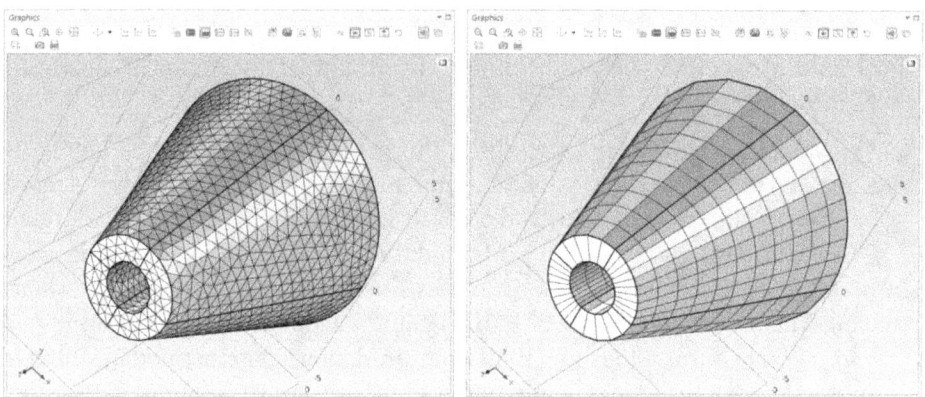

FIGURE 2.9 Free mesh (left) and Mapped mesh (right) for a geometry in COMSOL 5.

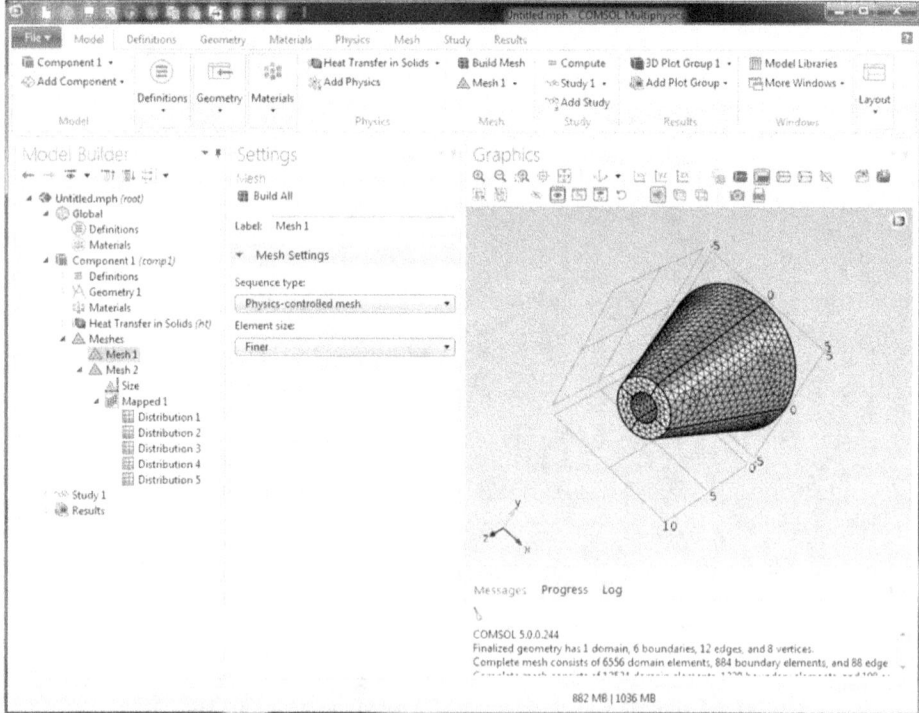

FIGURE 2.10: Mesh building tools and Settings in COMSOL 5.

can choose the resolution of the mesh and also manipulate the meshing parameters in this window by choosing options from Mesh Settings.

At this point the model is ready for analysis. Users can run their model by clicking on the Study button in the toolbar and choosing Compute. Additional studies can be implemented by clicking on the Add Study button.

After running the simulation for the selected Study, the results are displayed in the Graphics window. The default results vary according to the physics of the model. Users can add or modify these results by using post-processing tools, listed in the Results ribbon bar under the Result tab.

A detailed or summary report can be generated by clicking on the Report button, under the Results tab in the toolbar. This is a very useful feature for communicating and filing model results.

COMSOL 5 MODULES

COMSOL has many ready-to-use modules to handle modeling most, if not all, commonly occurring engineering problems. In addition, users can solve unconventional governing equations/PDEs using available Mathematics modules in COMSOL.

Following is a list of COMSOL physics/application modules available for purchase. Additional features and modules are released with newer versions of the software. For an updated and complete list, check the COMSOL website (www.comsol.com).

- CAD Import Module
- Design Module
- CFD Module
- Pipe Flow Module
- Structural Mechanics Module
- Non-Linear Structural Materials Module
- Fatigue Module
- Multibody Dynamics Module
- Heat Transfer Module
- Optimization Module
- AC/DC Module

- Mathematics Module

- Chemical Transport Module

- Mixer Module

- Microfluidics Module

- Molecular flow

- Electrodeposition and Corrosion Modules

- Acoustics Module

- Batteries & Fuel Cells Module

- Geomechanics Module

- MEMS Module

- RF Module

- Wave Optics

- Plasma and Semiconductor Modules

- Subsurface Flow Module

- Particle Tracing Module

- Ray Optics Module

COMSOL 5 MODEL AND APPLICATION LIBRARIES AND TUTORIALS

After installing COMSOL, many other resources become available to the users to support their modeling tasks at hand. One of these resources is the Model Libraries, which offers solved models for training, teaching, or modification. Registered users can download these models and supporting documents (usually in PDF format). Models available in the libraries are useful in order to start a model with similar or closely related physics and modify them according to a specific/desired modeling problem. A new Application library is available in version 5, which could be useful for Applications built based on COMSOL models. The Application library is currently small in volume, but it is gradually populated with new Application model files. Similar to a COMSOL model, an Application model can be loaded and used for learning from its features and user interface layout.

There are also two types of Help documents available for users under the Help button in the toolbar: Documentation and Dynamic Help. The Help Documentation offers users access to an extensive, searchable list of documents that explain interface icons and keys, as well as details of modules, physics, meshing, geometry, post-processing, and more. Users at varying levels of expertise can refer to the documentation to find more details about COMSOL features as well as answers to their specific questions. The Dynamic Help feature opens a specific section of the Help Documents relevant to the section or feature in use at hand.

APPLICATION BUILDER AND COMSOL SERVER™

A major upgrade in version 5 is the Application Builder tool. Application Builder provides excellent flexibility for communicating model results to other users such that they can change desired/designed parameters/data in the app to evaluate and examine the relevant results. It is like having the built COMSOL model as an "engine" in the background of the app and using it "implicitly" to run the model for different values of model parameters/data. The final result works like an "app" that is commonly used with mobile media. Application Builder enables users to build and edit (currently in Windows® environment) a "custom design" GUI for any model built in COMSOL Multiphysics. The model Application, which can be password protected, can then run independently from (or jointly with) the original model. The Application Builder tool provides two Editors to build and edit the user interface of the Model Application; (1) Form Editor (referred to a New Form, in version 5.1) and (2) Method Editor (referred to a New Method, in version 5.1). Form Editor is a sketch-type editor with quick-and-fast, yet practical, facilities and tools, whereas the Form Editor is more comprehensive and has programming capabilities. Users are referred to the COMSOL Application builder manual for further details.

We will use Application Builder and Form Editor for selected model examples presented in this book, along with relevant instructions. Applications can be executed on any operating system either by using a web browser or COMSOL Client for Windows®. The latter is a free download piece of software.

Another new service from COMSOL is the COMSOL Server, which requires a license for users who would like to have their own server set up. COMSOL license holders could also benefit from the Amazon Cloud server and have access to the COMSOL Server, through the World Wide Web. Once a user has a network connection to the COMSOL Server, from any desktop, laptop, or mobile device, they can run COMSOL Applications, which are accessible to the user. The combination of COMSOL Application Builder and COMSOL Server has democratized the modeling technology by providing cloud-type platform facilities. Currently only model Application files can be uploaded to the COMSOL Server.

Following instructions can lead a user to the COMSOL Server for uploading/running a Model Application:

1. Sign in to COMSOL Server: www.comsol.com/try-comsol-server

2. Follow the instructions given, until you reach to the page where you are logged in to the COMSOL Server page.

3. Run an application model from the Application library, or Upload a model Application and then run it.

Users can share their model Applications by clicking on the Administration tab, available in the COMSOL Server web page.

Example 2.1: Building an app for Andrew's squeezing mechanism

In this example we build an app for a model, i.e. Andrew's squeezing mechanism. The model is available from the COMSOL Model Library (Multibody_Dynamics_Module/Verification_Models/andrews_mechanism).[2] Andrew's mechanism consists of 7 rigid bodies linked at frictionless hinges, a motor, and a spring, as shown in Figure 2.11. Points A, B, C, and O are fixed in their plane coordinates. The governing differential-algebraic equations for the dynamics of the mechanisms consist of 14 differential and 13 algebraic equations.

The model parameters that we use as inputs for this app are the spring stiffness, initial spring length between points C and D, and the applied moment/torque at point O.

---

[2]Model made using COMSOL Multiphysics® and is provided courtesy of COMSOL. COMSOL materials are provided "as is" without any representations or warranties of any kind including, but not limited to, any implied warranties of merchantability, fitness for a particular purpose, or noninfringement.

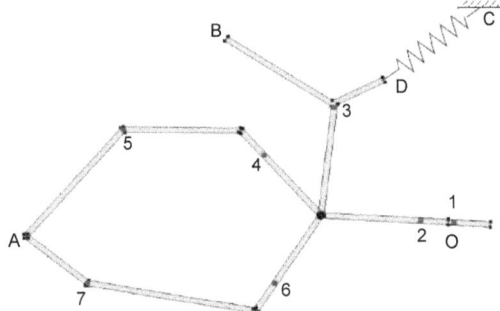

FIGURE 2.11 Schematic of Andrew's squeezing mechanism, with links' center of masses highlighted.

Instructions for building the app:

1. Launch COMSOL 5 and click on Application Wizard button, in the New window. The Select Model for Application window appears. In the window you should see a list of COMSOL Modules for which you have purchased licenses. Under this list, click on the Browse icon. Locate the file andrews_mechanism.mph, usually located in the folder where your COMSOL files are saved. Click on the file and open it. Users should note that sometimes Model files are not saved with their corresponding results, in the Model Library. You may need to open this file, as a model, in COMSOL and run it before opening it for building the app.

2. After opening the Model file, the New Form window should appears, as shown in Figure 2.12. This window has three tabs: Inputs/outputs, Graphics, and Buttons. Under each tab the relevant model features are listed. From the list in Inputs/outputs tab under Parameters, select Spring coefficient (c0), Unstretched length of the spring (l0), and Applied moment (M0) and move them (by clicking on the arrow) to the Selected window, on the right. Once these Parameters are moved to the Selected window, their default values, as defined in the original model, show up in the Preview section of the window. Similarly, click on the Graphics tab and from the items listed in the window, select Displacement (mbd) and move it to the Selected window. A Graphics window appears in the Preview section. Users have the option of adding more Graphics windows, as well. However, we recommend minimizing the creation of these Objects in order to optimize the layout of the app interface. Now click on the Buttons tab and select Compute Study 1 item and move it to the Selected window. Figure 2.13 shows the result.

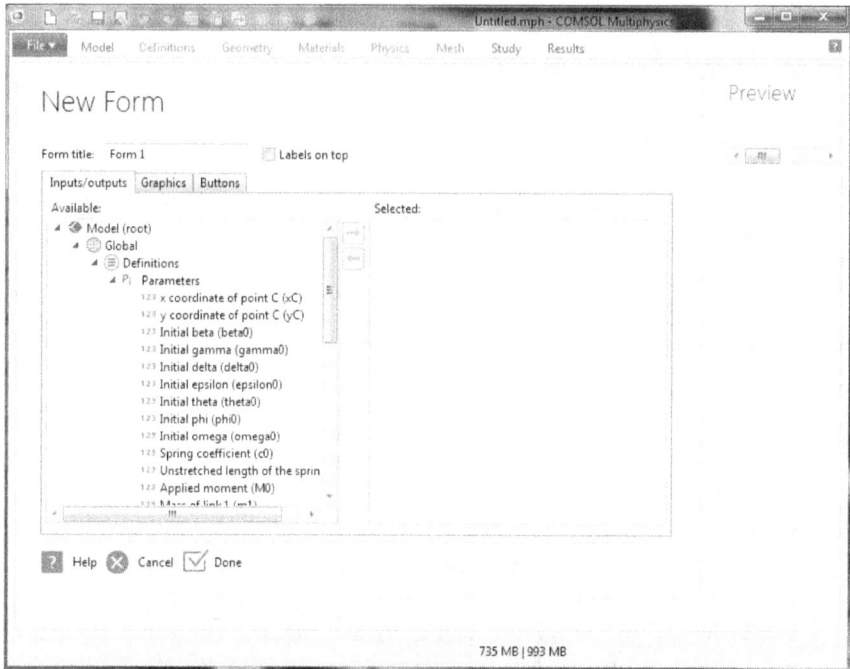

FIGURE 2.12 The New Form window for building model Application.

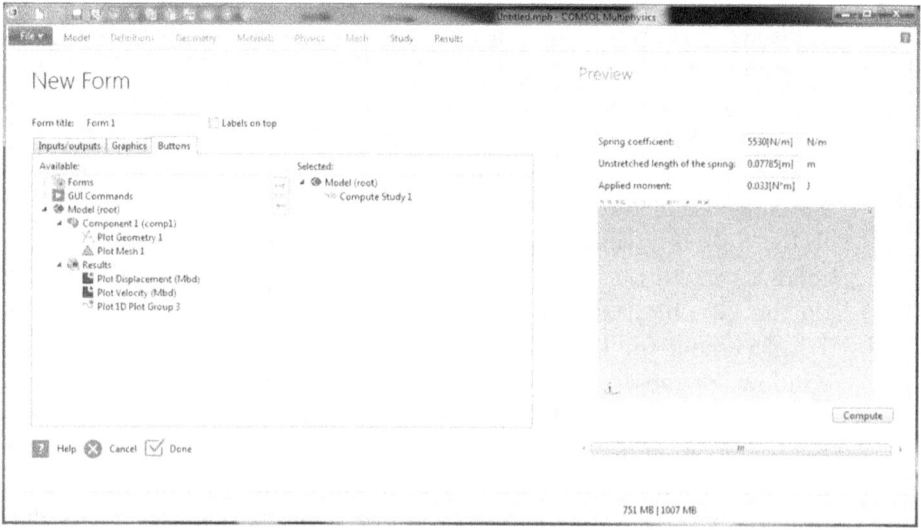

FIGURE 2.13 Model App's default interface as appears in New Form window.

3. Click on the Done button. The Form Editor desktop window appears, as shown in Figure 2.14. Click on File from the menu toolbar and save the file as App_Example 2.1.mphapp. The Form Editor provides the tools for editing the Form1 or the layout of the App's GUI. Objects in the Form1 window can be dragged and moved around, as desired. Also the Test Application button could be used during building the app interface. The Test Form tool could be used as well to check the layout of the interface.

4. Drag the Graphics window, the Form1 window, and move it to the right hand side of the Form1 window. Also drag the Compute button and move under the area where the parameters are listed. In order to organize the layout of the objects more precisely, click on Grid, located in the ribbon bar. This tool is very useful to know the location of each Object in the layout. Right-click on the highlighted cell/row on the top of the Graphics Object and select Grow from the list. Similarly, do the same by right-clicking on the highlighted bottom cell/column on the left side

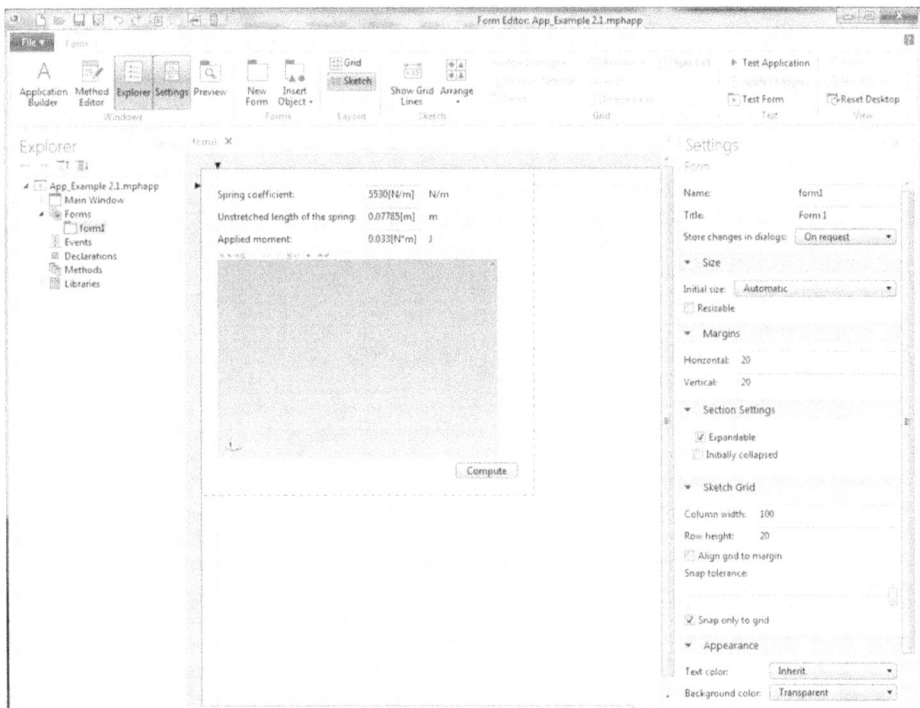

FIGURE 2.14 The Form Editor window interface with Form1 for the Model App.

and selecting Grow. We change the appearance of the button Compute and add more buttons to the Form1. Click on the Compute button. In the Settings window select compute_32.png from the list for Picture and Large for Size. Note that in the Choose Command to Run section, relevant operational commands are assigned to this button. Click on an empty cell to highlight it, maybe the one to the left of the Compute button, and then click on Insert Object button, from the ribbon toolbar, and select Button from the Input section. In the corresponding Settings window change the Text to Geometry, select Geometry_32.png for Picture and Large for Size. To assign an operation to this button, Locate the Choose Commands to Run section and select Model > Component 1 > Geometry 1. Click on the Run button, located underneath the list. The list under Command should show Build Geometry 1. Click on the Plot icon and then again in the Arguments cell area, use Edit Argument button to assign form1/graphics1 to the Arguments cell. This way we will use the same Graphics window and do not need to have several ones, which crowd the layout, otherwise. Similarly, create another button (name it Animation) and assign operations Export > Player 1 and Displacement (mbd) to it. You may want to add more buttons, for example, plotting more results. Use alignment for the Objects in the cells, as well. The trial Form1 should be similar to or closely look like the one shown in Figure 2.15.

FIGURE 2.15 Form1 layout for the Model App.

5. Save the file and click on Test Application button, from the ribbon toolbar. A new window appears. This window is the App GUI. When you resize it, you will notice that the window size changes but the Graphics window does not! Close this window, and in the in Form1, in the Form Editor window, right-click on the Graphics Object, two times, and choose Fill Horizontally and Fill Vertically, respectively. Now again click on Test Application. The APP GUI appears and by resizing it, you would have the Graphics window resized, accordingly (see Figure 2.16). Users may want to fine-tune the layout of the App GUI and edit further. Test the App by clicking on the Geometry button; the mechanism geometry should show in the Graphics window. Click on the Animation button. The App will use the Model to compute the displacements and the Graphics window should show the animation, after computation is finished. Users may want to change the parameters, as well, and run the App by clicking on the Compute button. The App could be used to study the mechanism dynamical behavior.

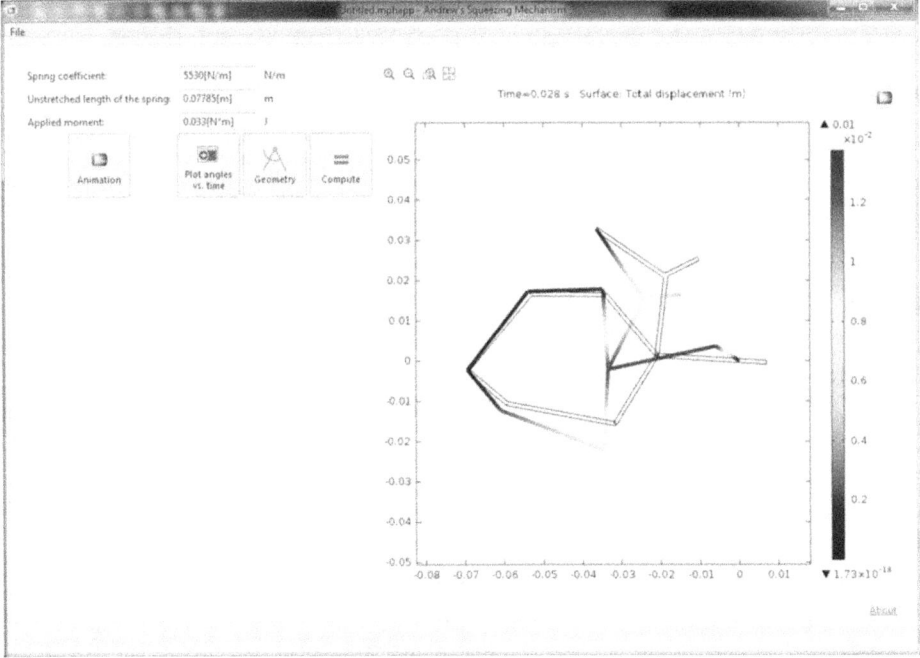

FIGURE 2.16 App GUI for Andrew's squeezing mechanism.

This concludes this example. However, more tools are available in the Form Editor and Method Editor facilities that enable interested users to build quite sophisticated apps.

GENERAL GUIDELINES FOR BUILDING A MODEL WITH COMSOL

The major sequential steps for building a model using COMSOL for a given problem are as follows:

1. Define the problem, including physics and materials involved.

2. Identify the governing equations, boundary conditions, and main physics involved to have a clear understanding of the scope of the problem's solution.

3. Possibly perform a rough hand calculation or search for similar models or experimental results. These could be used for verification and validation of the model.

4. Launch COMSOL.

5. Use COMSOL features to assign the dimension (1D, 2D, 3D, etc.), the physics involved, and the temporal (steady or transient, etc.) of the problem.

6. Build the geometry of the problem (if required), import your CAD file, or use LiveLink to access your model geometry.

7. Assign material properties to the built geometry blocks of the problem.

8. Add physics according to Steps 1 and 2.

9. Create a mesh or finite elements for the built geometry.

10. Solve/run the model.

11. Visualize the results and validate them, either using hand calculations or comparing with similar known results (Step 3).

12. Build an app, using Application Builder. This step is optional but highly recommended.

13. Create a report of the model and its specifications.

Many engineering problems involve either one type of physics or, if they involve multiphysics, can be simplified to a dominant one. For example, consider the laminar flow of a fluid at a high value of Reynolds number.

Since the Reynolds number may be interpreted [15] as the ratio of inertia over viscous forces exerted on the fluid, then for a fluid with a high-value Reynolds number we can neglect viscous forces and consider the fluid as an ideal frictionless one. When turbulence effects are present, flow becomes more complicated and equivalent turbulence-induced viscous stresses (so-called Reynolds stress) should be included. Or in the case of a steel beam, for example, we can neglect the effects of deformation due to shear stresses at the cross section of the beam and assume that the beam cross section remains perpendicular to the axis of the beam about which it is bent. These types of approximations in engineering are very common and sometimes are matters of "engineering art" or technical judgment. It takes much experience to simplify a problem without losing the dominant physics and obtain results that are useful in practice and have applications.

In the following chapters we use COMSOL to model example problems covering a wide range of concepts, including stationary equilibrium, dynamic equilibrium, buckling, time-dependent, fluid flow, heat transfer, electrical circuits, and transport phenomena. The main objective is to provide, for users and readers, a collection of solved examples that could be applied directly or lead to further solutions of similar or more complex problems using COMSOL. It is assumed that readers are familiar with relevant engineering principles and governing equations. Nevertheless, in each example we offer brief explanations of physics involved along with governing equations and phenomena, as applicable. It is recommended that readers cover Chapters 1 and 2 before attempting examples in these chapters. As mentioned previously, the examples are solved using COMSOL 5. Readers could also open the files from the accompanying CD in newer version (when available) and study or modify the examples pertinent to their modeling needs.

CHAPTER 3

MODEL EXAMPLES FOR FLEXIBLE STRUCTURES, PARTS, AND ASSEMBLY

In this chapter, we use COMSOL features to model some problems in structural mechanics, including mechanical parts and assembly, static, dynamic, parametric study, and buckling of two- and three-dimensional structures.

For analyzing stresses within a 3D solid body, we use the general equilibrium equation (in tensor notation):

$$\frac{\partial \sigma_{ij}}{\partial x_j} + \chi_i = \rho \frac{\partial^2 u_i}{\partial t^2}$$

Where, σ_{ij} is the stress tensor that has nine elements in 3D space (i and j take values of $1, 2, 3$). For most cases in practice, we end up with a symmetric stress tensor and hence we would have six independent elements. The stress tensor can be related to the strain tensor and consequently to the displacement vector u_i, using constitutive equations (like Hooke's law) and kinematic compatibility of strains constraint. The body loads per unit volume χ_i, together with the divergence of stresses, should balance external loads, which could include inertia forces of the continuum with density ρ, and boundary conditions to have an equilibrium state for a given structure/machine. Applied loads may include static loads or dynamic ones. The equation of motion for the continuum should be solved for dynamic analysis of a structure. Each structure has a fundamental natural frequency under which it starts to vibrate. The fundamental natural frequency is a very important characteristic of a structure; if applied loads have the same frequency, then resonance will occur, displacements will become very large, and failure may occur as a result. For failure criteria, we use von Mises stress, which is basically the failure criterion for or limit for stresses based on maximum distortional energy. It assumes that yielding will occur when the distortional strain energy reaches that value which causes yielding in a simple tension test [16].

Example 3.1: Stress analysis for a thin plate under stationary loads

A plate is a structural member that can carry loads acting normal to or along its plane. The resulting stresses due to bending and normal and shear forces applied cause deformation. For a thin plate the deformation due to shear force along the thickness is ignored, whereas for a thick plate it is, usually, taken into account. A plate is considered thin when its thickness over a typical horizontal dimension of the plate is smaller than or equal to 0.1 in. Thick plate theory is that of Mindlin and Reissner, see [4]. Governing equations are the equilibrium equations, which are available in the COMSOL manual. We will refer to these equations in the solution steps.

As an example we consider a thin plate (12 in. by 8 in.), as shown in Figure 3.1. We would like to calculate the displacement and stress field in the plate resulting from the applied loads. Plate thickness is 0.4 in., and applied tension stress load is 500 psi at the right-middle edge. Modulus of elasticity of the plate material is 3E7 psi, and Poisson's ratio is 0.3. The boundaries on the left side of the plate are constrained, while the other plate sides are let free.

Solution:

1. Launch COMSOL and in the New window click on Model Wizard. Click on File in the toolbar and save the new file as Example 3.1 by clicking on File > Save As.

2. In the Select Space Dimension window choose the geometry dimension by clicking on 2D icon.

3. In the Select Physics window, choose the Physics by clicking on Structural Mechanics > Plate (plate). Click on the Add button to add to Added physics interfaces list, as shown in Figure 3.2. Also notice that the dependent variables are listed, as displacements (u, v, w), in the right side of the Select Physics window. Next, click on the arrow icon to move to the Study.

4. In the Select Study window, choose the type of analysis or study by clicking on Stationary (in the Study list) and click on the Done icon to

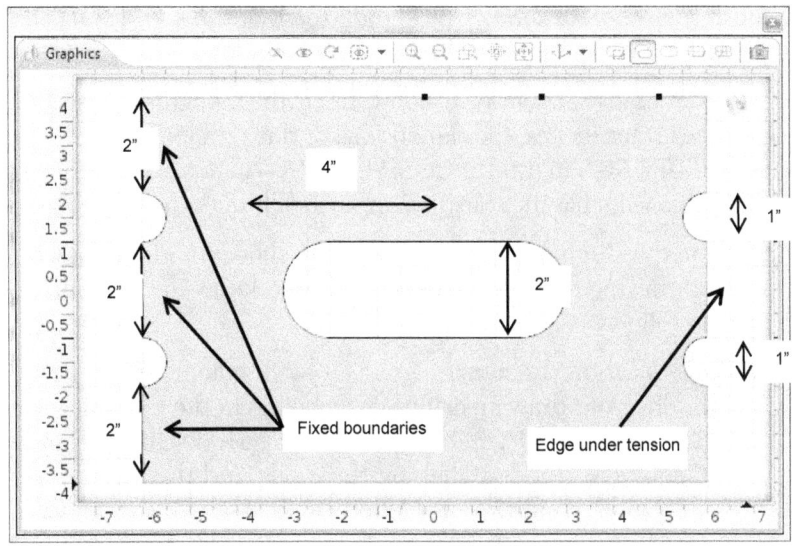

FIGURE 3.1 Geometry, dimensions, and boundary conditions.

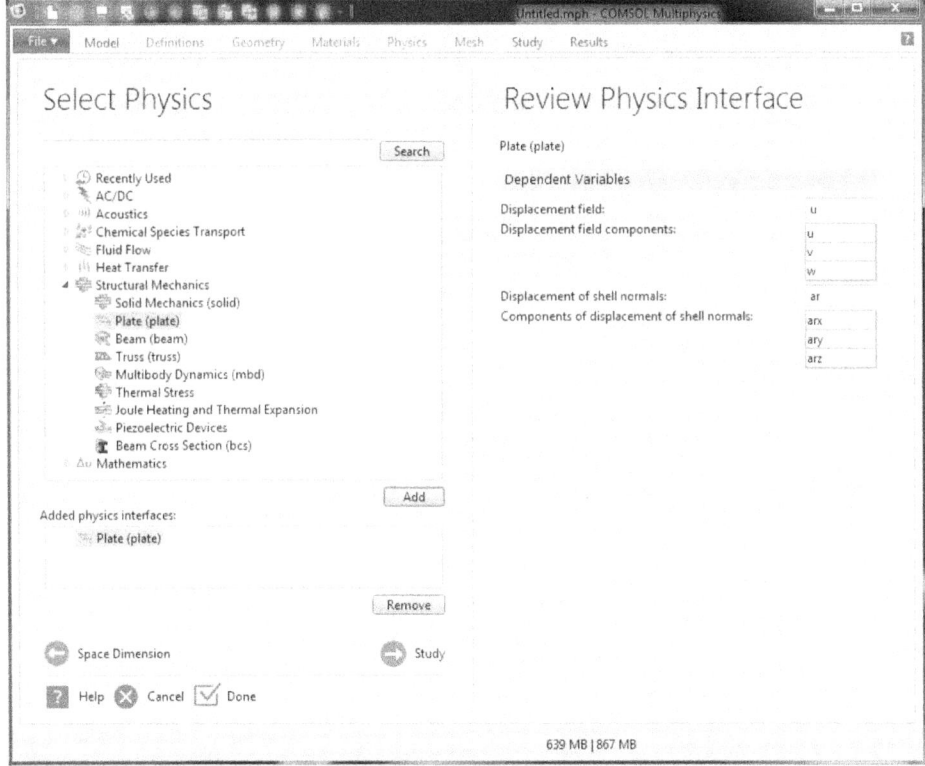

FIGURE 3.2 COMSOL 5 window for selecting and adding physics.

finish the problem setup. You should see that the COMSOL Desktop window opens, along with the Geometry window.

5. In the Settings Geometry window, open the Length unit section and choose (in) for inches, the dimensions of this problem. (Notice that the default Unit System used in COMSOL is SI—in this example, we keep it "as is" knowing that the software automatically does the unit conversion.)

6. Click on the Geometry tab in the toolbar, shown in Figure 3.3, to open a list of drawing tools in the toolbar ribbon. Draw the geometry of the plate as follows:

    6.1. Click on the Rectangle icon (Draw Rectangle (Center)) on the toolbar and draw an arbitrary rectangle in the Graphics window. The Rectangle 1 will be created in the Model Builder under the Geometry 1 node. Click on this node and then in the Settings Rectangle window enter Width = 12, Height = 8, and Position for Center is (0, 0). Then click on the Build Selected icon located at the top of the Settings window. Click on the Zoom Extents icon to scale the Graphics window to display the entire rectangle, as shown in Figure 3.4.

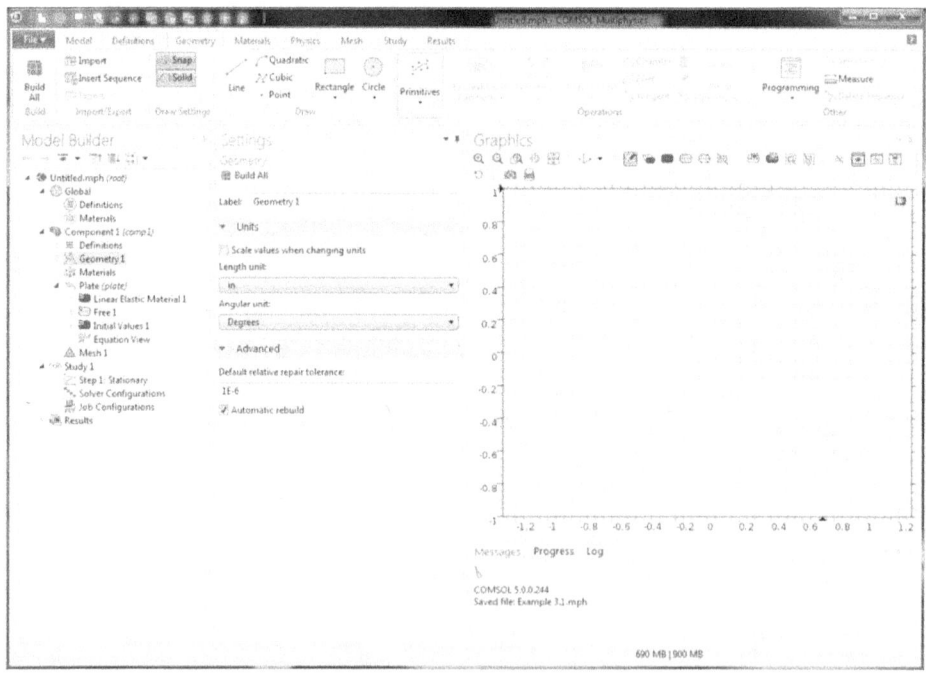

FIGURE 3.3 COMSOL Desktop with Settings and Graphics windows for geometry entries.

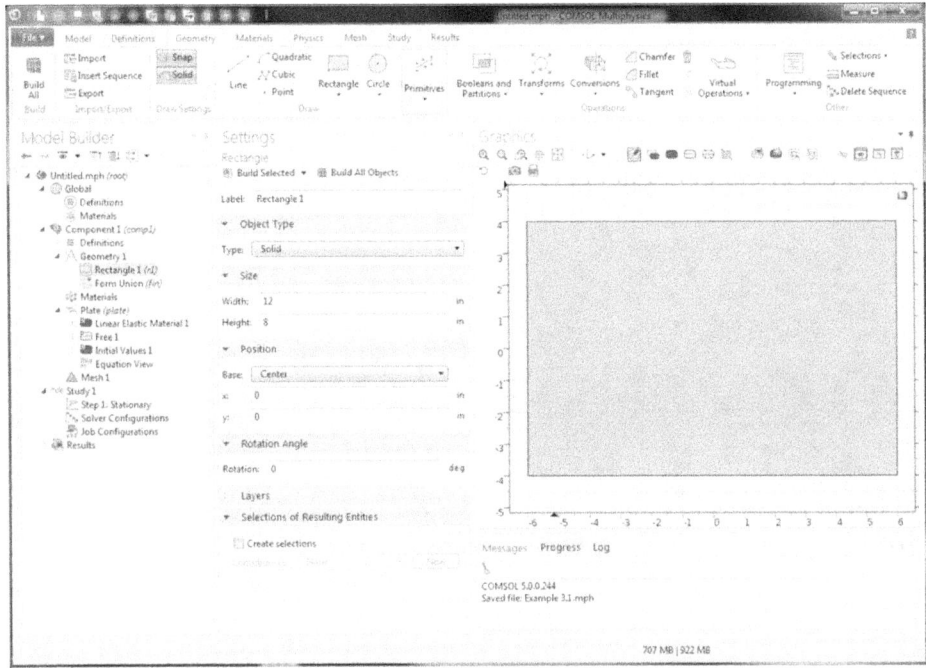

FIGURE 3.4 Graphics window with a rectangle and settings.

6.2. Draw four circles using the circle drawing tool, each with radius of 0.5 in. and their centers (Circle 1: x = −6, y = 1.5 in.), (Circle 2: x = −6, y = −1.5 in.), (Circle 3: x = 6, y = 1.5 in.), (Circle 4: x = 6, y = −1.5 in.). Results are shown in Figure 3.5.

6.3. Click on the Geometry tab in the toolbar. From the ribbon, open the list under Boolean and Partitions and select Difference. In the Settings window, select and add rectangle to the list under Objects to add. Activate the button under Objects to subtract and add the four circles to the list. Click on Build Selected icon. A Boolean operation is performed that subtracts the circles from the rectangle. Results are shown in Figure 3.6.

6.4. Similarly, draw a rectangle (Width = 4 in., Height = 2 in., Center at x = 0, y = 0) and subtract it from the resulting plate geometry. Again, draw two more circles with radii 1 in. (Circle 5: x = −2, y = 0 in.), (Circle 6: x = 2, y = 0 in.) and subtract them to create the resulting plate geometry shown in Figure 3.1.

6.5. Click on the Plate node in the Model Builder window, and then open the Thickness section and enter 0.4 [in] to assign

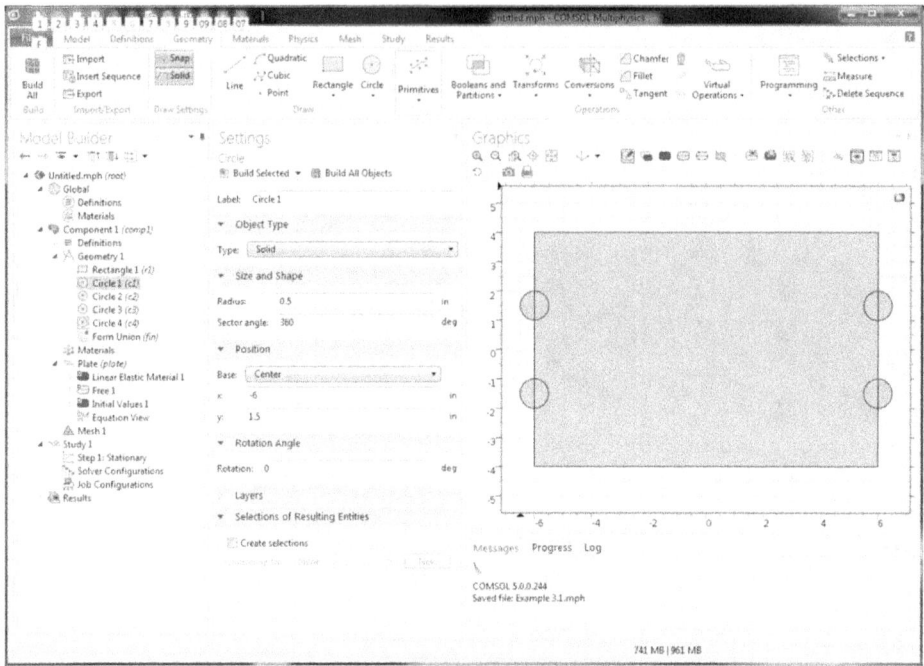

FIGURE 3.5 Graphics window with a rectangle and circles settings.

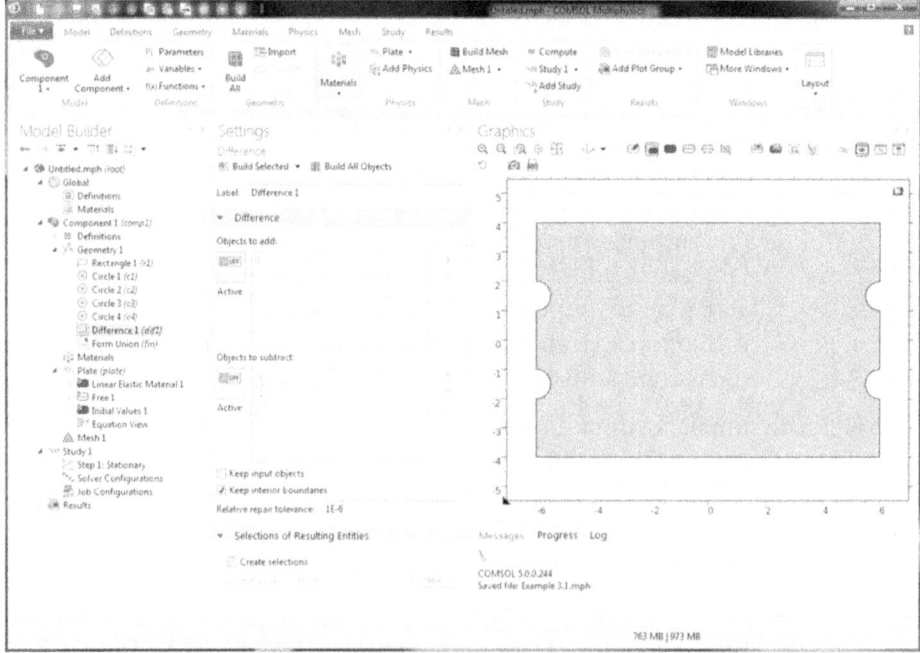

FIGURE 3.6 Graphics window with a rectangle and circles subtracted and settings.

the thickness of the plate. Notice the default is the SI unit (i.e. meters). It is optional but recommended to open the list under the Equation section to see the governing equations for the plate to be solved using FEM.

7. The next step is assigning the material properties. Move the pointer to the Plate node in the Model Builder window and open the list by clicking the Triangle icon. Click on the Linear Elastic Material 1 node to open the Settings window. Locate the Linear Elastic Material section. Under Young's modulus select User defined and enter 3e7 [psi], and under Poisson's ratio select User defined and enter 0.3, as shown in Figure 3.7. Notice that the default SI unit (Pa) is set for Young's modulus. COMSOL automatically converts the data from psi to Pa.

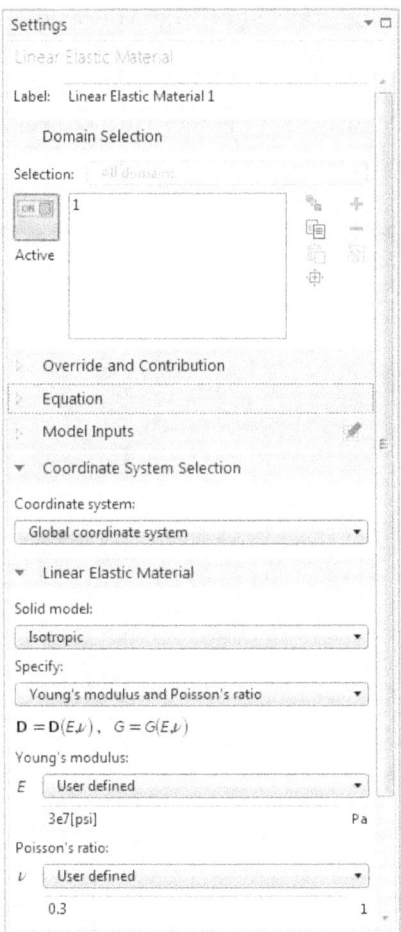

FIGURE 3.7 Data entry window for Linear Elastic Material Settings.

8. To assign the boundary conditions, click on the Physics tab from the toolbar and choose Fixed Constraint under the Boundaries button. In the Settings window, click on the left straight edges of the plate and add them to the Boundary Selection list. Results are shown in Figure 3.8.

9. Set the applied load by right-clicking on the Plate node in the Model Builder window and choosing Edge Load. The Edge Load 1 will be created under the Plate node. In Graphics window click on the middle-right side edge of the plate in the Graphics window, and add it to the Boundary Selection list in the Settings window. Locate Force section and choose the Load defined as force per unit area and enter 1500 [psi] in the x-component of the Edge load, since the load is applied as a tension on the edge of the plate. See Figure 3.9.

10. Create a mesh by clicking on the Mesh node in the Model Builder window, and in the Settings Mesh choose Physics-controlled mesh for Sequence type and Normal for Element size. Then click Build All to create the mesh. Statistics for the mesh are shown in the Messages window (usually located under/close to the Graphics window). A mesh with 1466 elements for this example was created. Results are shown in Figure 3.10.

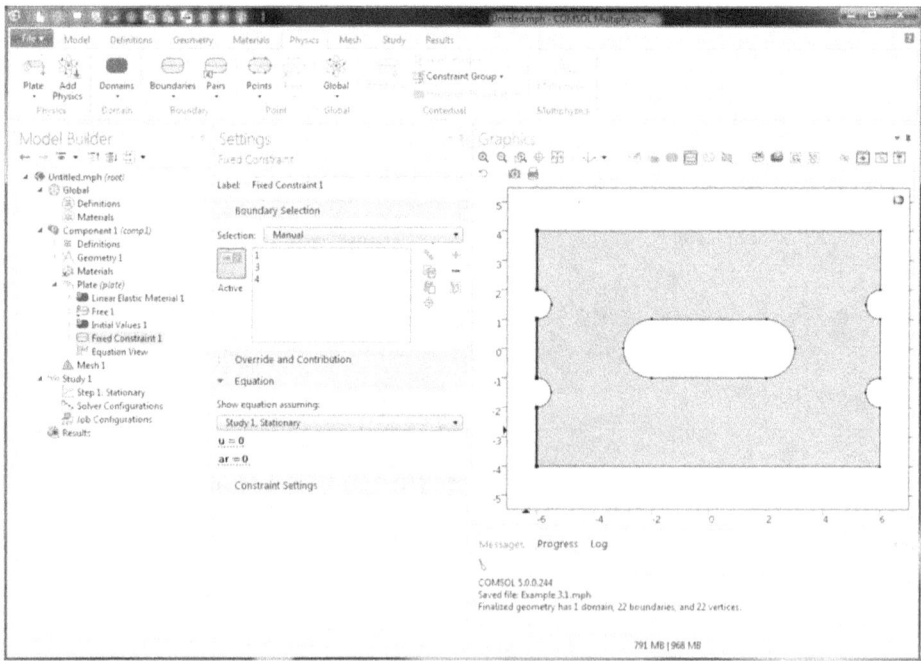

FIGURE 3.8 Fixed boundary conditions data entry window.

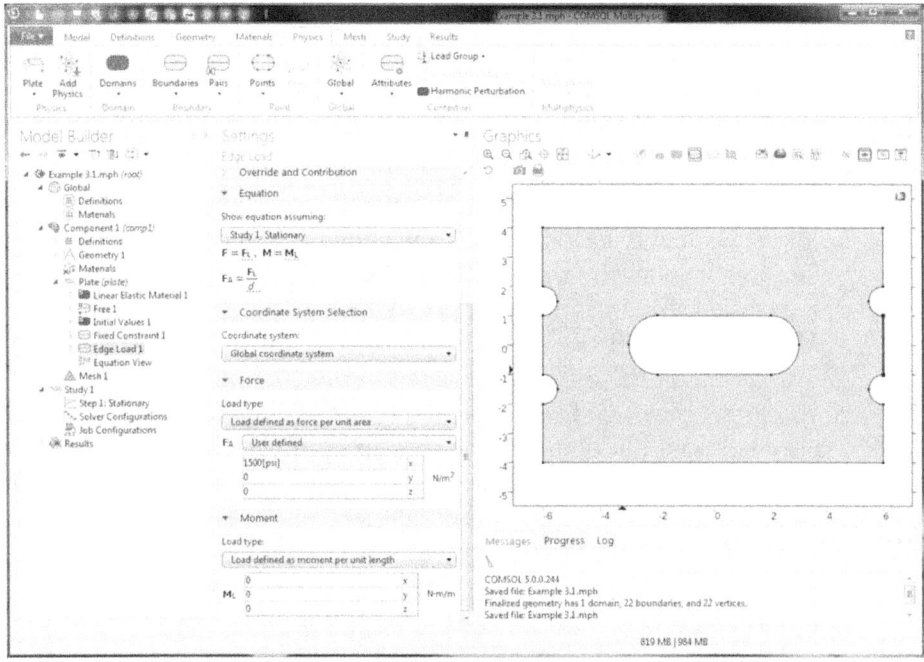

FIGURE 3.9 Load boundary conditions data entry window.

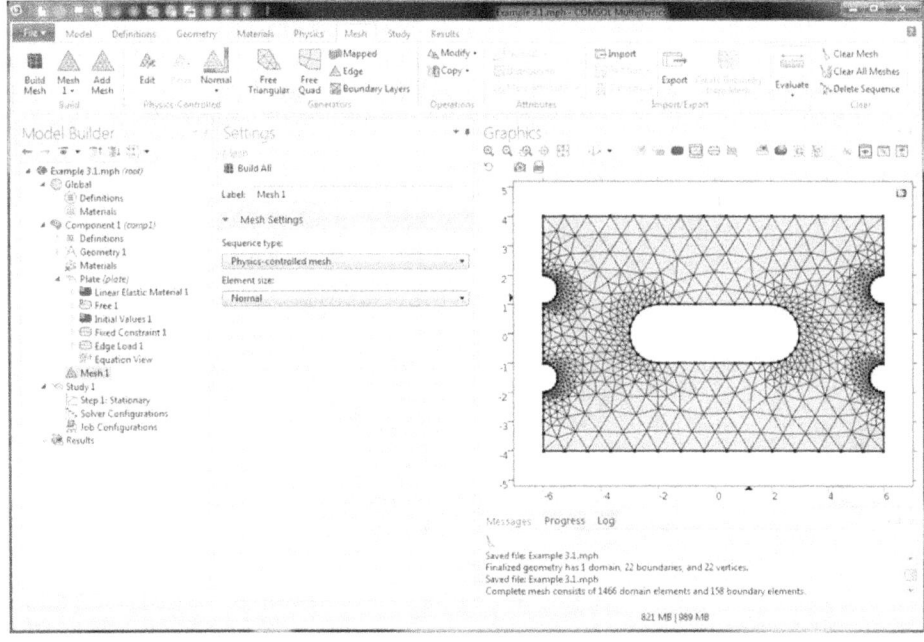

FIGURE 3.10 Mesh data entry window and resulting mesh.

11. To run the model, click on the Study tab from the toolbar and select Compute. The COMSOL default result for von Mises stresses shows up in the Graphics window. For this model, it takes about 6 seconds to run the finite element model with total 18540 d.o.f. (as registered in the Messages window) on a typical laptop computer.

12. To manipulate the results, click on Surface 1 node under the Stress Top (plate) node in the Model Builder window. This will open the Settings Surface window. In this window, open the list under Unit in the Expression section and choose psi, then open the Title section, choose Manual from the Title type, and enter von Mises stress for Example 3.1 (psi). Click on the (x–y) icon located in the Graphics window toolbar once to set the view. Then click on the Plot icon. Occasionally, you may need to Zoom In/Zoom Out the view to fit the results in the Graphics window. To show the displacement results, click on the Replace Expression icon (located on the right side of the Expression section) and double click on Plate > Displacement > plate.disp-Total displacement. To show the deformation, right-click on the Surface 1 node and select Deformation from the list. Change the Title type to Automatic, and then click on the Plot icon. Results are shown in Figure 3.11.

13. To create a report for this model, right-click on the Reports node in the Model Builder window and choose the level of the report (e.g. Brief Report). This will create Report 1 under the Reports. Rename Report 1 to Brief. In the Settings window, as shown in Figure 3.12, choose the desired report format (e.g. Microsoft Word) under Output format, choose the report Filename, and size and type (e.g. small and JPEG) under Images. A Preview of the final report can be viewed before writing it by clicking the Preview Selected icon button.

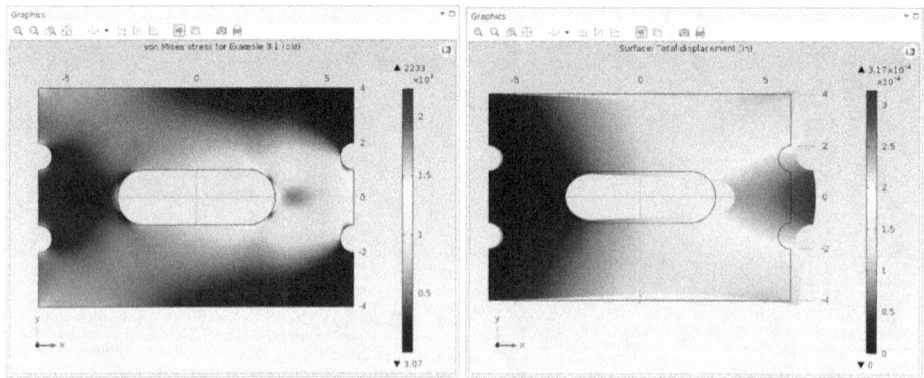

FIGURE 3.11 Results for von Mises stress and displacements.

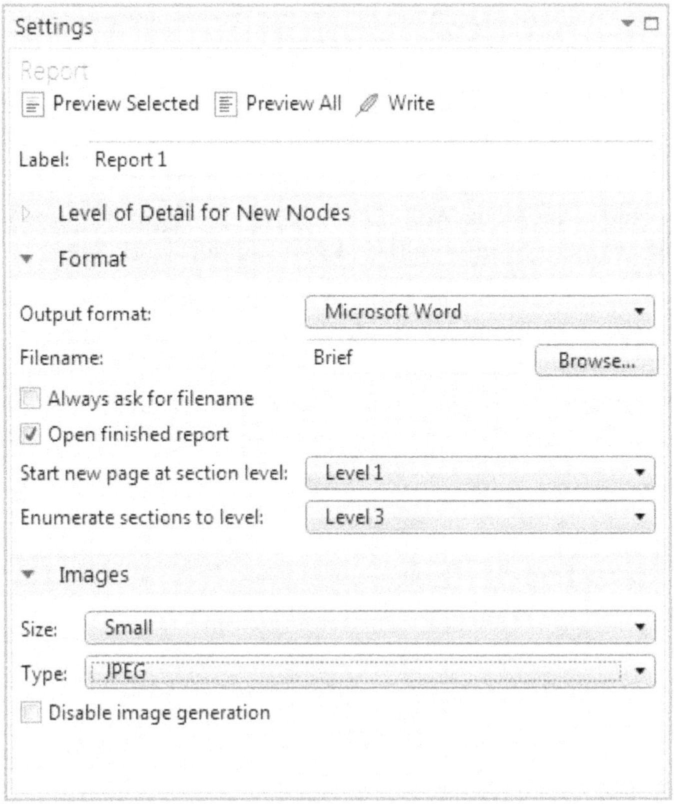

FIGURE 3.12 Data entry interface for creating a Report document.

Example 3.2: Dynamic analysis for a thin plate: Eigenvalues and modal shapes

For the plate given in Example 3.1, we would like to calculate the eigenvalues (i.e. natural frequencies). Natural frequencies for a structure/machine are important characteristics in terms of its vibration analysis. For a given structure, in this example a plate, there exists a set of frequencies at which it can vibrate. The smallest value of this set is called the fundamental natural frequency of the structure. Natural frequencies are the solution of the homogeneous governing equations, and applied loads usually do not have an effect (or a very small effect on the global stiffness matrix) on the solution. However, boundary and initial conditions do affect the value of natural frequencies [17].

In this example, we would like to calculate the natural frequencies of the plate given in Example 3.1 when initially at rest. Also we would like to calculate the corresponding modal shapes associated with natural frequencies.

Solution: We use the same model that we developed for Example 3.1, and add a Study scenario to it.

1. Open the model file Example 3.1. Save it as a new file with the name Example 3.2.

2. Click on the Study tab from the toolbar and then open the Study Steps list and select Eigenfrequency. Step 2: Eigenfrequency node will appear under Study 1 in the Model Builder window.

3. To run the model only for dynamic analysis, click on the Step 1: Stationary node under Study 1 in the Model Builder window. The Settings Stationary window will open. Under the Physics and Variables Selection section, click on the check mark in the Solve for column in the table. An X will appear, as shown in Figure 3.13.

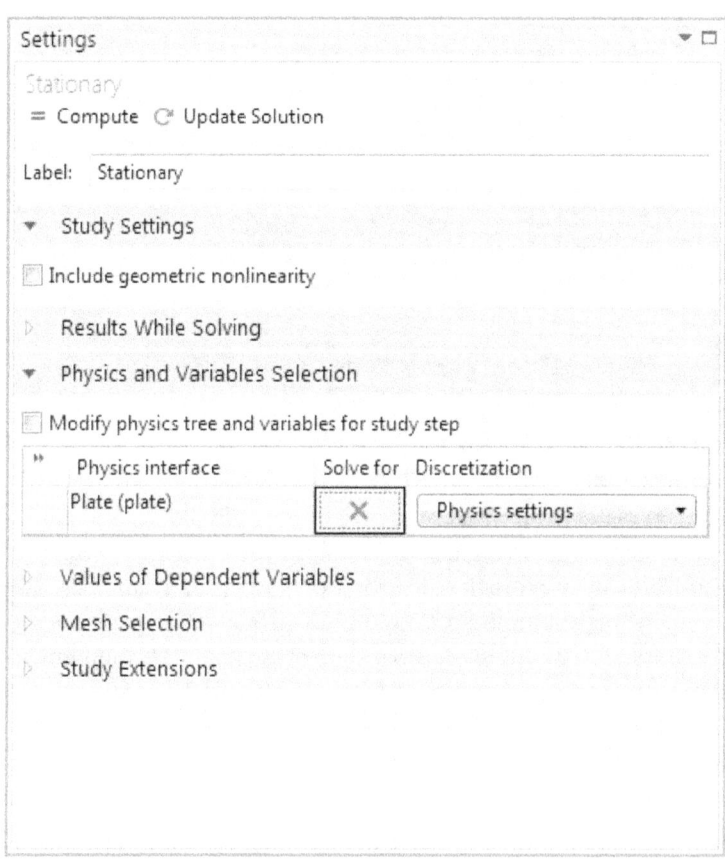

FIGURE 3.13 Settings window entries for Study Steps.

4. For dynamic analysis, the density of the material is required. Open the list in Model 1 (by clicking on the triangle icon on its left) and click on Linear Elastic Material node under Plate node. In the Linear Elastic Material section, under Density, open the list, choose User defined, and enter 7850. The unit used for density is kg/m^3, as shown in Figure 3.14.

5. Click on the Study tab from the toolbar and run the model by clicking on the Compute button. Wait until the run is finished.

6. Right-click on Stress Top (plate) node in the Model Builder, choose Rename, and rename it to 2D Plot Group. In the Settings window, open the Eigenfrequency list in the Data section to see the calculated plate fundamental natural frequency (89.636 Hz) and five higher natural frequencies, as shown in Figure 3.15.

7. Since we have added a new study to the model, we should modify the Graphics. Click on the Surface 1 node in the Model Builder to open the Surface window. In the Expression section type in plate.disp, and under Unit open the list and chose mm. Check the box for Description. Under the Title section, open the Title type list and choose Automatic.

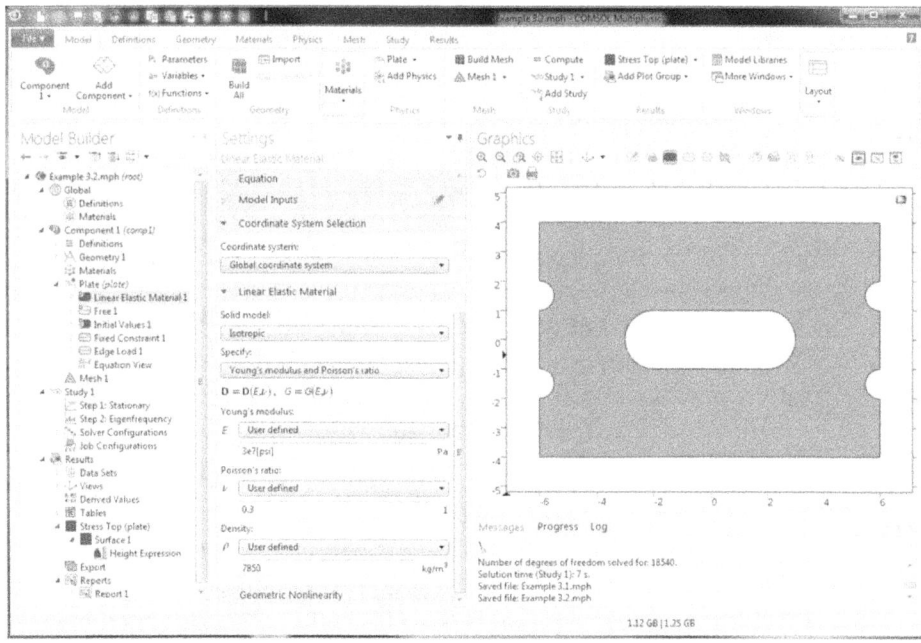

FIGURE 3.14 Material data entry Settings window.

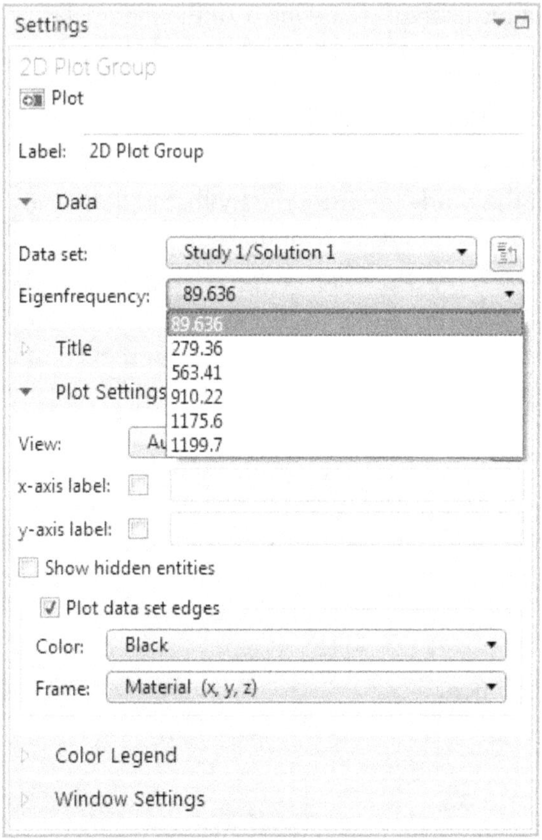

FIGURE 3.15 2D Plot Group settings and calculated Eigenfrequency.

In the Model Builder window, click on Height Expression and in the corresponding Settings window, under Unit choose mm. Click on the Plot icon to draw the graphics. By default COMSOL shows the fundamental natural frequency (the lower value of the frequencies of the plate), which is 89.636 Hz. The modal shape can be scaled relatively and arbitrarily. The result is shown in Figure 3.16.

8. To see an animation of modal shapes, we can create a movie. Right-click on the Export node in the Model Builder window and select Player from the list. A Player 1 node will be created and the Player window opens. In the Settings window, from Eigenfrequency selection list make sure All is selected. In the same window, open the list in front of Frame selection and choose All. In the Playing section, check the Repeat box. Click the Player icon to create the Player 1. Result is shown in Figure 3.17.

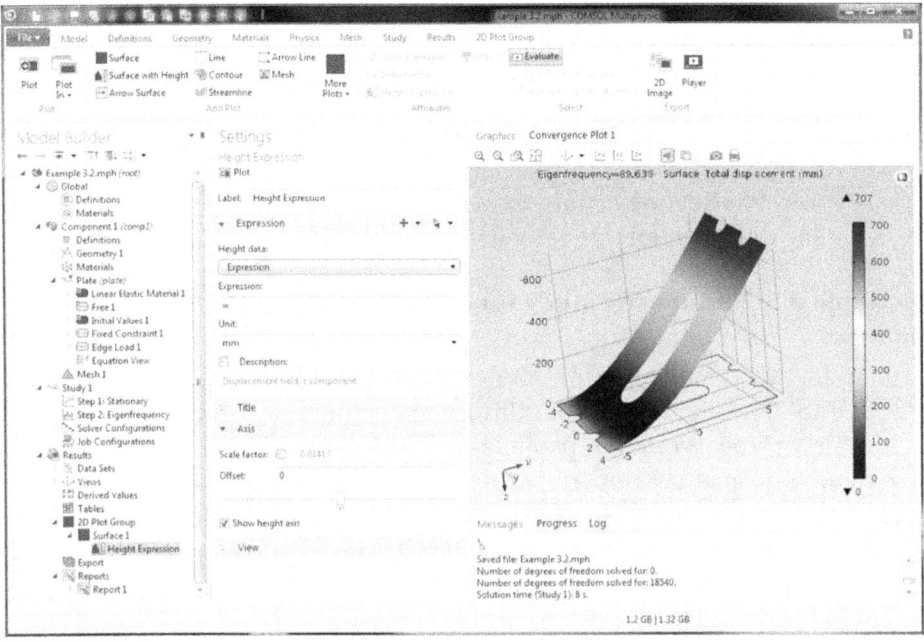

FIGURE 3.16 2D plot for fundamental eigenfrequency and modal shape displacement.

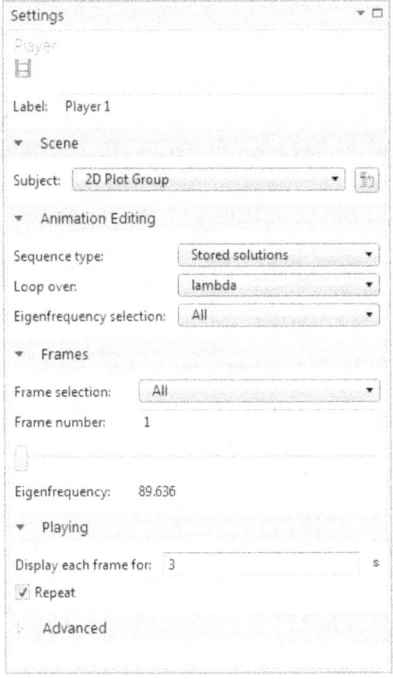

FIGURE 3.17 Data entry window for animation/Player for eigenfrequencies.

9. To see the movie, click on the Play icon located in the toolbar of the Graphics window. To stop the movie, click on the Stop icon. It is recommended to stop the movie occasionally to see the mode shapes at different modes. Alternatively, a specific frame can be shown by moving the slider bar in the Frames section.

10. Use the Reports feature to generate a report for Example 3.2, following the procedure described in Example 3.1.

Example 3.3: Parametric study for a bracket assembly: 3D stress analysis

In this example, we introduce Model Library, a useful resource in COMSOL. An example from this library (Structural_Mechanics_Module/Tutorial_Models/bracket_parametric[3]) will be rebuilt using COMSOL 5. Parametric analysis, importing existing CAD files into a model and setting up global parameters also will be explained. The model also demonstrates 3D stress analysis available through COMSOL tools applications. This model has been treated extensively for different loading and analysis types in Model Library.

For this example, we consider a bracket assembly made out of steel and held with bolts and that carries external loads applied through two pins, as shown in Figure 3.18.

External loads P applied on the inner surface of the two holes in the bracket arms have a sinusoidal distribution $P = P_0 \cos (\alpha - \theta_0)$, where P_0 is the maximum stress, θ_0 is the angle from the y-axis defining the orientation of the load in the plane of the holes, with $-\dfrac{\pi}{2} < \alpha - \theta_0 < \dfrac{\pi}{2}$. We will find the resulting von Mises stress distribution, and also build an App for this model.

Solution:

1. Launch COMSOL and open a new file, and save it as Example 3.3.

2. Click on the Model Wizard button and choose 3D from the Select Space Dimension window. In the Select Physics window, open the

---

[3] Model made using COMSOL Multiphysics® and is provided courtesy of COMSOL. COMSOL materials are provided "as is" without any representations or warranties of any kind including, but not limited to, any implied warranties of merchantability, fitness for a particular purpose, or noninfringement.

list under Structural Mechanics, click on Solid Mechanics (solid), and click on Add. Click on Study arrow button, select the Stationary node, and click on Done icon. Double-check to make sure that the file is saved.

3. To set up global parameters, click on Parameters, under the Model tab from the toolbar. The Settings Parameters window will open and display a table. Enter the table values as shown in Figure 3.19.

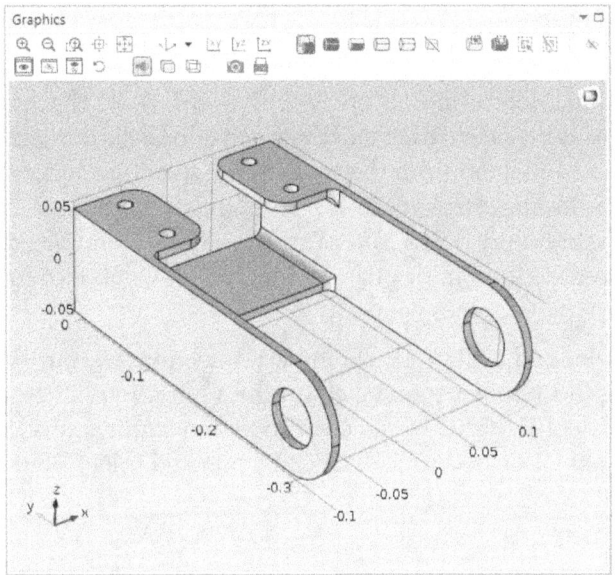

FIGURE 3.18 Geometry of bracket and applied-load pinholes.

Settings

Parameters

▼ Parameters

" Name	Expression	Value	Description
theta0	0[deg]	0 rad	load direction angle
R	25[mm]	0.025 m	Hole radius
YC	-300[mm]	-0.3 m	Y-coordinate of the hole center
P0	2.5[MPa]	2.5000E6 Pa	peak load/stress

FIGURE 3.19 Parameters data entry for Example 3.3.

Information listed under Description column is optional. Alternatively, these data could be uploaded from the companion CD.

4. To import the Geometry file, click on the Geometry tab from the toolbar and select Import from the ribbon list. The Import Settings window will open. In this window, click Browse and browse to the folder Structural_Mechanics_Module\Tutorial_Models. In the Tutorial_Models folder, double-click on the file bracket.mphbin, then click Import. See Figure 3.20. You should have the bracket geometry, as shown in Figure 3.18, in the Graphics window.

5. This model has complex loading on the bracket holes, specifically sinusoidal loads on half of the inner surface of the holes. To set up the loads we need to set variables and a local coordinate system. Click on the Model tab from the toolbar and select Functions; from the list select Local > Analytic. In the Settings window, enter the variables as shown in Figure 3.21. (*Note*: Variables used should be exactly the same as those defined in the Parameters table.) Click Plot to see the function in the Graphics window.

6. In order to facilitate the boundary selection, we put them in groups using the Explicit tool. Click on the Definition tab from the toolbar and select Explicit. In the corresponding settings window type Bolt 1 for Label. Locate Input Entities section and select Boundary from the

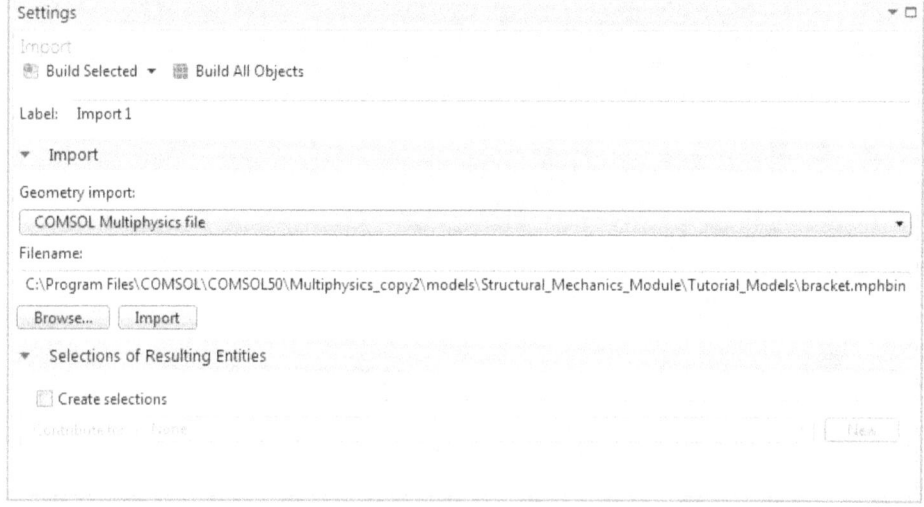

FIGURE 3.20 Importing a CAD file.

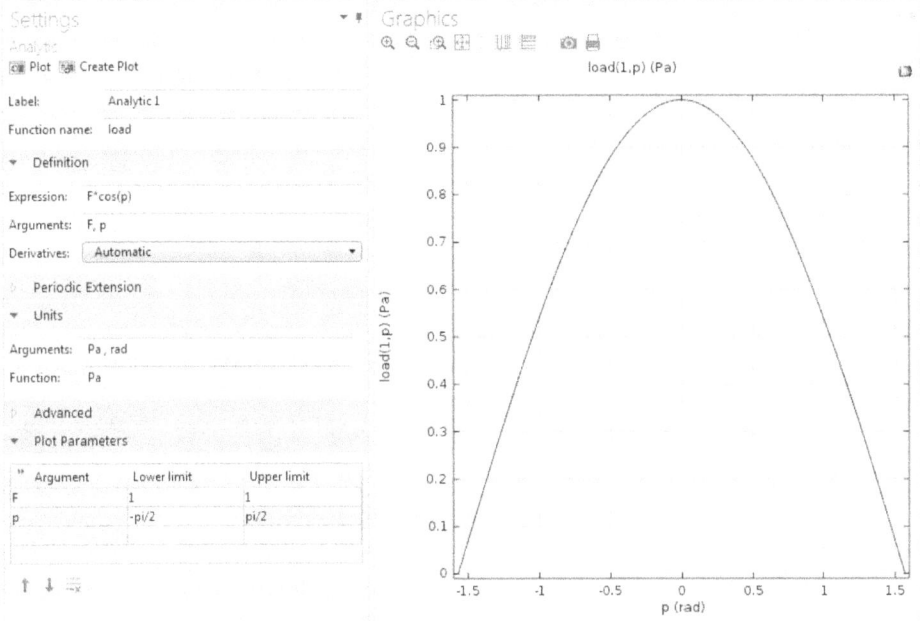

FIGURE 3.21 Data entry for Load function.

list for Geometric entity level. Select boundaries 18, 19 and add them to the list. Similarly, Create Bolt 2 (with boundaries 20, 21), Bolt 3 (with boundaries 31, 32), Bolt 4 (with boundaries 33, 34). Now we make a union of the bolts. Click on the Union in the Definition ribbon bar. In the Union settings window, enter Bolt holes for Label. Select Boundary for Level. Under the Selection to add window, click on add (i.e. + icon) and select Bolt 1–Bolt 4 from the list in the Add window, click OK.

7. To set up the rotating local coordinate systems for applied load on the holes, right-click on the Definitions node under Component 1 (comp1) in the Model Builder window and select Coordinate Systems > Cylindrical System. The Cylindrical System settings window will open. In this window enter data, as shown in Figure 3.22.

8. To assign materials to the bracket geometry, click on the Materials tab in the toolbar and then click on Add Material button. From the Add Materials window open the list under Built-In, click on Structural Steel, and select Add Material to Component. Note that check marks will appear in the Material Contents section list for Density, Young's

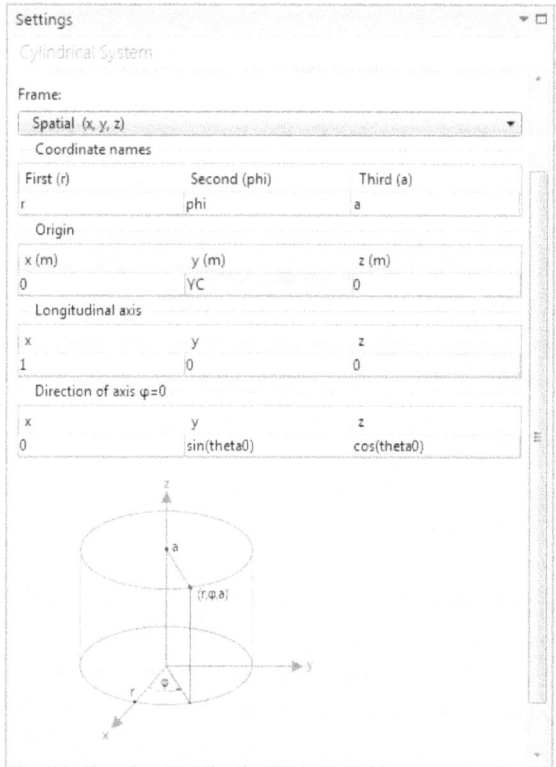

FIGURE 3.22 Data entry for local Cylindrical coordinates.

modulus, and Poisson's ratio (you may have to open this section by clicking on the pointer arrow) in the Settings window. Although other material properties are listed, only these three are required for the type of physics involved in this model. Click on Add Material button to close it.

9. To assign the boundary conditions, click on Physics tab from the toolbar; then click on Boundaries button and select Fixed Constraint. The Fixed Constraint settings window will open. In the Graphics window, locate Boundary Selection section and select Bolt holes from the list for Selection, as shown in Figure 3.23.

10. To assign the applied load, right-click on the Solid Mechanics (*solid*) node in the Model Builder window and select Boundary Load. The Boundary Load settings window will open. Click on boundaries 4 and 5 (only) of the inner surfaces of the right holes of the bracket (the

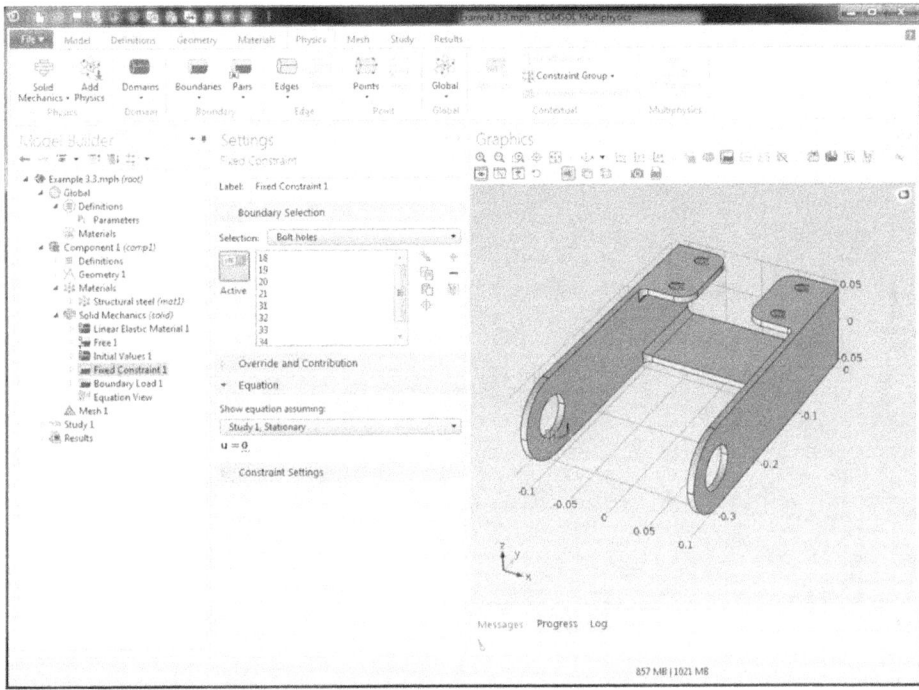

FIGURE 3.23 Data entry for Bolt holes as fixed boundary conditions.

one with negative x-coordinate value) in the Graphics window and add them to the Selection list. Under Coordinate System Selection section, open the list to see the options and choose Cylindrical System 2. In the Force section, type pressure (case-sensitive) for the r component, as load (−P0, sys 2.phi)*(abs(sys 2.phi) > pi/2). The symbols for applied forces should appear in the Graphics window. (Otherwise, to turn the physics symbols on, from the main toolbar click on File and select Preferences from the list. In the Preferences window click on the Graphics and Plot Windows and select the Show physics symbols check box, then click OK.) Similarly, create Boundary Load 2 and assign boundaries 42, 43 to it. For the force enter load (P0, sys 2.phi)*(abs(sys 2.phi) < pi/2). Results are shown in Figure 3.24.

11. Click on the Mesh node in the Model Builder. The Mesh window will open, as shown in Figure 3.25. In this window, select Normal for the Element size and click on the Build All icon. Wait until the mesh is created and appears in the Graphics window.

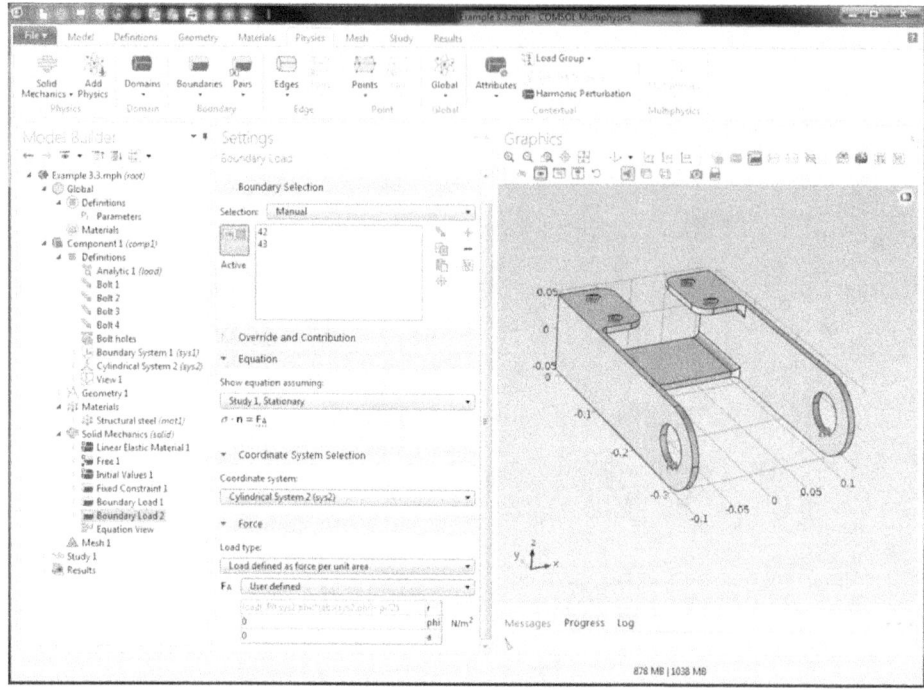

FIGURE 3.24 Data entry for Boundary Loads 1&2.

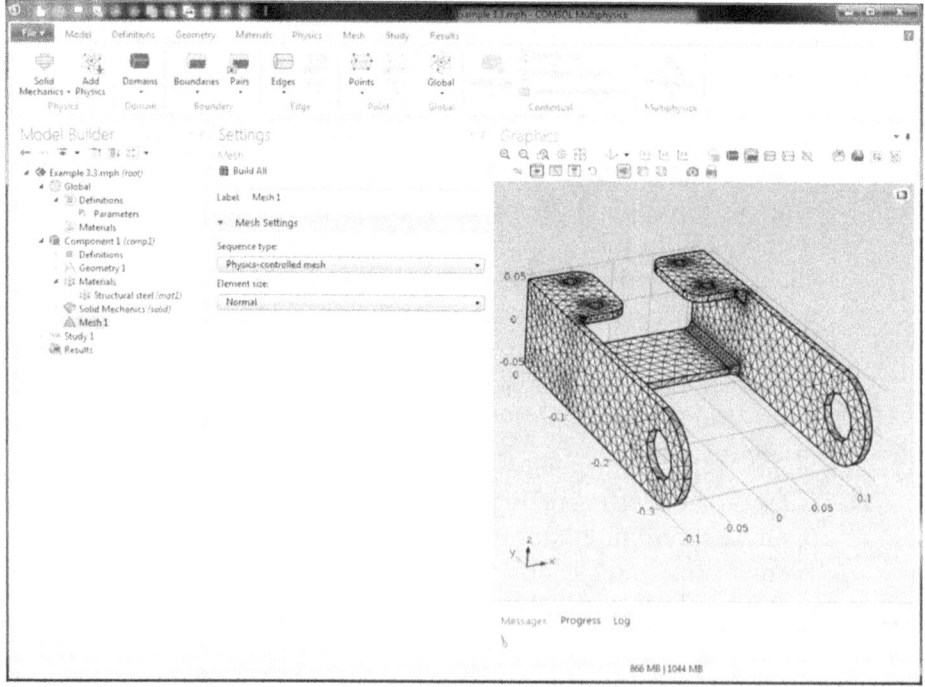

FIGURE 3.25 Mesh data entry window and the resulted mesh.

12. To run the model, we use a parametric analysis based on the load direction designated by angle theta 0. Click on Study 1: Stationary node. In the corresponding settings window locate Study Extensions section and expand it. Check the Auxiliary sweep box and click on Add (+) icon, from the list select theta 0 (load direction angle). Enter range (0, 10, 160) under the Parameter value list, and deg under Parameter unit. Select No Parameter from the list for Run continuation for. Click on Compute button.

13. The Default result von Mises stress distribution, for theta 0 = 160°, mapped on the deformed bracket, will appear in the Graphics window after the model run is finished, as shown in Figure 3.26. Maximum value is about 200 Mpa, which is less than yield stress value for structural steel (260 Mpa). This validates the choice of a linear elastic material and model to analyze this structure. Results for other values of theta 0 can be plot by choosing the desired value from the settings window for Stress (solid).

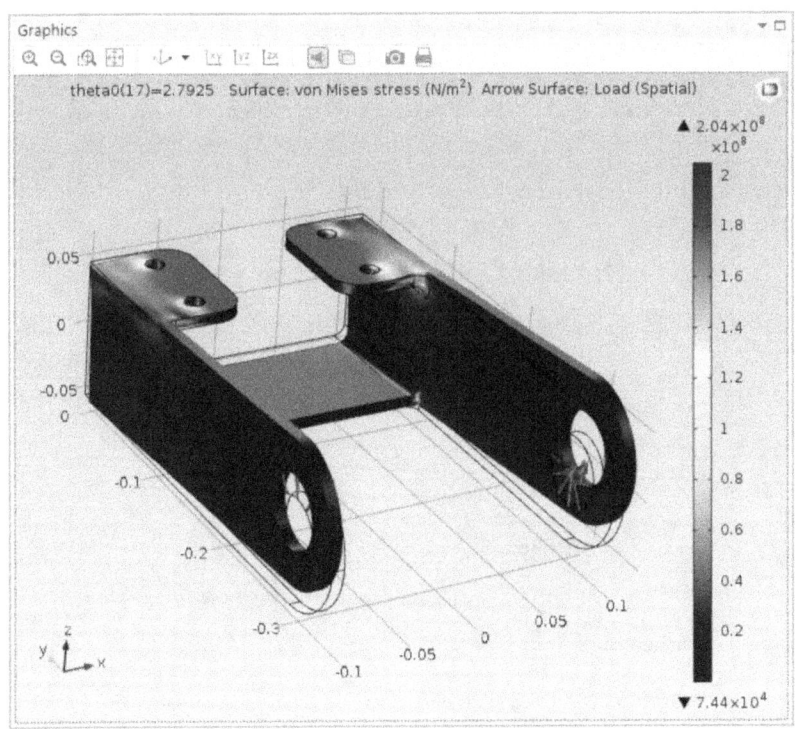

FIGURE 3.26 von Mises stress distribution window.

14. To show the direction of applied forces, right-click on the Stress (*solid*) node located under Results and select Arrow Surface. The Arrow Surface settings window will open. In this window, click on the Replace Expression icon and select Solid Mechanics > Load > Load (solid.FperAreax, ..., solid.FperAreaz) (Spatial) from the list. Click the Plot icon to graph the results. See Figure 3.27.

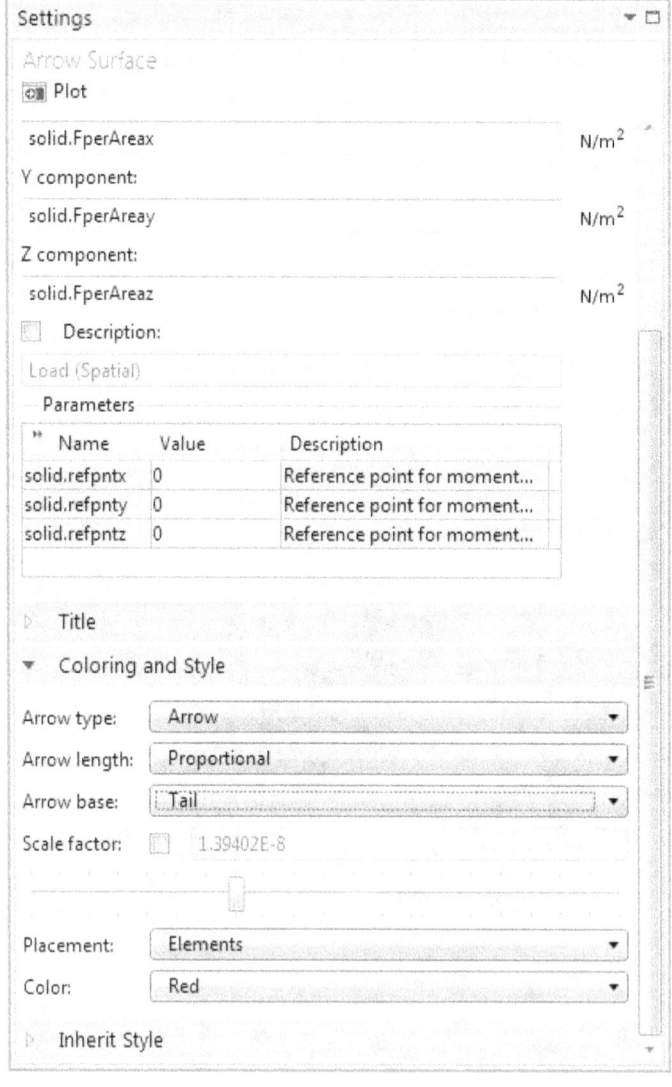

FIGURE 3.27 Applied loads direction settings for Arrow Surface Plot.

15. von Mises (σ_v) and principal stresses $(\sigma_i,\ i = 1, 2, 3)$ are related as $\sigma_v = \sqrt{\dfrac{(\sigma_1 - \sigma_2)^2 + (\sigma_2 - \sigma_3)^2 + (\sigma_1 - \sigma_3)^2}{2}}$. To show the principal stresses, right-click on Results and select 3D Plot Group. Right-click on the newly created 3D Plot Group 2 node in the Model Builder and select More Plots >Principal Stress Volume. The Principal Stress Volume settings window will open. In this window, under Positioning section enter the following data as shown in Figure 3.28 and then click the Plot icon.

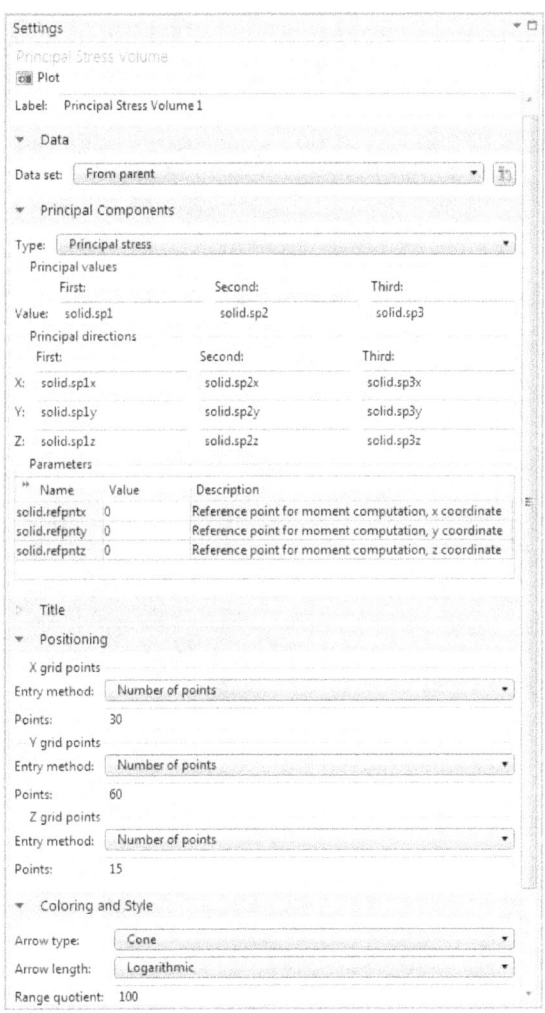

FIGURE 3.28 Principal Stress Volume data entry window.

16. In the Graphics window, click on the Zoom Extents icon. Principal stresses are shown with arrows (red the largest, green the medium, and blue the smallest—consistent with the coordinate axes). Similarly, principal stresses can be shown for other sections of the bracket.

17. It is useful to show the values of the reaction forces exerted by bolts. Right-click on the Derived Values node under Results in the Model Builder window and select Integration > Surface Integration. The Surface Integration settings window will open. In the Graphics window, click all bolts holes surfaces to select and add them to the list in the Selection section in the Surface Integration window. Under Expression, click on the Replace Expression icon and select Solid Mechanics > Reactions > Reaction force (Spatial) > solid.RFx-Reaction force, x component. Click the Evaluate icon. In the Expression section, type solid.Rfy and click Evaluate. Again, type solid.RFz in the Expression section and click Evaluate. The Values of the components of reaction force vector appear in a Table, under the Graphics window, as shown in Figure 3.29.

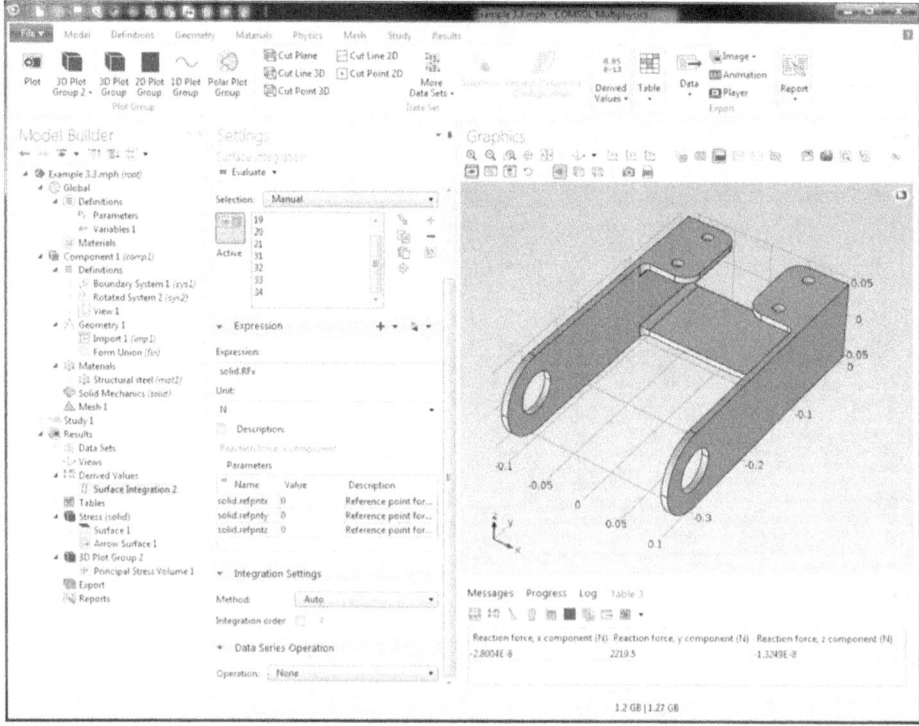

FIGURE 3.29 Surface Integration data entry window and reaction forces results.

18. To study the effect of the orientation of the applied load, click Step 1: Stationary node (under Study 1 in the Model Builder window). The Stationary settings window will open. In this window, click on the Study Extension section to expand it. Under Parameter value list, type range (0, pi/4, pi) This will change the value of parameter theta 0 from 0 to degrees with increments of $\dfrac{\pi}{4}$ degrees. See Figure 3.30.

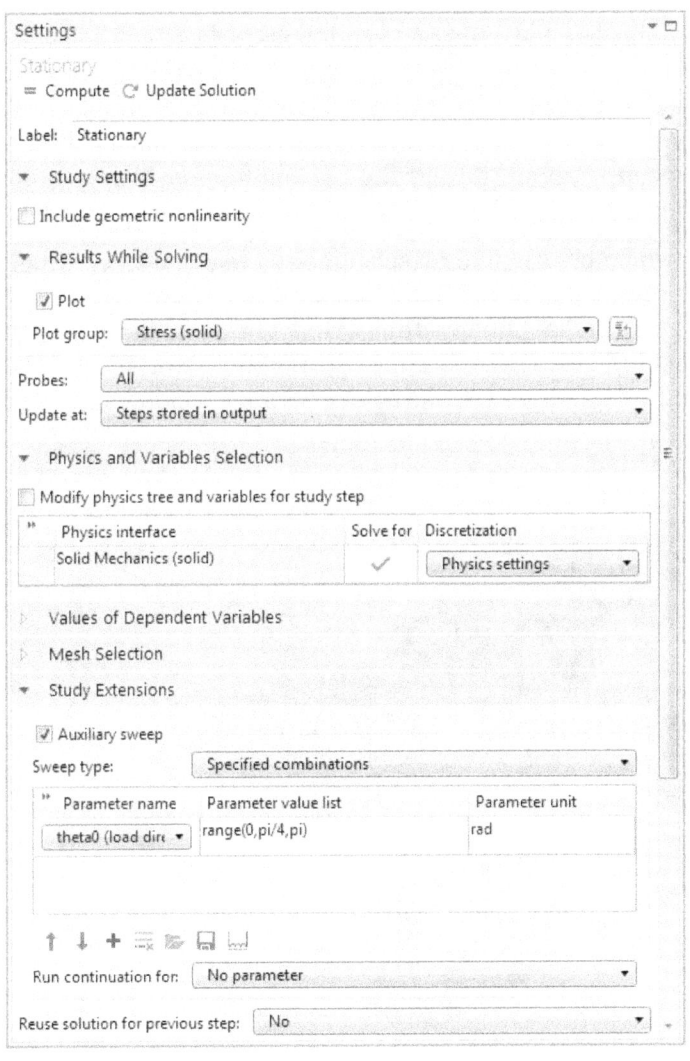

FIGURE 3.30 Study Extension data entry window for parametric sweep.

19. Click on the Study tab in the toolbar and select Compute. Wait for the model run to finish the calculations. The results will appear in the Graphics window. Click Stress (solid) node in the Model Builder and, from the corresponding window, under Data section, select 1.5708 from the list of Parameter values (theta 0) and click on the Plot icon. Note that the direction of applied forces (shown by red arrows) changes. Similarly, choose another value for the list (e.g. 45) for theta 0 and plot it. The result is shown in Figure 3.31.

20. To calculate the reaction (for all bolts, select Bolt holes) for different orientations of the applied load, right-click the Derived Values node (under Results in the Model Builder window) and select Integration > Surface Integration. Repeat the operations similar to those explained in Step 17 above. Results for reaction forces will appear in a Table, as shown in Figure 3.32.

21. It would be useful to make a graph of reaction forces. Right-click the Results node (in the Model Builder window) and select 1D Plot Group. In the 1D Plot Group settings window, expand the Legend section and choose lower right from the list. Right-click on the 1D Plot Group 4 and select Table Graph. In the Settings window under Data, locate

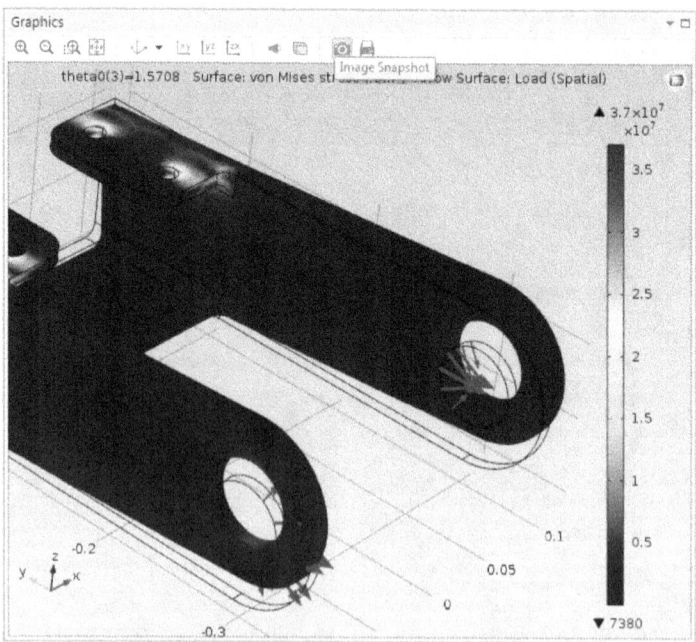

FIGURE 3.31 von Mises stress, applied loads, and deformed bracket geometry.

Table 2

theta0 (rad)	Reaction force, x component (N)	Reaction force, y component (N)	Reaction force, z component (N)
0.0000	7.3429E-8	6.8840E-7	-2.0330E-6
0.78540	4.0844E-8	-5.3961E-5	4.7472E-5
1.5708	-1.5642E-8	-1.9699E-5	7.0027E-8
2.3562	-6.2807E-8	-6.5320E-5	-4.4702E-5
3.1416	-7.3304E-8	-3.0114E-7	-3.5077E-6

FIGURE 3.32 Results for derived reaction forces at the Bolt holes.

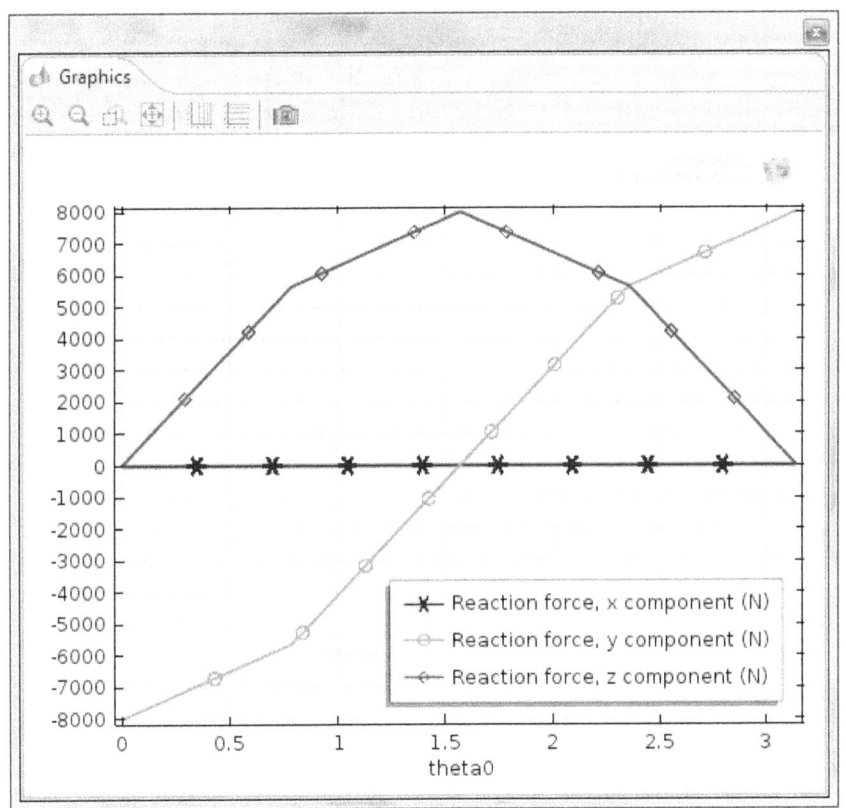

FIGURE 3.33 Reaction forces for different theta 0, orientation angle.

Table, open the list, and select Table 2. Under the Legends section, check the Show Legends box (you may need to expand this section to view the options). Click Plot. A graph showing the values of reaction forces (in N) versus theta 0 (in rad) will appear in the Graphics window. Results are shown in Figure 3.33.

22. Perform an approximate hand calculation and compare obtained results versus those. This verification/validation is left as an exercise for readers.

23. To build an app for this model, launch COMSOL if not already running. From File menu select New. In the New window select Applications Wizard. In the Select Model for Application window click on Browse, locate and open Example 3.3.mph. The New Form window will appear. In this window, select entries Heat transfer coefficient, Frequency, and ambient temperature under Parameters, from the Inputs/outputs tab and move them to Selected window. Click on Graphics tab; select and move Flow and Stress (fsi) to Selected window. Click on the Buttons tab; select and move Compute Study 1 to Selected window. See Figure 3.34. Click on Done. Form Editor desktop window appears. Save the file as App_ Example 3.3.

24. Several editing tools are available in the Form Editor window. We use some of these tools in order to lay out the app interface. Click on Grid from the ribbon bar to relocate the objects in the Form 1. Use the Settings for each object to modify the size, affiliate a picture, and specify the function. Refer to detailed instructions given for Example 2.1, for a guide. Final results for the app interface are shown in Figure 3.35.

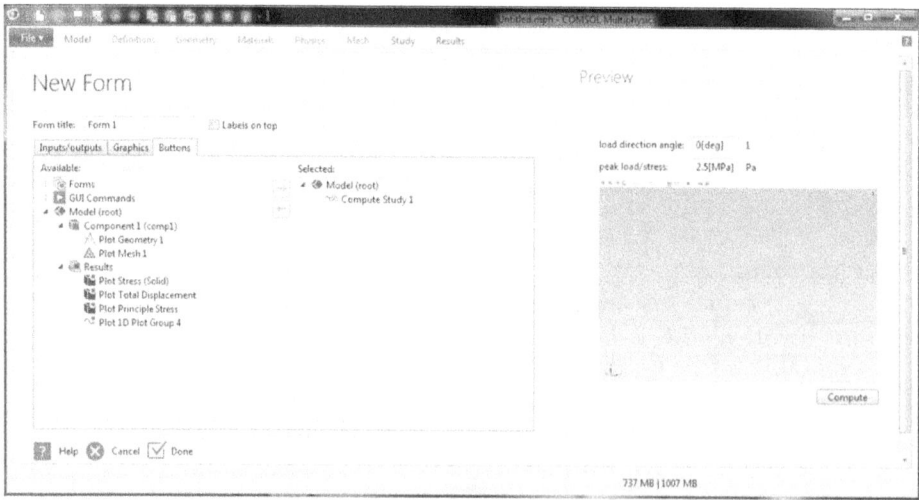

FIGURE 3.34 Application settings for Example 3.3.

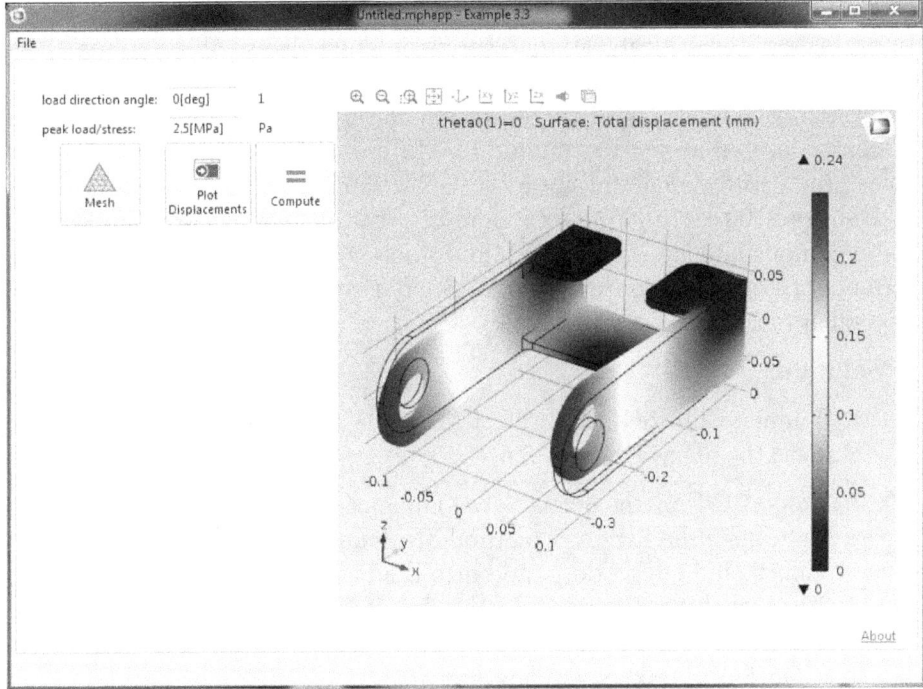

FIGURE 3.35 Application GUI for Example 3.3.

Example 3.4: Buckling of a column with triangular cross section: Linearized buckling analysis

A column is a structural member that supports applied load, mainly in compression. For certain values of an applied load and boundary conditions, the column may fail and exhibits very large deformations. The failure situations are examples of instability from neutral equilibrium condition for the given column geometry, material, and boundary conditions.

Buckling is categorized as a bifurcation problem from a mathematical point of view since it has more than one equilibrium situation when and after the column becomes unstable. One important point to emphasize here is that buckling is not a material failure type; it is the result of the column becoming unstable under a given load. A column could buckle for discrete values of applied loads. The smallest value—clearly of interest to engineers—is called the critical load (Euler formula: for a simply supported column with length L and moment of inertia I, $P_{cr} = \pi^2 \, EI/L^2$). Buckling could also be considered an eigenvalue/eigenvector problem and is one of the approaches available in COMSOL. Readers interested in more in-depth discussions on buckling are referred to the COMSOL help manual and [16].

In this example, we use a linearized buckling analysis that provides an estimate of the critical load that causes sudden collapse of the column. We calculate the critical load for a column with an equilateral triangular cross section (side length is 30 cm) and a circular hole with radius 4 cm with its center located at x = 15 cm, and y = 12 cm. For material properties, we use the existing material library in COMSOL, Aluminum 6063-T83 with Density 2700 kg/m^3, Young's modulus 69E9 Pa, and Poisson's ratio 0.33. To demonstrate alternative COMSOL tools available, we mainly work with model tree nodes from the Model Builder window, instead of menu and toolbar tabs.

Solution:

1. Launch COMSOL and select Model Wizard from the New window. Save the file as Example 3.4.

2. Select 3D from the Select Space Dimension. In the Select Physics window, open the list for Structural Mechanics, click on Solid Mechanics (*solid*), and click Add to add it to the Selected physics interfaces list. Then click Study arrow. In the Select Study window (under Preset Studies list) select Linear Buckling and click on the Done icon. The COMSOL Desktop window appears.

3. Click on Parameters, under the Model tab from the toolbar. Enter col_height under Name in the Parameters window/table and 200 [cm] under Expression. Similarly, enter col_load and R_hole under Name, and 100 [Pa] and 4 [cm], respectively, under Expression.

4. Geometry of the column can be built in COMSOL or alternatively imported as a file. In this example we build it using 3D CAD tools available in COMSOL.

    4.1. Right-click on the Geometry 1 node in the Model Builder window and select Work Plane from the list. The Work Plane settings window will open. Click on the Plane Geometry node (under Work Plane 1). Click on the Draw Line icon in the toolbar and draw a horizontal line in the Graphics window. (Click at a point in the drawing area, move the mouse pointer to a second point, and click. Right-click to finish.) Under Plane Geometry node, click on Bezier Polygon 1 node (you may have to open the list). The Bezier Polygon window will open. In this window, click on Segment 1(*linear*) in the Polygon Segments section. Change the coordinates of the line to (0, 0) and (30 [cm], 0), as shown in Figure 3.36, and click the Build Selected icon.

4.2. In the same window, click on Add Linear and enter the coordinates (30 [cm], 0) and (15 [cm], 0.15*3^0.5). Click the Build Selected icon. Click on the Zoom Extents icon to see all the geometry in the Graphics window. Result is shown in Figure 3.37.

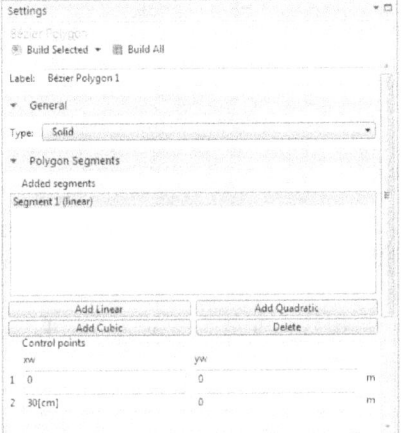

FIGURE 3.36 Line point coordinates data entry.

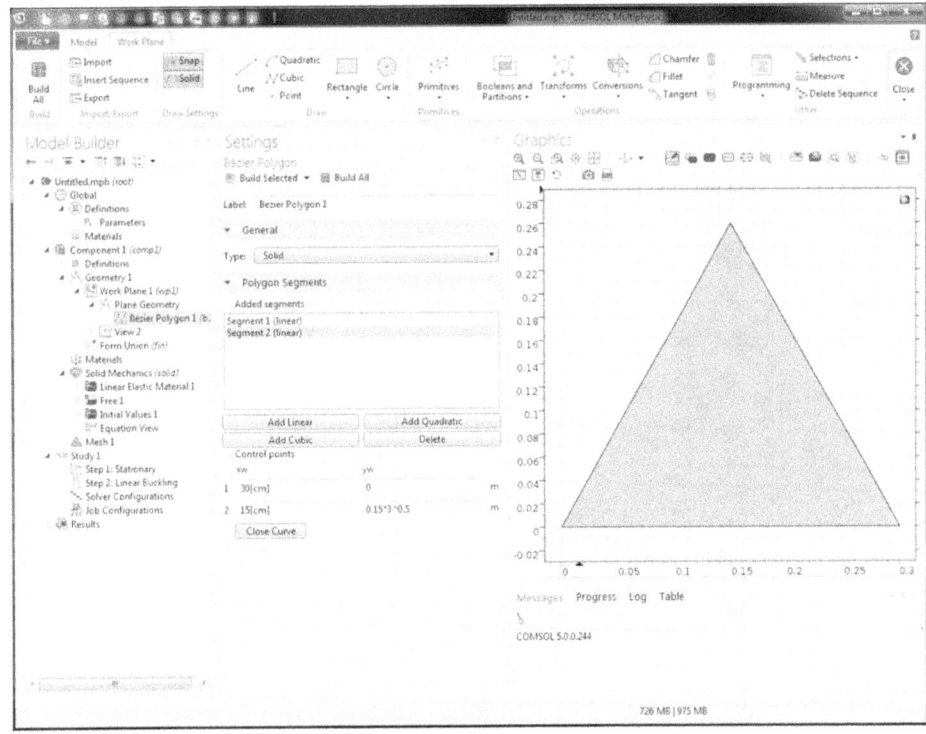

FIGURE 3.37 Control points data entry and resulting polygon geometry.

4.3. To add fillets, in the Model Builder window right-click on the Plane Geometry node and select Fillet. In the Fillet settings window, add the three vertices of triangle to the Vertices to fillet list by clicking on each one and then right-clicking. Enter 1 [cm] for the Radius and click Build Selected. The filleted triangle will appear in the Graphics window, as shown in Figure 3.38.

4.4. To create the circle, draw a circle inside the triangle by right-clicking on the Plane Geometry node and selecting Circle from the toolbar. In the Circle settings window, enter 4 [cm] for the Radius, 15 [cm] for xw, and 12 [cm] for yw. Click on the Build Selected icon. To create the hole inside the triangle, click on the Select Objects icon (located on the Graphics window toolbar) and Alt + Click at any point in the Graphics area. The entire geometry is selected. Right-click on the Plane Geometry node and select

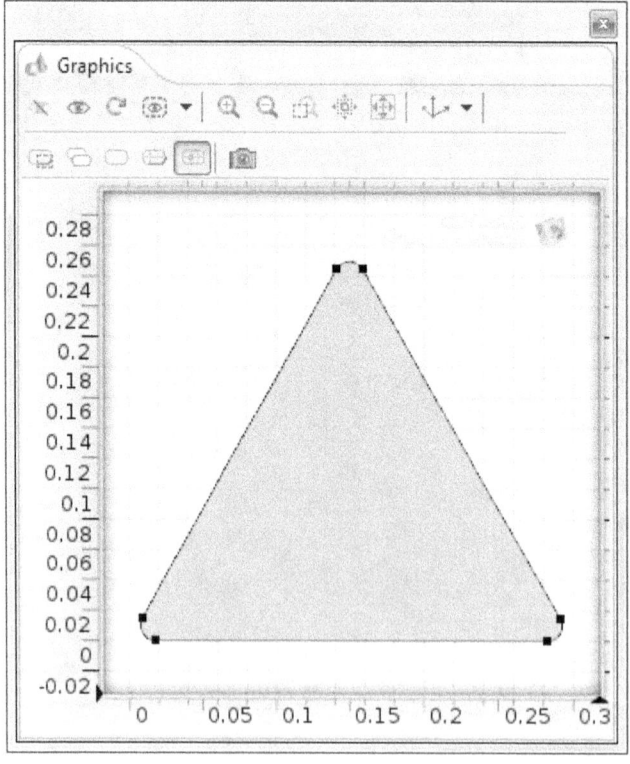

FIGURE 3.38 Column cross-section geometry.

Booleans and Partitions > Difference. The triangle with a circular hole appears in the Graphics window, as shown in Figure 3.39.

4.5. To extrude the cross section, right-click on Work Plane 1 and select Extrude. The Extrude settings window opens. In this window under Distance, enter col_height and click on the Build Selected icon. Click on Zoom Extents in the Graphics window to see the entire Column geometry. See Figure 3.40.

5. To assign materials to the geometry, right-click on the Materials node in the Model Builder and select Add Material. In the Add Material window, open the list for Built-In, right-click Aluminum 6063-T83, and select Add to Component 1. Close Add Material window. See Figure 3.41.

6. Define a global parameter for the applied load on the column. Click on the Parameters node in the Model Builder window. In the Parameters window table, type col_load under Name and 100 [Pa] under Expression.

7. To define the boundary conditions, right-click on the Solid Mechanics (*solid*) node in the Model Builder window and select Fixed Constraint from the list. Assign this constraint to the base of column by clicking

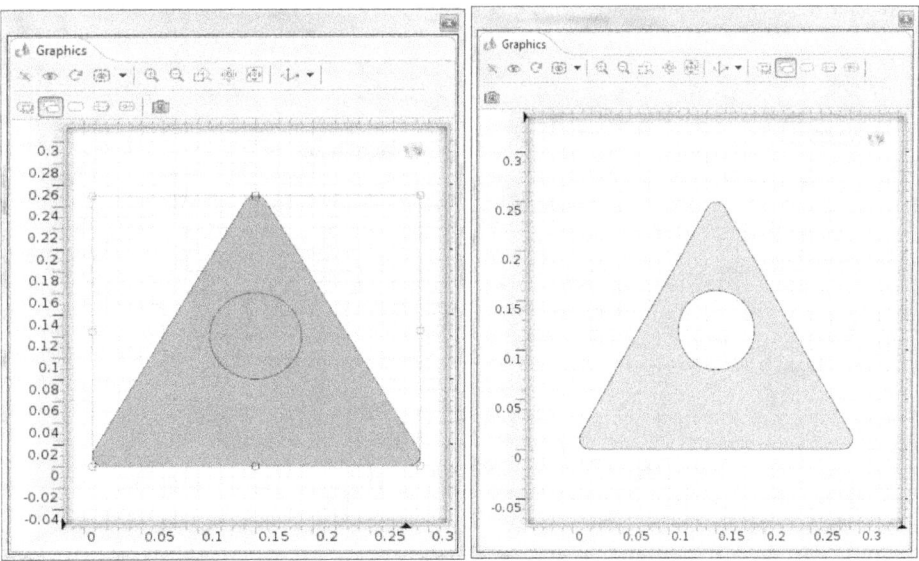

FIGURE 3.39 Column cross-section geometry with subtracted circular hole.

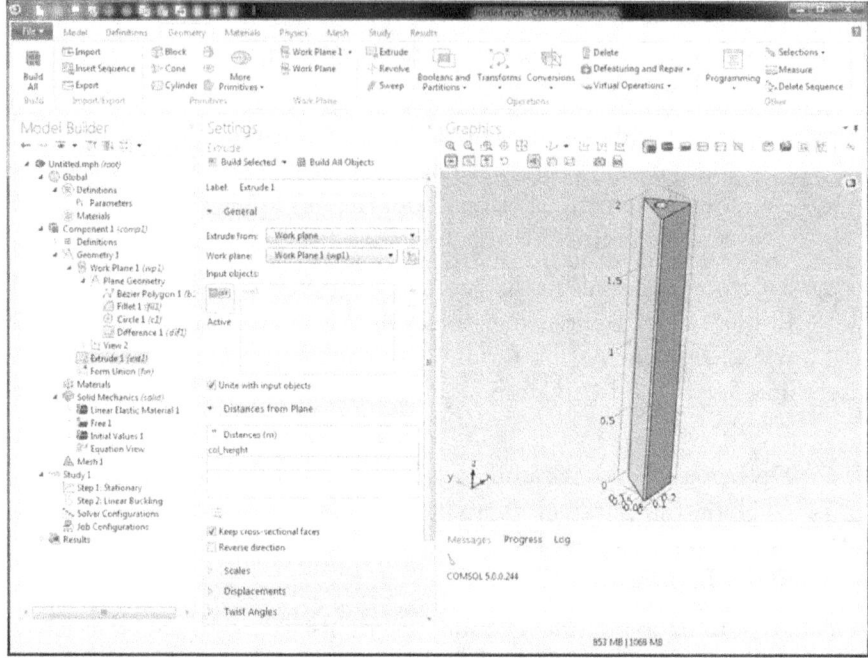

FIGURE 3.40 Graphics window showing column geometry.

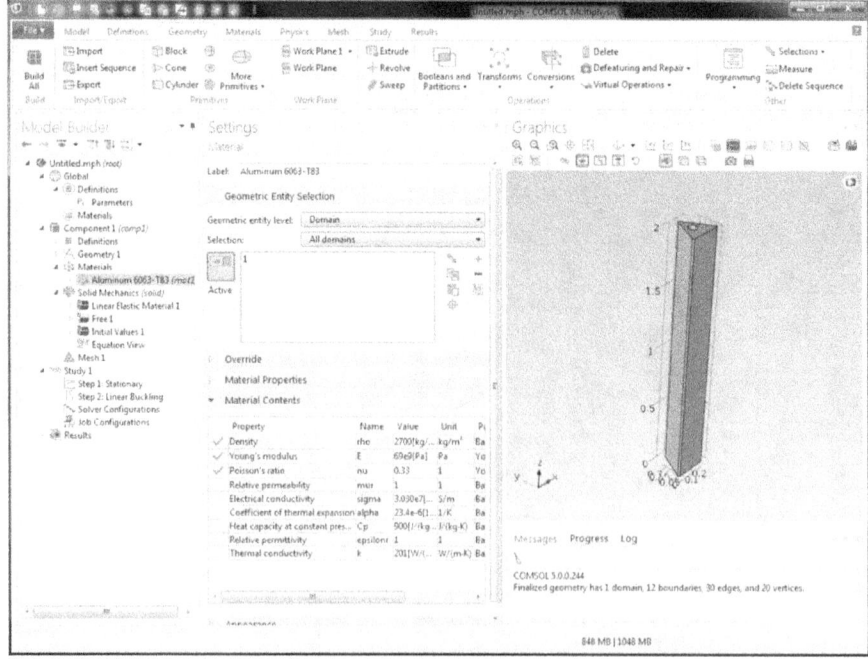

FIGURE 3.41 Material data entry for the Column.

on the base surface (located in the x-y plane) of the column, and add it to the Selection list in the Settings window. Users may need to rotate the column geometry to see the base. To define the applied load on the top surface of the column, right-click on the Solid Mechanics (*solid*) node and select Boundary Load from the list. Assign this load to the top of the column by clicking on the top surface (located in the z = col_height plane) of the column, and add it to the Selection list. In the Boundary Load settings window, locate Force section and enter – col_load (the minus sign is needed since load is applied in the negative z-direction) for the z component, as shown in Figure 3.42.

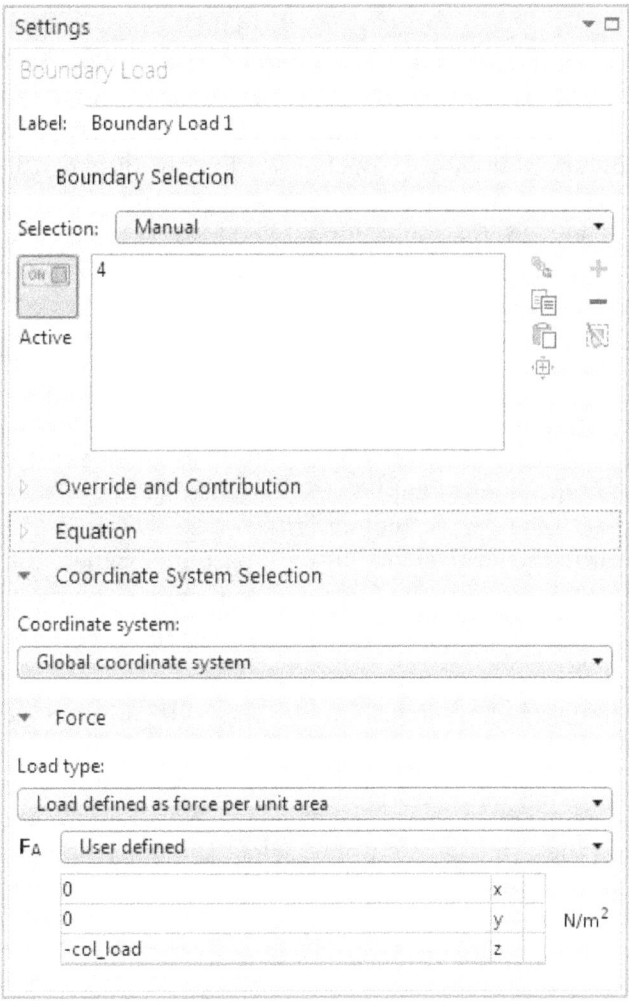

FIGURE 3.42 Boundary Load data entry.

8. Click on Mesh 1 node. In the Settings window click on Build All, with accepting all default options.

9. To run the model, right-click on the Study 1 node and select Compute. The solver will first solve the equilibrium equations and then find the first eigenvalue or the critical load factor. The result for first mode shape of buckling and value of critical load factor (1.6862E6) is shown in the Graphics window.

10. To calculate the critical load of buckling for this column with assigned load factor and boundary condition, use $P_{cr} = \lambda * col\_load$, where λ is the critical load factor. This gives $P_{cr} = (1.6862\ E6 \times 100\ Pa) = 168.62\ Mpa$.

11. To perform a parametric analysis for the applied loads, right-click on the Study 1 node in the Model Builder and select Parametric Sweep. The Parametric Sweep window opens. In this window under Parameter names, click on the (+) icon to display a list and select col_load. In the space under Parameter value list, enter 100, 200, 300. These are the values for applied load on the column. To run the model, right-click on the Study1 node and select Compute. Wait for computations to finish.

12. The Results appear in the Graphics window. Click on the Mode Shape (*solid*) 1 node in the Model Builder under Results node. In the corresponding settings window, type in Mode Shape (solid)-parametric load for the Label. Open the Parameter value (col_load) list (the three values of applied load are listed here). Choose, for example, 200. The corresponding critical load factor (i.e. 8.4312E5) will appear in the Critical load factor (space under the Parameter value). To plot this result in the Graphics window, click on the Plot, as shown in Figure 3.43. Similarly, the results for other loads can be obtained and plotted.

13. Now we build an Application for this mode; save the file again. Click on File from the File menu and select New. In the New window click on Application Wizard button. Click on Browse and locate the file Example 3.4.mph; click Open. The New Form window will open. Under the Tab Inputs/outputs, select and move parameters Column height (col_height), Column load (col_load), and Column hole

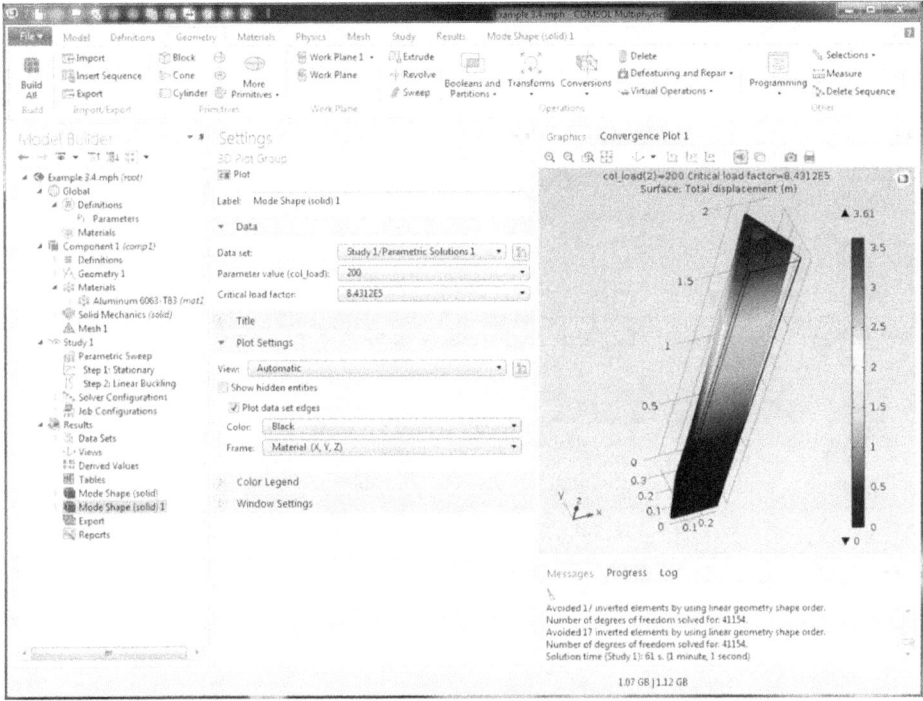

FIGURE 3.43 Results for buckled column displacement and critical load factor.

radius (R_hole) from the list in Available window to the Selected window.

14. Click on Graphics tab. Select and move Geometry 1 and Model Shape (solid) from the Available list to the Selected area. Click on the Buttons tab; select and move Plot Geometry 1 and Compute Study 1 from the Available list to the Selected area. Results are shown in Figure 3.44. Click on Done button; Form Editor Desktop window will open.

15. In the Form Editor window, rearrange the Plot Geometry and Compute buttons and bring them to the top of the window. Also resize the Graphics window, shown. Click on Test Form button, located in the ribbon toolbar, to see the preview of the Form1. Click on Test Application button; to see a preview run of the Application, change some of the parameters and run the Application to make

FIGURE 3.44 New Form windows for Selected items for Input/outputs, Graphics, and Buttons (clockwise from top left).

sure that it works as per design. When satisfied with the final layout of the Form1, save it as App-Example 3.4. Users may want to use the tools available in Form Editor, to change the fonts, button sizes, etc.

16. Now we open and run the Application independently. Launch COMSOL 5 and click on File >Run Application. Locate and open App_Example 3.4.mphapp file. The Application will open, as shown in Figure 3.45. Change the parameters, for example type in 180 [cm] for Column height, 3.5 [cm] for Column hole radius. Click on Plot Geometry to see the new geometry in the graphics window. Run the Application by clicking on the Compute button. The Application will run using the new parameters. After computation is done, the new load factor is given as 6.82E5.

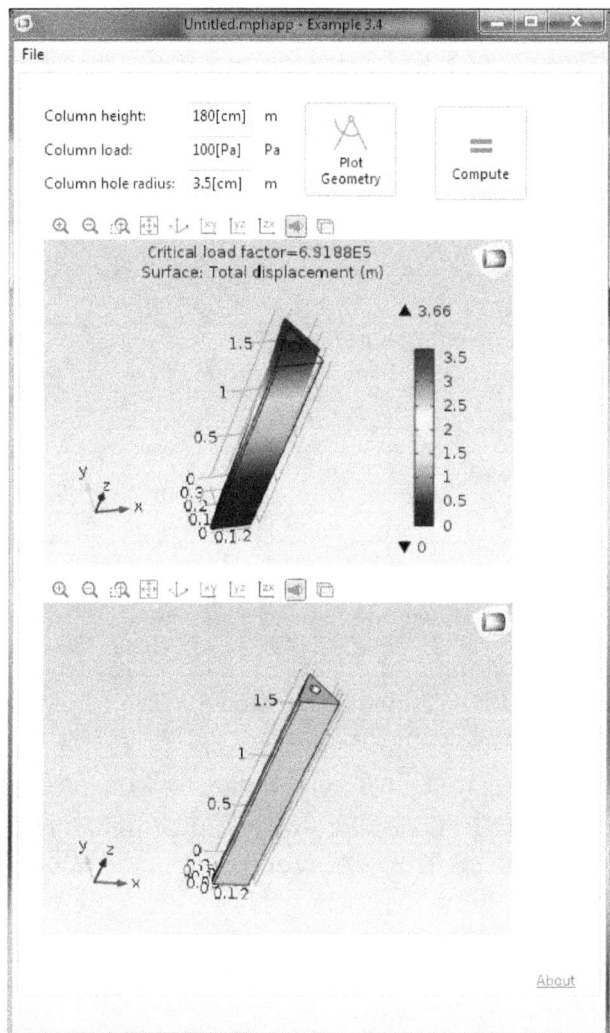

FIGURE 3.45 App GUI for model Example 3.4.

Example 3.5: Static and dynamic analysis for a 2D bridge-support truss

A truss is a structure that carries loads applied to it and results in compression or tension in its members. In this example, we use tools available in COMSOL to analyze a typical bridge structure under static and dynamic loads, including calculating its fundamental natural frequency. Truss dimensions (in feet) are given in the Figure 3.46.

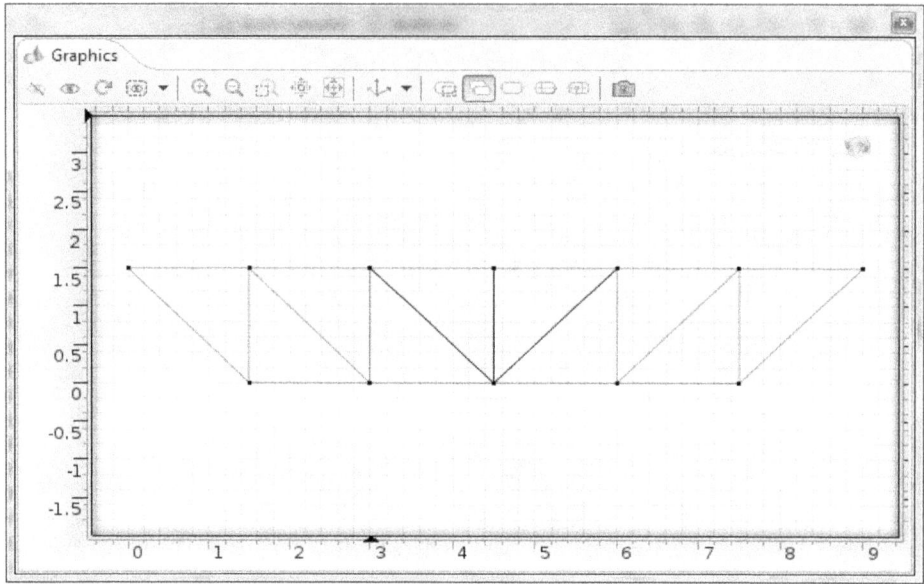

FIGURE 3.46 Geometry of the truss.

Solution:

1. Launch COMSOL and click on Model Wizard in the New window. Save the file as Example 3.5.

2. In the Select Space Dimension window, click on the 2D button.

3. From the Select Physics list, expand Structural Mechanics node and from the list select Truss (truss) and click on the Add button. Click on Study arrow button.

4. From the Select Study window, select Stationary and click Done. At this stage the physics, type of study, and level of dimension (2D) of the model are set for the truss.

5. Click on the Model tab from the toolbar and select Parameters. In the corresponding settings window enter the data as shown below:

Settings				▾ ☐
Parameters				
▾ Parameters				
" Name	Expression	Value	Description	
point_loadDy	-8e3[N]	-8000 N	vertical load at point D	
xsection_area	2*3e-3[m^2]	0.006 m²	top and bottom members-x-sectional area	
diag_xsection_area	3e-3	0.003	diagonal members-x-sectional area	

Now we draw the geometry of the truss in the Graphics window using CAD tools available in COMSOL.

6. Since the dimensions are given in feet, change the Length unit to ft, from the list in the Units section in the Geometry settings window. Note that all units will be automatically converted to the SI units. Click on the Geometry tab from the toolbar, select Line located in the ribbon, and draw a horizontal line in the Graphics window. To draw a line, click on a point, then drag the mouse to another point and then click again (to release the draw line tool, right-click). In the Model Builder window, expand the Geometry node and click on the Bezier Polygon 1(*b1*). In the corresponding window, click on Segment 1 (linear) and enter coordinates for x: 0 and y: 1.5, x: 9 and y: 1.5. Click the Build All Objects icon to draw the line segment AG.

7. Using the same tool, draw the rest of truss members, as shown in Figure 3.47, using the following data:

Node	A	B	C	D	E	F	G	H	I	J	K	L
x: (ft)	0	1.5	3	4.5	6	7.5	9	1.5	3	4.5	6	7.5
y: (ft)	1.5	1.5	1.5	1.5	1.5	1.5	1.5	0	0	0	0	0

8. To assign material to the model, click on the Materials tab from the toolbar and click on the Add Material button. In the Add Material window expand Built-In and select Structural steel from the list. Click on Add to Component. Click on the Add Material button, in the ribbon bar, to close it. Numerical values of material properties will be listed in the Material Contents section, as shown in Figure 3.48. These values can be modified, if needed. In this example, we accept the default values. In the Settings window select Boundary from the list for Geometric entity level, to add all truss members to the list.

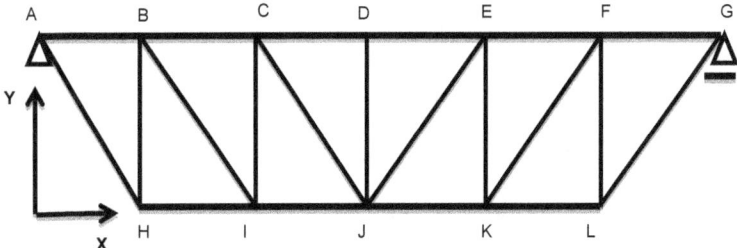

FIGURE 3.47 Geometry of the truss and its joints coordinates.

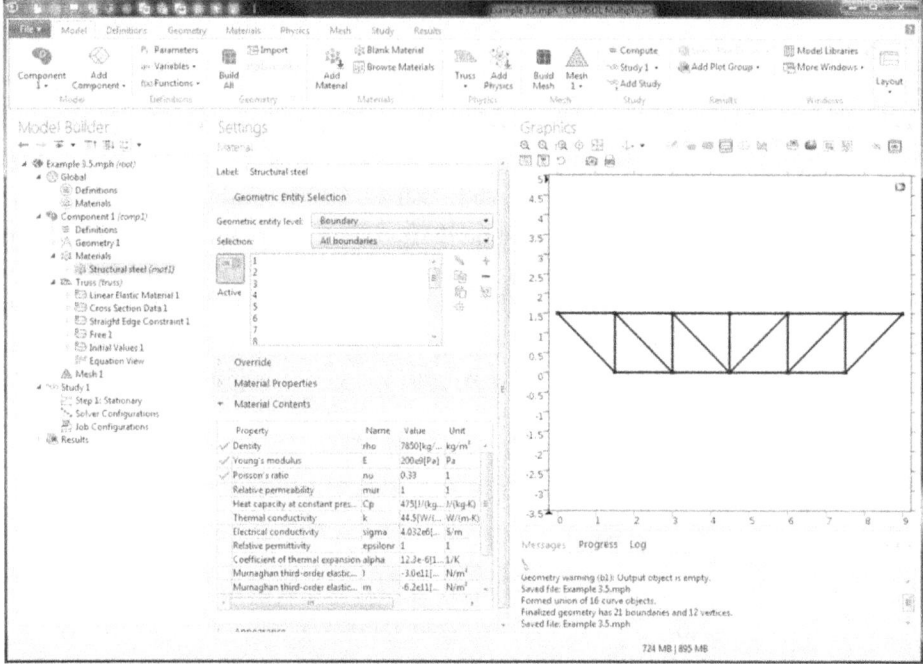

FIGURE 3.48 Material properties data entry for the truss members.

We have built the truss geometry and assigned material properties. We now define boundary conditions, applied loads, and truss members shape properties. We refer to the truss sketch for nodes.

9. Click on the Cross Section Data node located under Truss (*truss*) node. In the corresponding settings window, locate Cross Section Data and enter 3e−3 for Area. This is equal to the area of two 4 × 4 × 5/8" in. angle shape steel members. Create another Cross Section Data node and enter xsection_area for its Area value. Select and add all truss members associated with the upper and lower horizontal sides. The diagonal members would then automatically assigned to Cross Section Data 1.

10. To add point loads, right-click on the Truss (*truss*) node and select Point Load from the list. In the corresponding settings window, as shown in Figure 3.49, select and add the top-middle node (node D, see Figure 3.47) to the Selection list. Locate Force section and enter 0 for x and point_load Dy for y. Similarly, create another Point Load and select nodes B and C to it. For the load values, enter 2e3 * 0.1 for x and −2e3 for y.

11. For boundary conditions, click on the Physics tab from the toolbar and from the list under Points select Pinned. In the Settings window, add

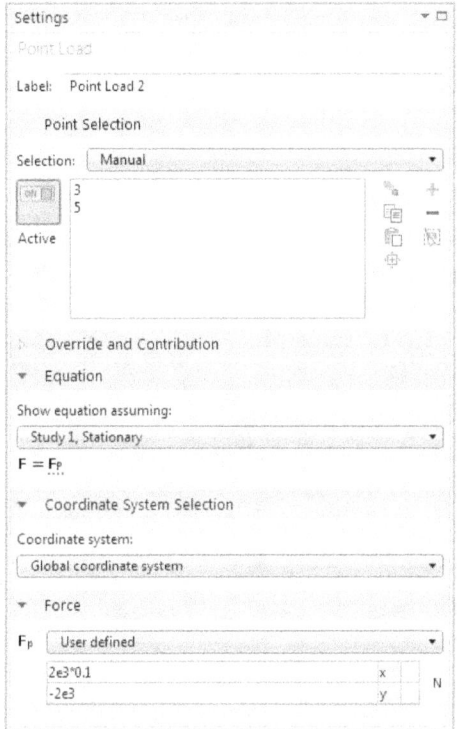

FIGURE 3.49 Point load data entries.

node A to the Selection list. This will assign a pinned support condition for this node that has x = y = 0 for its displacements. For the other support node G, we constrain displacement in y-direction to be zero, y = 0. Right-click on the Truss (*truss*) node and select Prescribed Displacement from the list. From the Graphics window, select the corresponding node for node G and add it to the Selection list in the Prescribed Displacement settings window under Point Selection section. In the same window, locate the Prescribed Displacement section and check the box for Prescribed in y direction only. Make sure the value for U_{0y} is set to 0, as shown in Figure 3.50.

At this stage, we have the model ready for meshing. For truss structures since each member can actually be considered as a finite element, the mesh resolution mesh is irrelevant. COMSOL considers each truss member as an element, regardless.

12. Click on the Mesh 1 node in the Model Builder window and in the corresponding settings window select Extra coarse from the list for

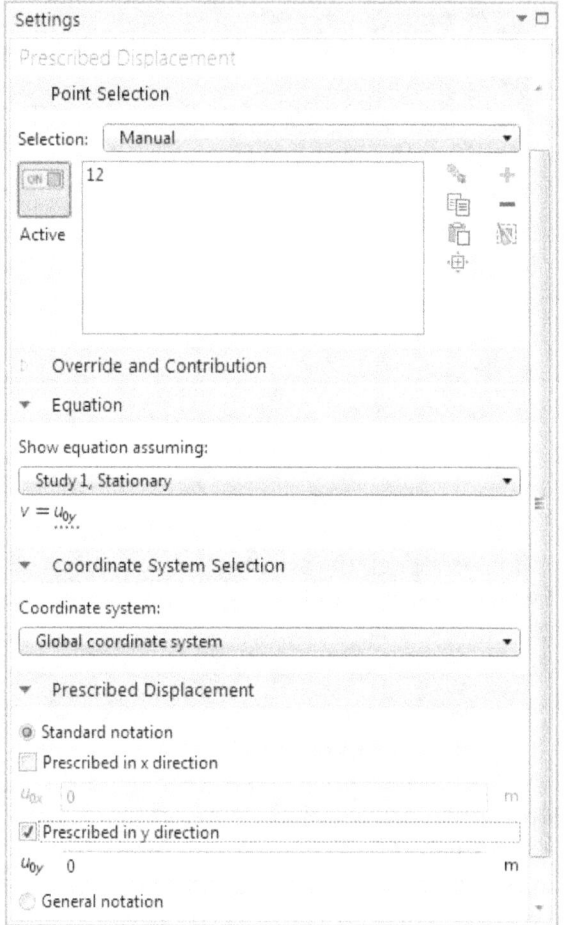

FIGURE 3.50 Boundary condition for prescribed displacement data entries.

Element size. Click Build All. Check the mesh built. Try this with a finer resolution and you will find out that again each truss member is considered as an element.

13. To run the model, right-click on the Study 1 node in the Model Build window and select Compute. Wait for the computations to finish.

14. Default results for normal forces for members will appear in the Graphics window, as shown in Figure 3.51, corresponding to Force (truss) node under the Results. Also, normal stresses for truss members can be shown by clicking on corresponding node Stress (truss).

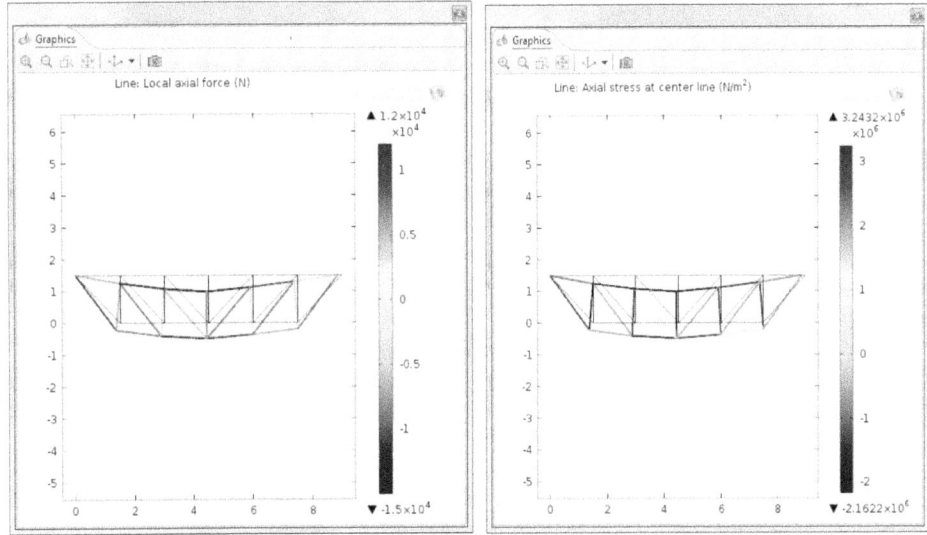

FIGURE 3.51 Results for truss members axial loads and stresses.

15. To show the displacement results, right-click on Results and select 2D Plot Group. A new 2D Plot Group node will appear under the Results tree. Rename the new node to Displacements. Right-click on Displacements and select Line. In the corresponding settings window, click on the Replace Expression icon and select Truss > Displacement > truss.disp-Total displacement. For the Unit select mm from the list. Click Plot. To show the deformed shape, right-click on Line 1 and select Deformation. Results are shown in Figure 3.52.

At this point static analysis is complete. For design purposes, a user can modify the loads or materials and design the truss to meet a design Code criterion. Building an app would be very useful for parametric analysis, for this model.

Now we perform dynamic analysis by adding studies to the existing model. To start, we calculate the natural frequencies of the truss. Usually the natural frequencies are calculated without any applied load; hence, the results are a solution to a homogeneous form of the governing equations. For some structures, compression or tension forces may affect the natural frequencies. For comparison, we calculate the natural frequencies with applied loads, as well.

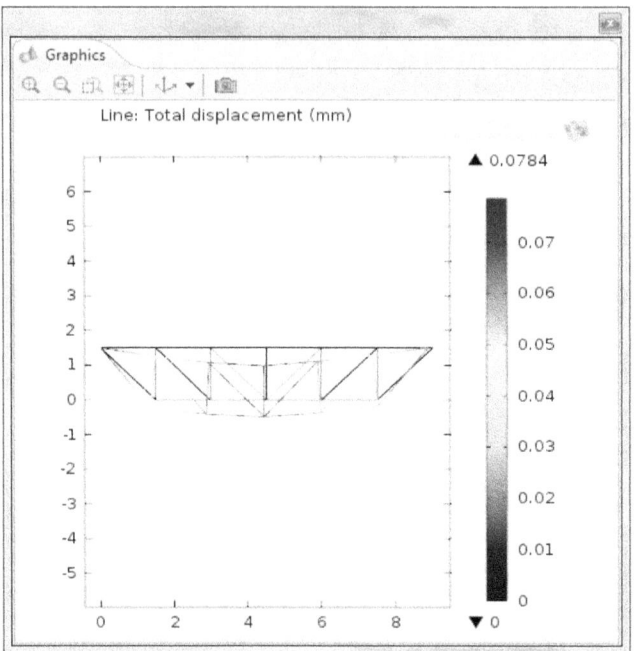

FIGURE 3.52 Results for truss joints displacements.

16. First rename the Static study case and its corresponding Solution 1 to Study1-Static and Solution 1-Static, respectively. Solution 1 node is under Solver Configurations.

17. Click on the Study tab from the toolbar and select Add Study. In the Add Study window, select Eigenfrequency and click on the Add Study icon. A new Study node will appear in the Model Builder window. Rename it Study 2-Eigenfrequency w/o load. Right-click on the Study 2-Eigenfrequency w/o load node and select Compute. After calculations are done, rename the corresponding Solution to Solution 2-Eigenfrequency w/o load.

18. Default results show the first eigenfrequency (132.63 Hz) and the corresponding displacements for modal shapes, as shown in Figure 3.53. Change the units for displacements to mm. To show the results for any one of the other five eigenfrequencies, simply click on Mode Shape (truss) node in the Model Builder window and from the Settings window locate Data section and select one from the list, click Plot. Results for first and sixth eigenfrequencies (132.63 and 929.71 Hz) are shown below.

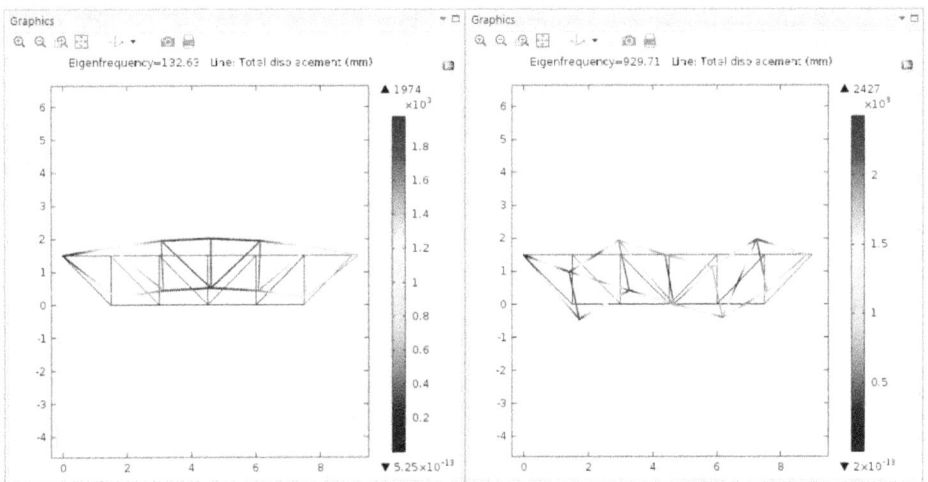

FIGURE 3.53 Results for truss modal shape displacement for 2 of its Eigenfrequencies.

19. To calculate the eigenfrequencies with loads, click on the Study tab from the toolbar and select Add Study. In the Add Study window, select Prestressed Analysis, Eigenfrequency, and click on the Add Study icon. A new Study node will appear in the Model Builder window. Rename it Study 3-Eigenfrequency with load. Click on the Compute button, under Study in the toolbar. After calculations are done, rename the corresponding Solution to Solution 3-Eigenfrequency with load.

20. The default results for first eigenfrequency (132.62 Hz) with corresponding displacements show up in the Graphics window. Since the difference between eigenfrequencies with and without loads is very small, we can ignore the effect of loading for calculating natural frequency of the truss.

The first dynamic excitation of this truss (fundamental modal shape) happens at the frequency of 132.63 Hz. In other words, if a dynamic load with the same frequency is applied to this truss, then resonance will happen and the truss will exhibit displacements with very large values. It would be useful to study the behavior of this truss for a range of harmonics, or applied loads with a range of frequencies including the fundamental natural one. We will do this in the following steps:

Click on the Study tab from the toolbar and select Add Study. In the Add Study window, select Frequency Domain and click on the Add

Study icon. A new Study node will appear in the Model Builder window. Rename it Study 4-frequency domain and expand it. Click on the Step1: Frequency Domain node. In the corresponding window, enter 20, 60, 99, range (100, 3, 140), 150, 200, 300, 320 for the Frequencies in the Study Settings section. These are harmonics for a series of applied dynamic loads. We have a higher resolution around 132.63 Hz to capture the resonance. Click on Compute. After calculations are done, rename its corresponding Solution to Solution 4-frequency domain.

21. The default results for member normal forces and stresses will appear in the Graphics window. Forces are shown in Figure 3.54, for the last frequency at 320 Hz. Results for harmonics can be shown by selecting the desired value from the list of Parameter values (freq (HZ)) in the Plot settings window.

22. It would be useful to draw the displacement of a point, say, at the middle of the truss (node D) for a range of harmonics. Right-click on the Results node in the Model Builder window and select 1D Plot Group. A new 1D Plot Group 1 will appear in the model tree. Rename

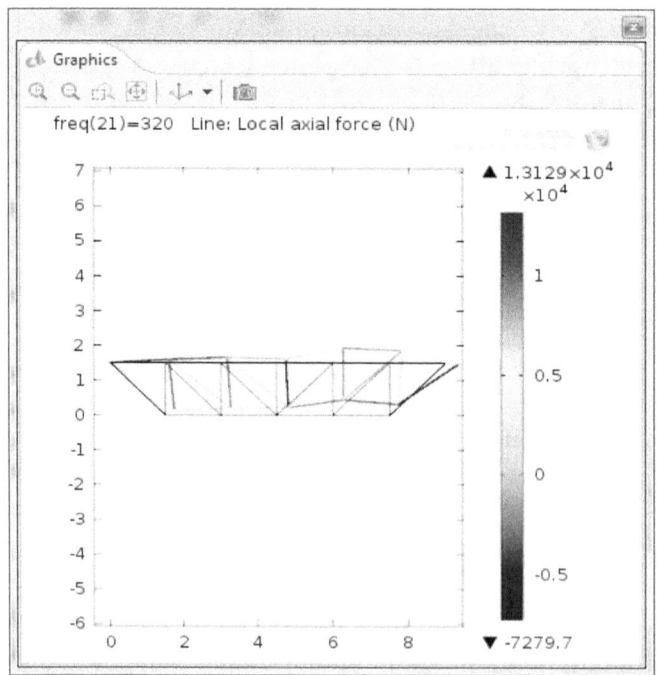

FIGURE 3.54 Results for truss members axial forces at 320 Hz.

it Displacements for harmonics. Right-click on this node and select Point Graph. In the corresponding settings window, select Study 4-frequency domain from Data set and nodes representing D, H, and F from the Graphics window and add them to the Selection list. Change the unit to mm under Unit. Click Plot. The results for displacements for the range of harmonics will appear in the Graphics window, as shown in Figure 3.55.

These results show the resonance close to 133 Hz. So far for this model we have not introduced damping to the vibration. In practice damping is the capacity of the structure to absorb and dissipate part of the energy applied to the structure. We add the damping either by adding extra dampers or it is provided by the material used as a result of intrinsic friction. In COMSOL, damping can be added to the model using Rayleigh Damping or Loss Factor Damping. By using an isotropic structural loss factor of 0.01, we introduce damping to the model.

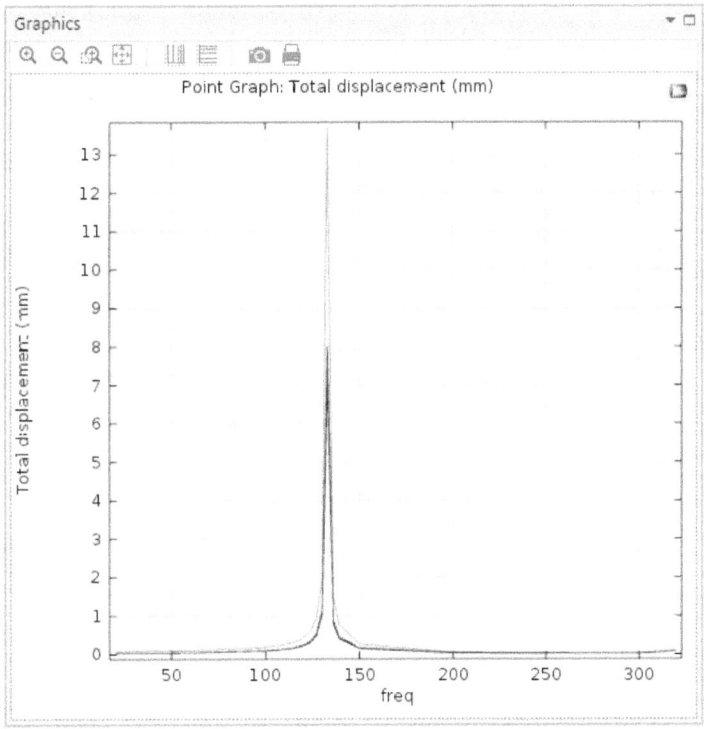

FIGURE 3.55 Total displacement at joint D, H, and F for different values of harmonic loads. The resonance is clearly shown.

23. Expand the Truss (*truss*) node in the Model Builder window, right-click on Linear Elastic Material 1, and select Damping. From the Damping window, locate Damping settings section and select Isotropic loss factor from the list for Damping type. Select From material for Isotropic loss factor. Now we need to add the loss factor to the list of material properties. Expand Materials node and click on Structural steel (*mat 1*). In the corresponding settings window, from the list under Material Contents, enter 0.01 for the value of eta_s, Isotropic structural loss factor, as shown in Figure 3.56.

24. To run the model, click Compute. Wait for computations to finish.

25. Click on Displacements For harmonics to see the results for damped values for displacements, as shown in Figure 3.57.

26. When damping is added to the mode, COMSOL automatically considers it for eigenfrequency calculations, as well. In order to eliminate this, expand Study 2-Eigenfrequency w/o load node and click on Step1: Eigenfrequency. In the corresponding settings window, expand the Physics and Variables Selection and check the box for Modify physics tree and variables for study step. From the list, locate Damping 1 and click on it. Then click the Disable icon, as shown in Figure 3.58.

27. Repeat Step 25 for Study 3-Eigenfrequency with load and save the model.

28. Now we build an Application for this mode; save the file again. Click on File from the File menu and select New. In the New window click on the Application Wizard button. Click on Browse and locate the file Example 3.5. mph; click Open. The New Form window will open. Under the Tab Inputs/outputs, select and move parameters from

▼ **Material Contents**

Property	Name	Value	Unit
✓ Density	rho	7850[k...	kg/...
✓ Isotropic structural loss factor	eta_s	0.01	1
✓ Young's modulus	E	200e9[...	Pa
✓ Poisson's ratio	nu	0.33	1

FIGURE 3.56 Window for adding damping as Isotropic loss factor.

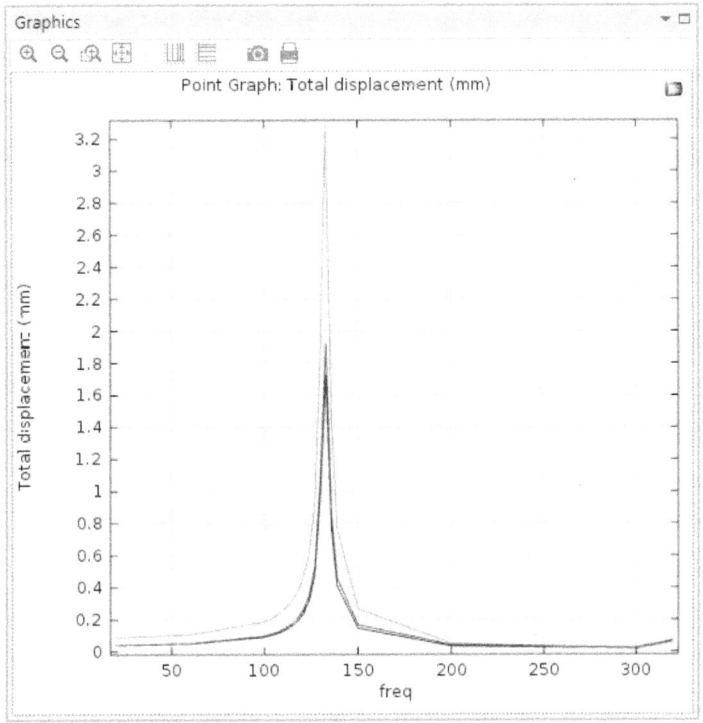

FIGURE 3.57 Total displacement at joint D, H, and F with damping for different values of harmonic loads.

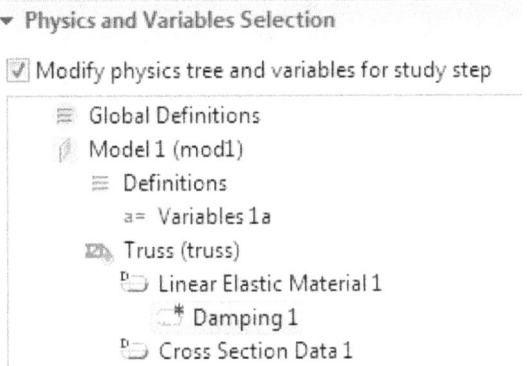

FIGURE 3.58 Window for modifying physics for damping.

the list in the Available window to the Selected window, as shown in Figure 3.59. Click on Graphics tab, and select and move Force (truss), Mode Shape (truss) w/o load, and Displacements for harmonics to the Selected window. Click on the Buttons tab; select and move Compute

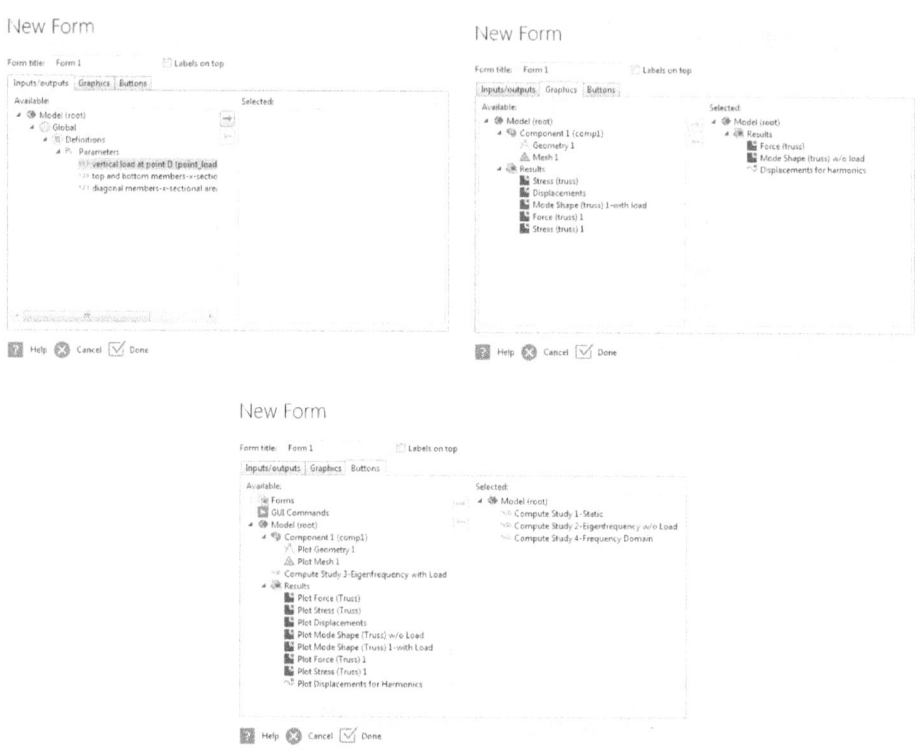

FIGURE 3.59 New Form windows for Selected items for Input/outputs, Graphics, and Buttons (clockwise from top left).

Study 1-Static, Compute Study 2-Eigenfrequency w/o Load, and Compute Study 4-Frequency Domain to the Selected window. Click on Done button; Form Editor Desktop window will open.

29. In the Form Editor window, rearrange the Plot Geometry and Compute buttons and bring them to the top of the window. Also resize the Graphics window, shown. Click on the Test Form button, located in the ribbon toolbar, to see the preview of the Form1. Click on the Test Application button to see a preview run of the Application; change some of the parameters and run the Application to make sure that it works as per design. When satisfied with the final layout of the Form1, save it as App-Example 3.5. Users may want to use the tools available in Form Editor, to change the fonts, button sizes, etc.

30. Now we open and run the Application independently. Launch COMSOL 5 and click on File > Run Application. Locate and open App_Example 3.5.mphapp file. The Application will open, as shown in

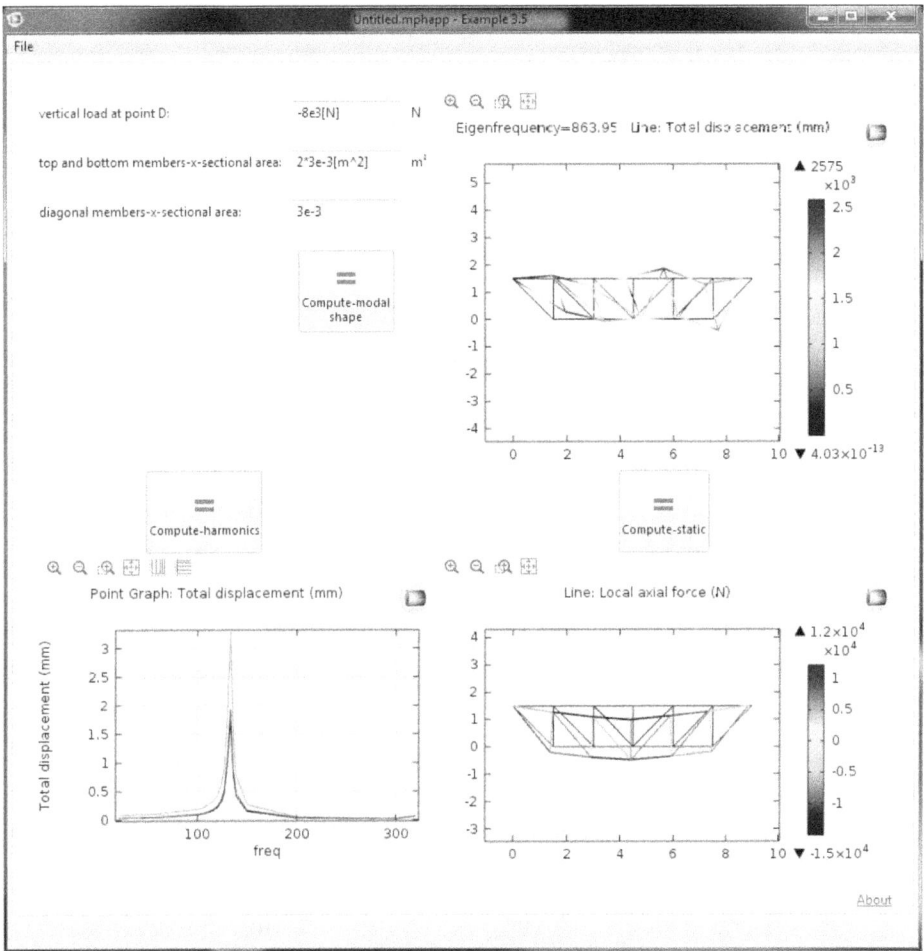

FIGURE 3.60 App GUI for model Example 3.5.

Figure 3.60. Change the parameters, for example, type in −10e3 [N] for vertical load at point D, 4e−3 for diagonal members-x-sectional area. Run the Application by clicking on Compute buttons. The Application will run using the new parameters.

Example 3.6: Static and dynamic analysis for a 3D truss tower

In this example, we use tools available in COMSOL to analyze a 3D structure under static load. We also calculate its natural frequencies. Shape and dimensions are given in Figure 3.61 and Table 3.1.

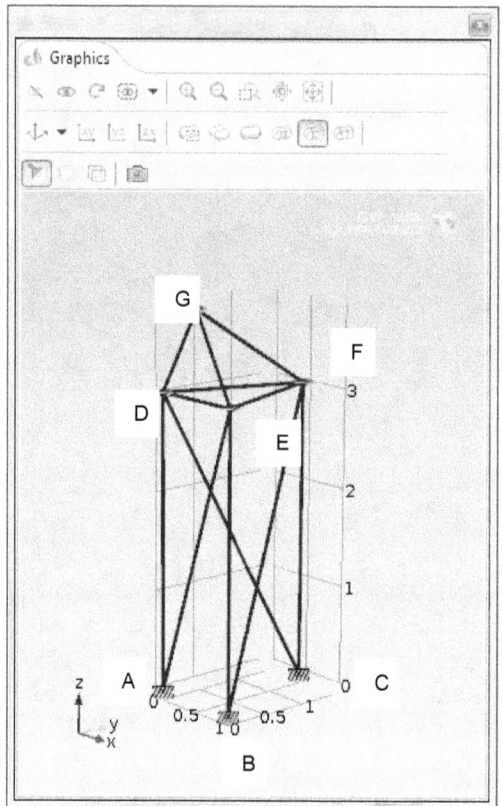

FIGURE 3.61 Geometry for the 3D truss.

Node	X (m)	Y (m)	Z (m)
A	0	0	0
B	1	0	0
C	0.5	$\sqrt{2}$	0
D	0	0	3
E	1	0	3
F	0.5	$\sqrt{2}$	3
G	0.5	0	$3+\sqrt{3}/2$

TABLE 3.1 Joint Coordinates for the 3D Truss

Solution:

1. Launch COMSOL and in the New window click on Model Wizard. Save it as Example 3.6.

2. In the Select Space Dimension window, click on the 3D button.

3. From the Select Physics list, expand the Structural Mechanics node and from the list select Truss (truss), then click on Add. Then Click on Study arrow icon.

4. In the Select Study window, select Stationary and click on Done.

   Now we draw the geometry of the truss in the Graphics window using CAD tools available in COMSOL.

5. Click on the Geometry tab from the toolbar and select More Primitives > Bezier Polygon. In the corresponding settings window, locate Polygon Segments section and click Add Linear. Enter the coordinates for joints A&D for member AD. Click Build Selected. Rename Bezier Polygon 1 to Bezier Polygon AD. Right-click on Bezier Polygon AD (b1) and select Duplicate. In the corresponding settings window,

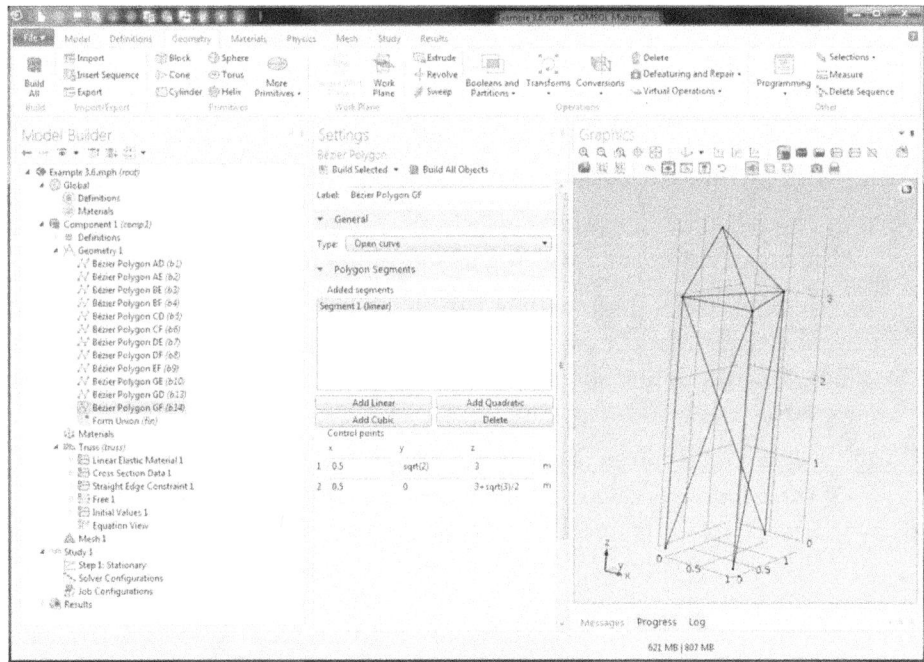

FIGURE 3.62 Geometry of truss in Graphics window.

enter the coordinates for nodes A&E. Similarly, draw the remaining truss members, total of 14 members. Results are shown in Figure 3.62.

6. Add materials to the model. Click on the Materials tab from the toolbar and then click on the Add Material button. In the Add Material window, expand Built-In and select Aluminum. Click on the Add to Component icon. In the Settings window add all truss members to the Selection list. Close Add Material window.

7. Click on the Physics tab from the toolbar and select Cross Section Data under Edges. A new node will be added to the model tree; rename it L2×2×3/8 in. In the corresponding settings window, change Selection to All edges and enter 8.8E−4 in the space provided for A, under Area.

8. To define the boundary conditions (support types) and point loads, click on Points and select Pinned. In the corresponding settings window, select joints representing A, B, and C from the Graphics window and add them to the Selection list. Similarly, select point Load and in the corresponding settings window select the node that represents D and add it to the Selection list. In the same window, enter 150 for the value of y in the Force section. Similarly, add three more point loads as follows: for point G, force components are (0, 350, −150); for point E, force components are (0, 200, 0); and for point F, force components are (−300, 0, 0).

9. Click on the Study tab from the toolbar and select Compute. Wait for computations to finish.

10. The default result will appear in the Graphics window showing truss members forces and stresses, as shown in Figure 3.63.

Although from these results we can read normal forces of the truss members using the legend, it would be useful to have the exact numerical values. We will define some variables for normal forces and extract their values from the results database.

11. Click on the Results tab from the toolbar and select Derived Values > More Derived Values > Line Maximum. A new node, Line Maximum 1, appears in the model tree window, under Derived Values node. In the corresponding settings window add select and add any truss member to the Selection list, for example, member BF. Locate Expression section and change the Expression to truss.Nxl. Click on the Evaluate icon. The axial force for member BF appears in the Table window, as 1.896 kN. Results are shown in Figure 3.64.

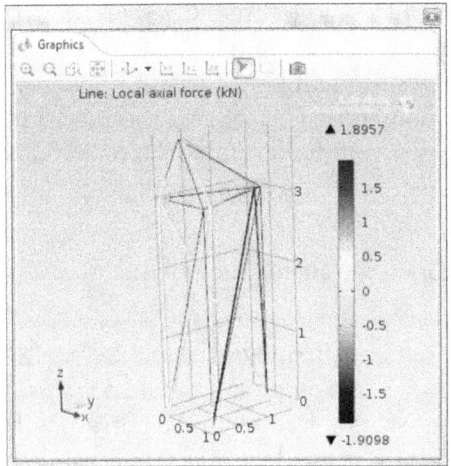 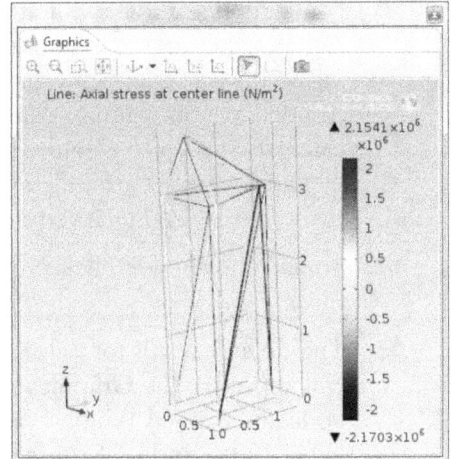

FIGURE 3.63 Results showing axial forces and stresses for truss members.

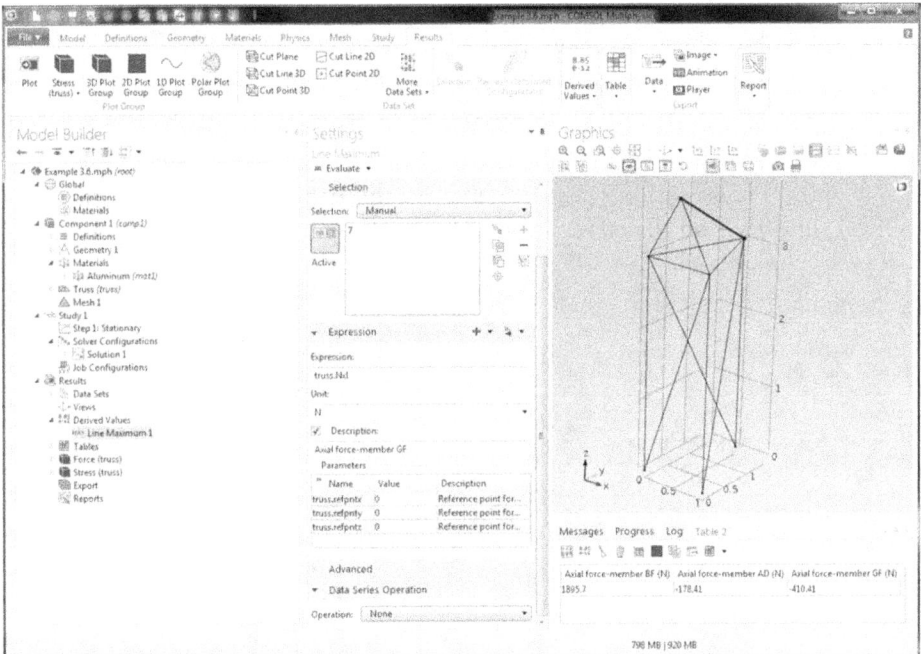

FIGURE 3.64 Evaluation of member BF, AD, GF axial forces.

12. Now we would like to calculate the natural frequencies of the space truss. Click on the Study tab from the toolbar and then click on Add Study. In the Add Study window click on Eigenfrequency from the list and click on Add Study.

13. A new node Study 2 will appear in the model tree. Click on Step 1: Eigenfrequency (located under Study 2 node) and in the corresponding settings window enter 6 for Desired number of eigenfrequencies (if needed, this number can be modified). This will set the model to calculate the first 6 eigenfrequencies. Another useful tool is Search for eigenfrequencies around, which can calculate eigenfrequencies close to any desired value, if required.

14. To run the model, click on Compute. Wait for the computations to finish.

15. The default result for natural frequency (first eigenvalue 32.878 HZ) will appear in the Graphics window. Click on the Mode Shape (*truss*) node and expand Window Settings section located in the corresponding settings window. From the list for Plot window, select New window and give the title Plot 1. This will create a new Graphics window for showing the results. Results for other values of eigenfrequencies can be plotted by choosing the desired value from the list. The results for first and sixth eigenfrequencies are shown in Figure 3.65.

16. To animate the modes of vibration for different eigenvalues, right-click on the Export node in the Model Builder window and select Player. In the Player settings window, modify the data as shown in Figure 3.66 and click on Play icon, located in the Graphics window toolbar. Modal

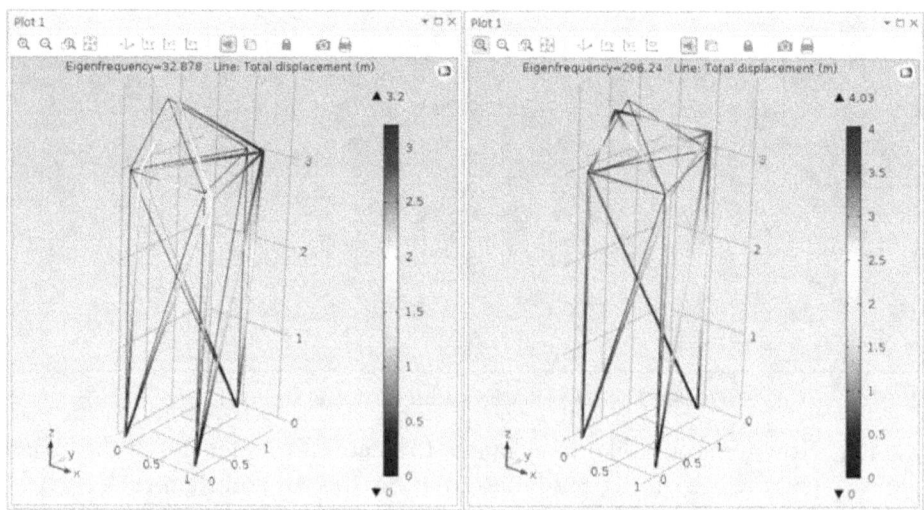

FIGURE 3.65 Results showing total modal shapes and displacements for two eigenfrequencies.

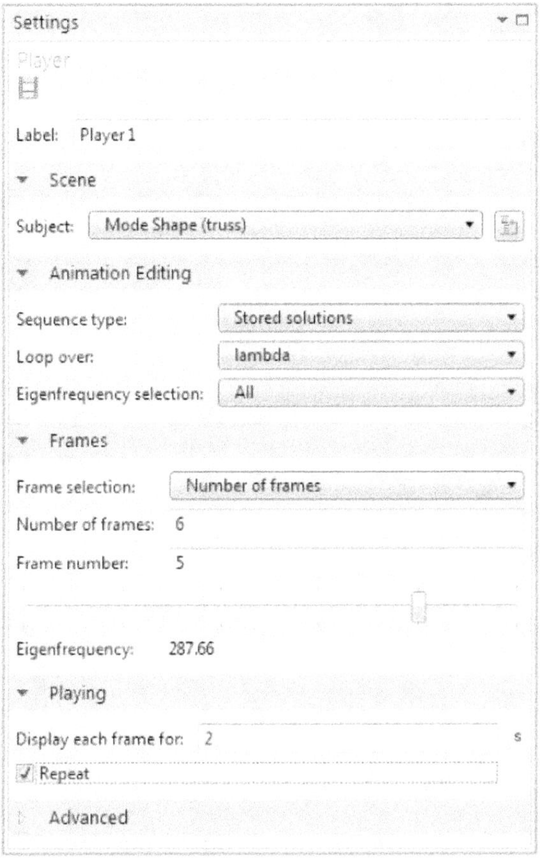

FIGURE 3.66 Results showing Player data entry for animation.

shapes animation show up in the Plot 1 window. To stop the animation, click on the Stop button, located on the toolbar in the Graphics window.

CHAPTER 4

MODEL EXAMPLES FOR INTERNAL FLUID FLOWS
Steady and Transient

In this chapter, we use COMSOL modules to model some examples in fluid mechanics. Models include dynamic, parametric study, swirling, and moving boundary conditions of two- and three-dimensional flows, but mainly internal flows. Modeling and analysis of fluids flow is more complex than that for linear solid mechanics. This is mainly because nonlinear governing equations (i.e. Navier-Stokes) for fluid flow are velocity vector components (u, v, w), pressure (p), and (F_x, F_y, F_z) component of body force. Density and dynamic viscosity of the fluid are ρ and μ, respectively.

$$\rho\left(u\frac{\partial u}{\partial x} + v\frac{\partial u}{\partial y} + w\frac{\partial u}{\partial z}\right) + \frac{\partial p}{\partial x} = \mu\left[\frac{\partial^2 u}{\partial x^2} + \frac{\partial^2 u}{\partial y^2} + \frac{\partial^2 u}{\partial z^2}\right] + F_x$$

$$\rho\left(u\frac{\partial v}{\partial x} + v\frac{\partial v}{\partial y} + w\frac{\partial v}{\partial z}\right) + \frac{\partial p}{\partial y} = \mu\left[\frac{\partial^2 v}{\partial x^2} + \frac{\partial^2 v}{\partial y^2} + \frac{\partial^2 v}{\partial z^2}\right] + F_y$$

$$\rho\left(u\frac{\partial w}{\partial x} + v\frac{\partial w}{\partial y} + w\frac{\partial w}{\partial z}\right) + \frac{\partial p}{\partial z} = \mu\left[\frac{\partial^2 w}{\partial x^2} + \frac{\partial^2 w}{\partial y^2} + \frac{\partial^2 w}{\partial z^2}\right] + F_z$$

$$\frac{\partial u}{\partial x} + \frac{\partial v}{\partial y} + \frac{\partial w}{\partial z} = 0$$

In engineering and industry we encounter many problems that require analysis and modeling of flow of a fluid and, in most cases, around or inside complex geometries. A fluid flow is categorized as laminar or turbulent when the Reynolds number is small or large, respectively, as compared to unity. The Reynolds number is a dimensionless number that measures inertia forces against viscous forces (in general) applied to a fluid point or particle. When the Reynolds number is large, flow becomes unstable and additional equations are needed to analyze the resulting turbulent flow. This adds to the complexity of flow analysis and modeling [18]. Whereas for low-Reynolds flows, the inertia force can be neglected and the governing equation is a linear version of the Navier-Stokes equation or so-called Stokes equation.

Our main objective is to provide, for users and readers, some solved examples that can be used directly or lead to further solutions of similar or more complex flows using COMSOL. It is assumed that readers are familiar with relevant engineering principles and governing equations of fluid mechanics. In each example, we provide brief explanations of physics involved along with governing equations and phenomena, as applicable. It is recommended that readers cover Chapters 1 and 2 before attempting examples in this chapter.

Example 4.1: Axisymmetric flow in a nozzle: Simplified water-jet

Axisymmetric flows may exist in many industrial and engineering problems. By definition an axisymmetric flow (or geometry) exists when the flow variables do not vary about the axis of symmetry involved. For example, in a tube for a fully developed flow the velocity along the axis of the cylinder at a given radius does not change with respect to the angular dimension. In other words, if an observer at a given radius moves around the axis of the tube then s/he will not observe any changes in the value of fluid velocity. The realization of axisymmetric flow, if it exists, is very important since it could reduce a 3D flow analysis to a 2D one.

In this example, we model flow in a nozzle that has an axis of symmetry. The cross section and dimensions of the nozzle in the r–z plane are shown in Figure 4.1.

Solution:

1. Launch COMSOL and in the New window click on Model Wizard. Save the file as Example 4.1.

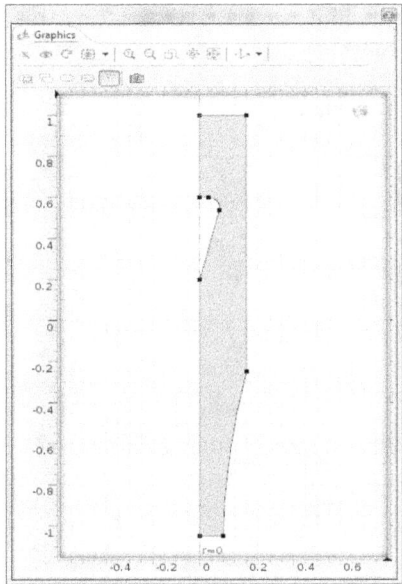

FIGURE 4.1 Geometry and dimensions of the Nozzle cross section in (r-z) plane.

2. In the Select Space Dimension window select 2D axisymmetric from the list.

3. In the Select Physics window, open the list under Fluid Flow and select Single-Phase Flow > Laminar Flow. Click on Add. Then click on the Study arrow icon to go to the next step.

4. In the Select Study window, click on Stationary and then click on Done. Make sure the file is saved.

5. Draw the geometry (or import a CAD file) of the nozzle cross section. Click on the Geometry tab from the toolbar and select Primitives > Bezier Polygon from the list. In the Bezier Polygon window, click on the Add Linear button and enter the following data:

	r:	z:	
1	0	1	m
2	0.2	1	m

6. Similarly, add another line segment by clicking on the Add Linear button and enter the following data:

	r:	z:	
1	0.2	1	m
2	0.2	-0.25	m

7. Add a quadratic line by clicking on the Add Quadratic button and enter the following data:

	r:	z:	
1	0.2	-0.25	m
2	0.1	-0.65	m
3	0.1	-1.05	m

8. Add two more lines with following data, and then click the Build Selected icon.

	r:	z:			r:	z:	
1	0	-1.05	m	1	0.1	-1.05	m
2	0	1	m	2	0	-1.05	m

See Figure 4.2.

9. Similarly, create another Bezier Polygon. In the Bezier Polygon window, click on Add Linear to create three line segments with the following data:

	r:	z:			r:	z:			r:	z:	
1	0	0.6	m	1	0.1	0.6	m	1	0	0.2	m
2	0.1	0.6	m	2	0	0.2	m	2	0	0.6	m

Click on the Build Selected icon to create the triangle. To remove the triangular shape, click anywhere in the Graphics window and press the CTRL + A keys. The geometry will change color. Click on the Booleans and Partitions > Difference icon, located in the toolbar. See Figure 4.3.

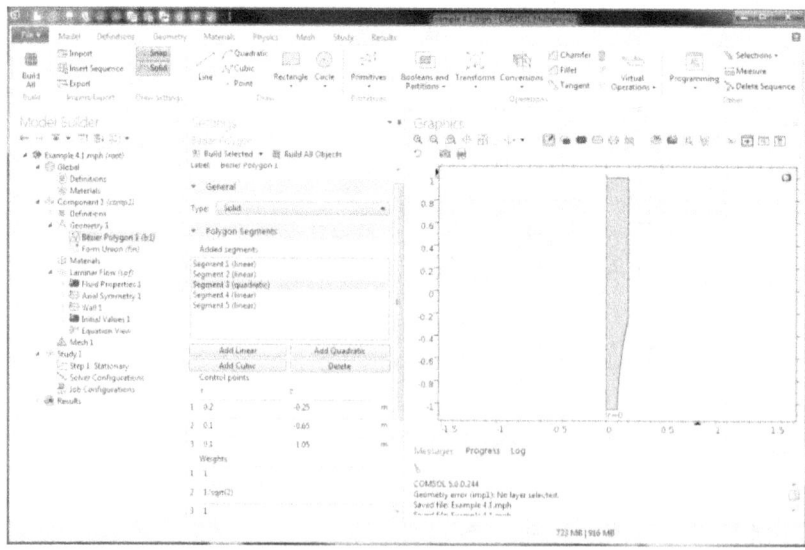

FIGURE 4.2 Geometry and line segments coordinates for the nozzle.

10. Add a fillet to the sharp corner of the triangular cut. Click on the Fillet from the toolbar. In the settings window, add the vertex of the sharp corner of the triangle to the list under Vertices to fillet. For Radius enter 0.03 (unit should read m) and click Build Selected icon. At this point the geometry of the nozzle cross section is complete, as shown in Figure 4.4.

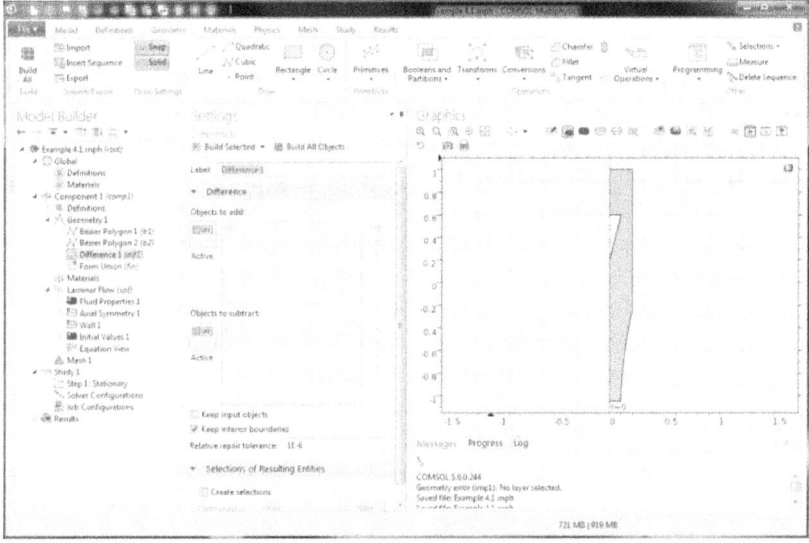

FIGURE 4.3 Geometry cross section with the cut.

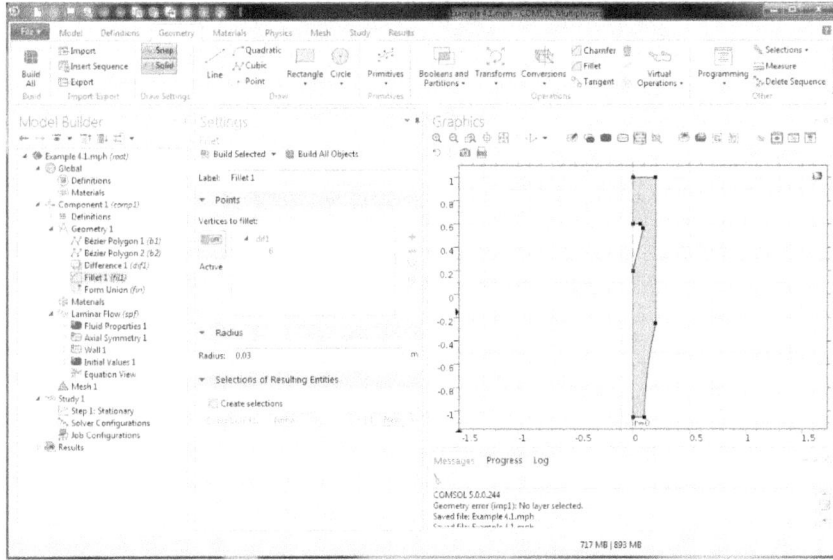

FIGURE 4.4 Geometry cross section with the cut and fillets.

11. To add material, right-click on the Materials node in the Model Builder and select Add Material. In the Add Material window, expand Liquid and Gases > Liquids and select Water from the list. Click on Add to Component. The properties of water are given as functions of its temperature as shown in Figure 4.5 under Material Contents section (expand this section if needed). Make sure the flow domain is selected.

12. This model is for an isothermal flow in the nozzle, but for calculating the properties of water the value of temperature is required. Click on the Model tab from the toolbar, and on Parameters. In the Parameters settings window, add the data as shown in Figure 4.6.

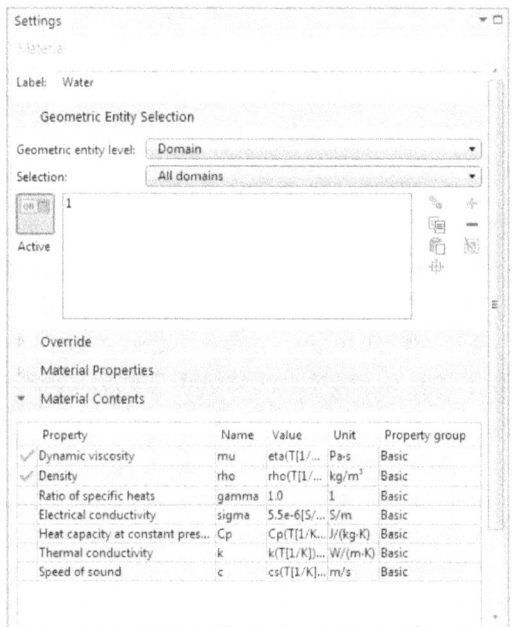

FIGURE 4.5 Material properties for nozzle.

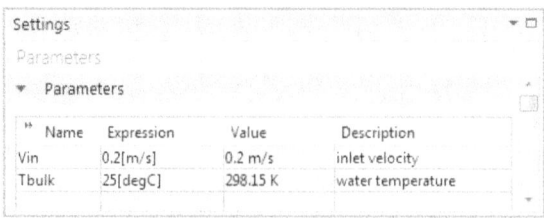

FIGURE 4.6 Parameters data entry.

Click on the Fluid Properties 1 node to open the Fluid Properties settings window and type in Tbulk for Temperature, located under Model Inputs section.

13. To define the boundary conditions, click on the Physics tab from the toolbar and select Inlet, located under Boundaries. In the Inlet settings window, add the edge at the top of geometry (z = 1) to the Selection list. In the Velocity section, enter Vin for U_0 (make sure that Normal inflow velocity is checked). This will assign the water velocity to the value set in the Parameters (Vin) entering into the nozzle from the top. Similarly, add Outlet boundary condition at the exit (the bottom edge) and set the Pressure value P_0 equal to 0. Double-check the Axial Symmetry 1 (automatically created) to have the vertical edges at $r = 0$ as axis of symmetry. See Figure 4.7.

14. To create a mesh, click on the Mesh 1 node in the Model Builder window. In the Mesh settings window select User-controlled mesh for Sequence type. Click on Size, in the Model Builder window and in the settings window select Extremely coarse from the Predefined list. Click on Build All. Figure 4.8 shows the mesh, created.

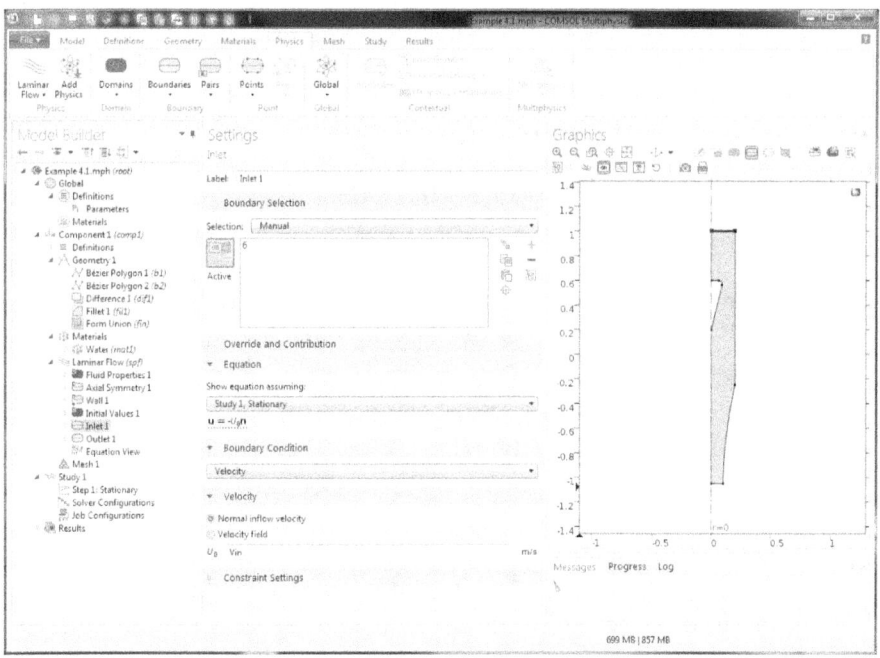

FIGURE 4.7 Inlet boundary condition data.

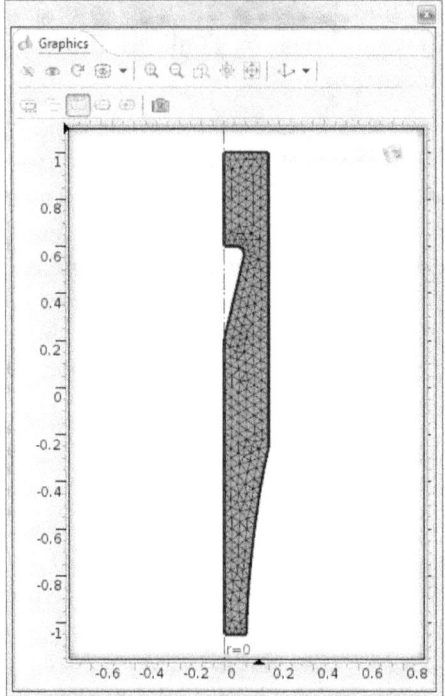

FIGURE 4.8 Mesh created for the Nozzle cross section.

15. To set up a parametric study based on inlet velocity, click on the Study tab from the toolbar and select Parametric Sweep. In the Parametric Sweep settings window under the Sweep type table, click on the plus sign (+) and open the list under Parameter names that appears in the table. From the list select Vin (inlet velocity) and type 0.5, 1, 3, 5 for Parameter value list, as shown in Figure 4.9. This sets the value of inlet velocity for a sequence of runs for the model. To run the model, click on Compute.

16. To show the results in 3D (recall that this model is axisymmetric), right-click on Velocity 3D (*spf*) node and select Surface. Repeat this operation and select Contour. Click on Contour 1 node and in the Contour window type p for the Expression. Click on the Plot icon. The velocity mapped by pressure contours will appear in the Graphics window. To see these values for different input velocity values, simply click on Velocity 3D (*spf*) node, and in the corresponding settings window select the desired value for Parameter value (Vin), under the Data section. Results for Vin = 5 m/s are shown in Figure 4.10.

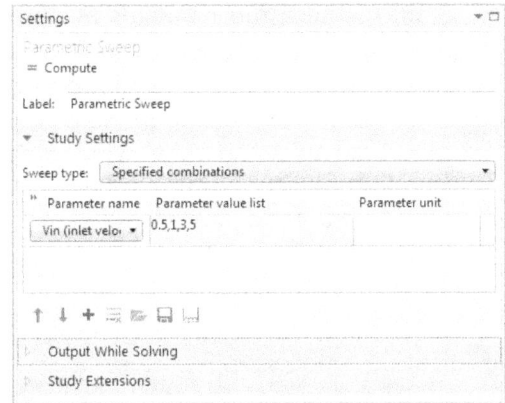

FIGURE 4.9 Parametric Sweep data for values of inlet velocity.

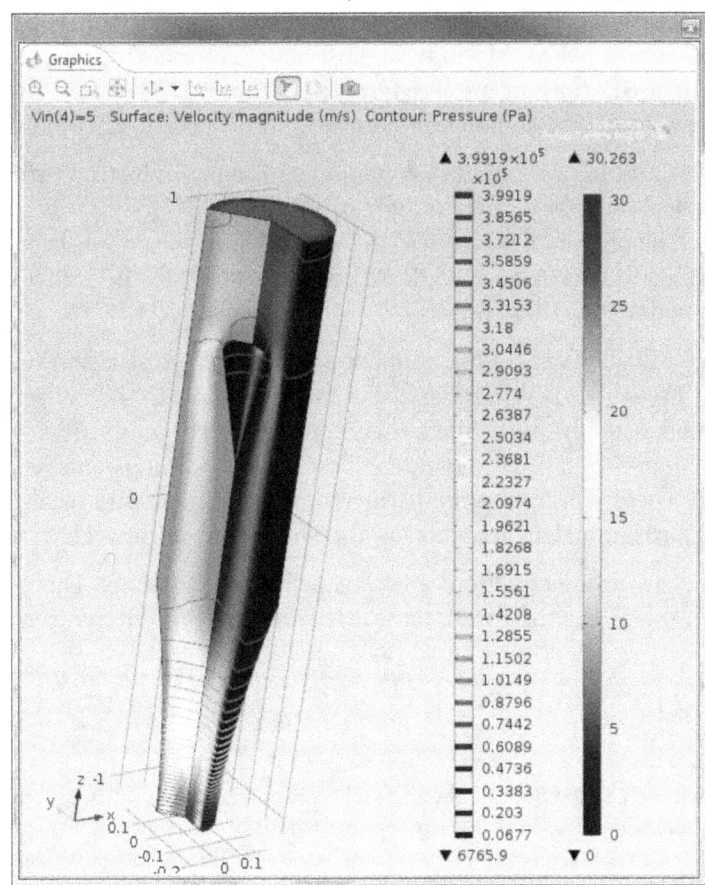

FIGURE 4.10 Results for surface velocity in the nozzle for inlet velocity of 5 m/s.

17. It would be useful to check the Reynolds number, based on nozzle diameter (i.e. 0.2 m) at the exit. Create a Line Average, by right-clicking on Derived Values and select Average >Line Average. In the Settings window select the exit boundary and click on Evaluate. The average velocities show up in the Table window; it is 21.534 m/s for Vin = 5 m/s. Therefore, the Reynolds number is about 4e6 (assuming water kinematic viscosity, 1e–6 m²/s). Similarly based on nozzle inlet (i.e. diameter of 0.4 m), the Reynolds number is about 2e6. These values could indicate a requirement for running this model with a turbulent model. We leave this as an exercise for users. See Tabatabaian [18], for turbulent flow modeling.

Example 4.2: Swirl flow around a rotating disk: Laminar flow

In this example, we rebuild in COMSOL 5 a model from the COMSOL Model Library (CFD_Module/Single-Phase_Tutorials/rotating_disk).[4] This model is a 3D flow in an axisymmetric geometry. We use this model as another example to demonstrate how to model a turbulence flow.

Flow around a rotating disk happens in many industrial processes and mechanical machines, such as high-speed fly wheels for energy storage and mixers. Calculation of stress field in the fluid and consequently on the rotating disk itself is required for design of the system for both low-speed and high-angular velocities cases.

The 3D geometry of the container and rotating disk is shown schematically in Figure 4.11. When the disk rotates, the flow in the tank will eventually reach a steady-state. Since the geometry is axisymmetric, we use a 2D axisymmetric model. However, the flow is three-dimensional and the fluid velocity vector has components in radial, rotational, and axial directions. The 2D axisymmetric geometry is also shown in Figure 4.11.

The governing equations are Navier-Stokes equations. The momentum transfer equations for a stationary, axisymmetric flow are written as:

$$\rho\left(u\frac{\partial u}{\partial r} - \frac{v^2}{r} + w\frac{\partial u}{\partial z} \right) + \frac{\partial p}{\partial r} = \mu\left[\frac{1}{r}\frac{\partial}{\partial r}\left(r\frac{\partial u}{\partial r} \right) - \frac{u}{r^2} + \frac{\partial^2 u}{\partial z^2} \right] + F_r$$

---

[4]Model made using COMSOL Multiphysics® and is provided courtesy of COMSOL. COMSOL materials are provided "as is" without any representations or warranties of any kind including, but not limited to, any implied warranties of merchantability, fitness for a particular purpose, or noninfringement.

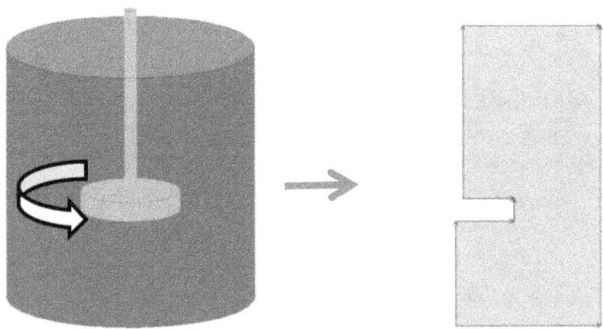

FIGURE 4.11 3D and 2D axisymmetric geometries for flow around a rotating disk.

$$\rho\left(u\frac{\partial v}{\partial r} - \frac{uv}{r} + w\frac{\partial v}{\partial z} \right) = \mu\left[\frac{1}{r}\frac{\partial}{\partial r}\left(r\frac{\partial v}{\partial r} \right) - \frac{v}{r^2} + \frac{\partial^2 v}{\partial z^2} \right] + F_\varphi$$

$$\rho\left(u\frac{\partial w}{\partial r} + w\frac{\partial w}{\partial z} \right) + \frac{\partial p}{\partial z} = \mu\left[\frac{1}{r}\frac{\partial}{\partial r}\left(r\frac{\partial w}{\partial r} \right) + \frac{\partial^2 w}{\partial z^2} \right] + F_z$$

where u is the radial velocity, v the rotational velocity, and w the axial velocity, μ viscosity, and p pressure. The body volumetric forces (F_r, F_φ, F_z) are all equal to zero in this model.

Solution:

1. Launch COMSOL and in the New window click on Model Wizard. Save the file as Example 4.2.

2. In the Select Space Dimension, select 2D axisymmetric from the list.

3. In the Select Physics window, open the list under Fluid Flow and select Single-Phase Flow > Laminar Flow. Click on Add and then on the Study arrow button.

4. In the Select Study window, select Stationary and click on Done.

5. To define the disk angular velocity as a parameter, click on the Parameters, from the ribbon toolbar. In the Parameters settings window, add the data Name: omega and Expression: 0.25*pi [rad/s], as shown in Figure 4.12.

6. Draw the 2D cross-section geometry with the dimensions as shown in Figure 4.13.

Edge	AB	BH	HG	GC	CD	DE	EF	FA
Length (m)	0.014	0.008	0.003	0.007	0.023	0.019	0.04	0.02

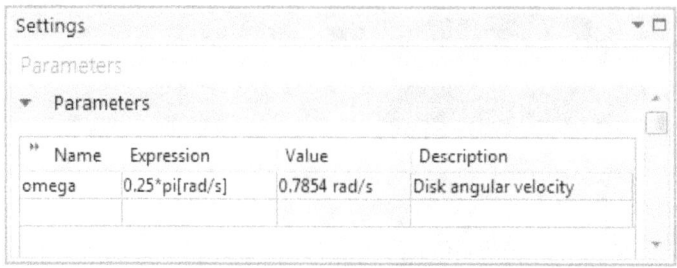

FIGURE 4.12 Data entry for Parameters.

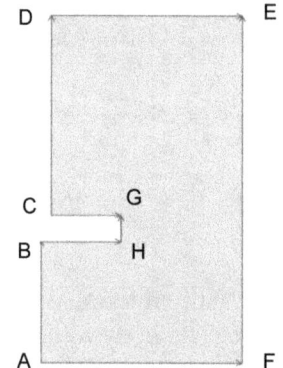

FIGURE 4.13 Axisymmetric geometry and dimensions for flow around a rotating disk.

Right-click on the Geometry 1 node in the Model Builder window and select Rectangle. In the Rectangle window (under the Size section), enter 0.02 for Width and 0.04 for Height. Under Position, select Corner for Base and enter 0 for r and 0 for z. Right-click on the Geometry 1 node in the Model Builder window and select Rectangle. In the Rectangle window (under the Size section), enter 0.008 for Width and 0.003 for Height. Under Position, select Corner for Base and enter 0 for r and 0.014 for z. Finally, right-click on the Geometry 1 node in the Model Builder window and select Rectangle. In the Rectangle window (under the Size section), enter 0.001 for Width and 0.023 for Height. Under Position, select Corner for Base and enter 0 for r and 0.017 for z. Click at any point in the Graphics window, press CTRL+A to highlight the geometry, then click on the Difference icon in the toolbar. Results are shown in Figure 4.14.

7. To add materials to the model, click on Materials tab from the toolbar and select Blank Material. The Material settings window, under the Material Contents section, enter the data for density and dynamic viscosity, as shown in Figure 4.15.

8. To define the details of the model physics and boundary conditions, click on the Laminar Flow node in the Model Builder window. Expand the

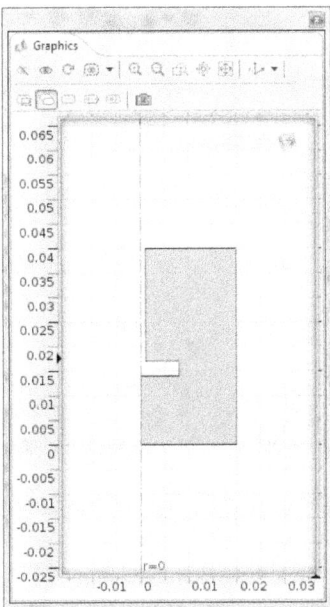

FIGURE 4.14 Axisymmetric geometry and dimensions.

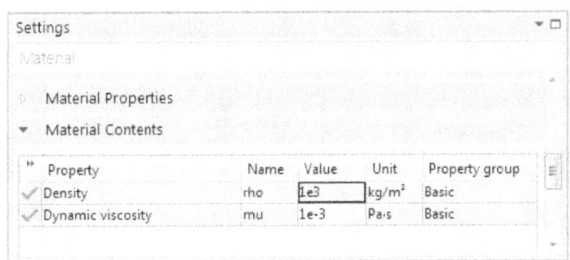

FIGURE 4.15 Material window showing fluid properties.

Physical Model section and select Incompressible flow from the list under Compressibility. Also check the box for Swirl flow, as shown in Figure 4.16.

9. Next we define the order of polynomial functions used for finite elements for velocity and pressure. The default is linear functions (P1 + P1); in this model we use a second order polynomial for velocity and a linear one for pressure (P2 + P1). This combination is recommended for Stokes and swirl flows (see Chapter 2 and the COMSOL manual for *The Laminar Flow Interface-Discretization*). Click on the Show icon located in the toolbar of Model Builder window and select Discretization. Then click on the Laminar Flow (*spf*), and in the Laminar Flow settings window select P2 + P1 from the list under Discretization of fluids, as shown in Figure 4.17.

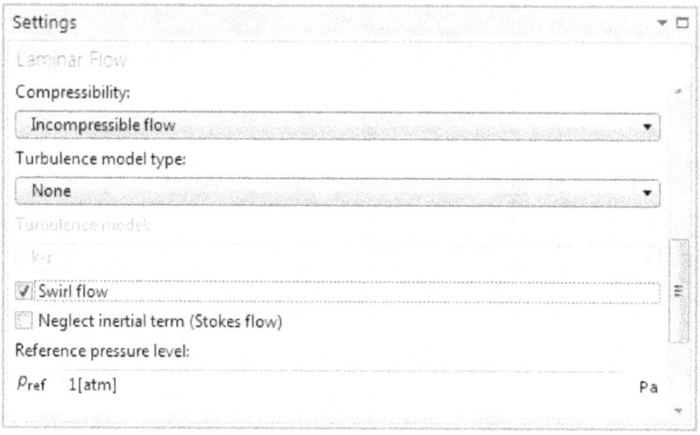

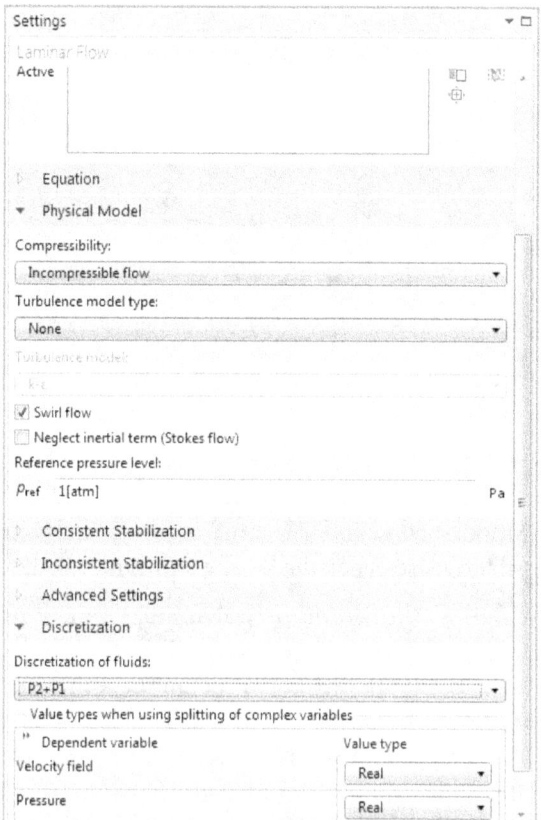

FIGURE 4.16 Laminar flow window showing fluid physics set up for swirling flow.

FIGURE 4.17 Laminar flow window showing discretization scheme setup.

Click the Physics tab from the toolbar and select Wall, under Boundaries. The Wall settings window will open. In this window, select and add the boundaries (3, 4, 5, 7) related to the shaft and disk by clicking on these boundaries in the Graphics window. In the same window under Boundary Condition, select Sliding wall from the list and type omega*r for V_w. This will set the rotational velocity of the disk as the boundary value for the fluid. See Figure 4.18.

For the boundary at the top surface of the fluid, we use symmetry boundary condition to allow for radial and rotational velocities and eliminate the velocity in the z-direction. Click on the Boundaries button and select Symmetry from the list. In the Symmetry settings window, add the edge on the top (which reps the free-surface, edge # 6) to the Selection by clicking on it in the Graphics window. To see the equations for this boundary condition, expand the Equation section.

Finally, we should set a reference point for the pressure since there is no output or exit point for the fluid. Click on the Points button

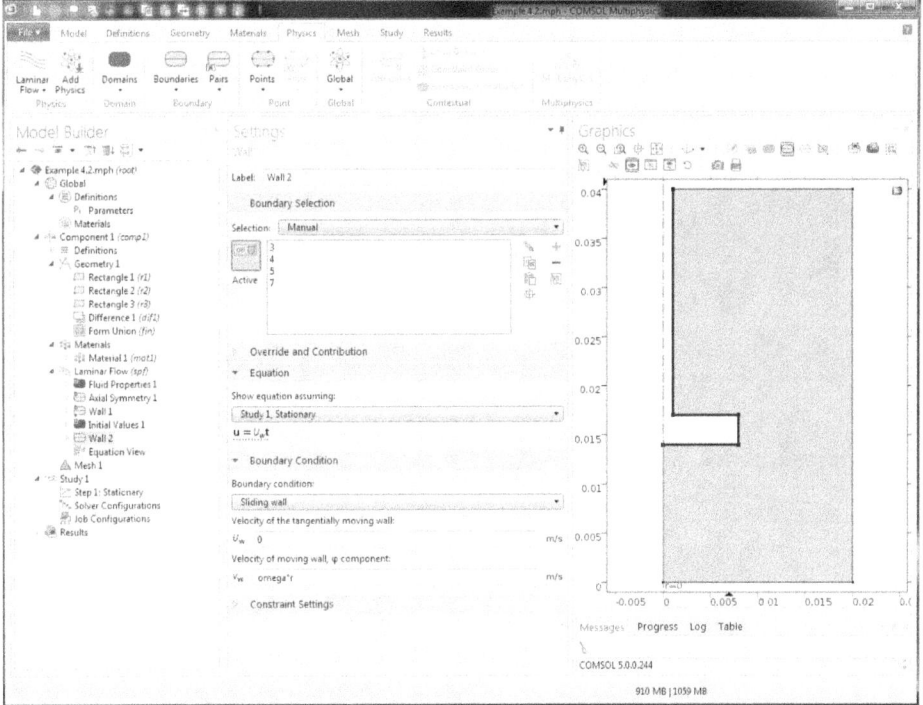

FIGURE 4.18 Wall window showing sliding wall boundary conditions setup.

and select Pressure Point Constraint. In the corresponding settings window, add vertex at the top-right corner of the geometry (# 8) to the list by clicking on the vertex.

The boundary conditions applied to this model are listed in Figure 4.19.

10. To run the model for a series of values of disk angular velocities omega, expand the Study 1 node in the Model Builder window and click on Step 1: Stationary. In the Stationary settings window, locate the Study Extensions section and expand it. Check the Auxiliary sweep box and click on the plus (+) icon. From the list, which appears under Continuation parameter, select omega (Disk angular velocity) and type 0.25*pi, 0.5*pi, 2*pi, 4*pi under the Parameter value list, as shown in Figure 4.20.

Edge	AB	BH	HG	GC	CD	DE	EF	FA
Boundary conditions	Axial symmetry	Sliding wall	Sliding wall	Sliding wall	Sliding wall	Symmetry	No-slip wall	No-slip wall

FIGURE 4.19 Boundary conditions applied to the walls of the rotating disc.

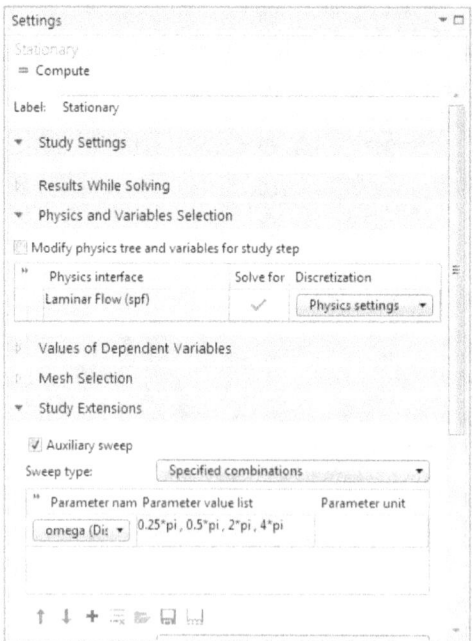

FIGURE 4.20 Stationary window showing parameter values set up for disk.

11. Run the model by clicking on the Study tab and then on Compute. After calculations are finished, the default result will appear in the Graphics window.

12. To manipulate the results, click on Velocity (*spf*) and in the 2D Plot Group select 0.7854 from the list for Parameter value (omega); then click the Plot icon. Add the streamlines to the 2D plot for velocity. Click on the Streamline button from the ribbon toolbar. In the Streamline settings window, set the expressions as shown in Figure 4.21; then click on Plot.

The model results for four values of omega are shown in Figure 4.22.

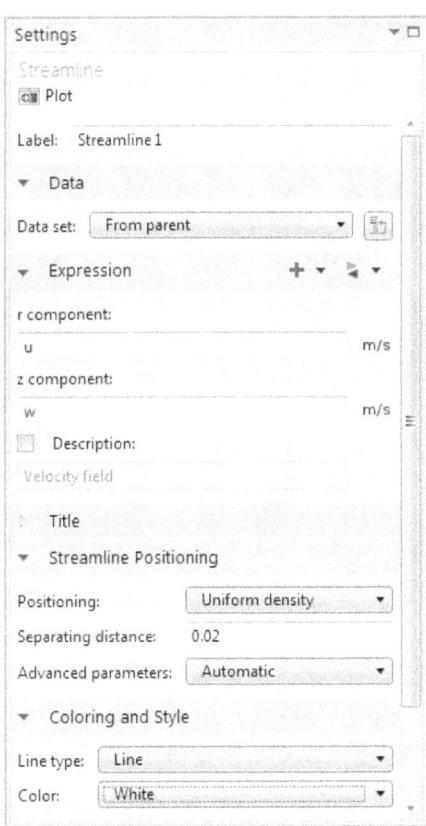

FIGURE 4.21 Stationary window showing parameter values set up for disk.

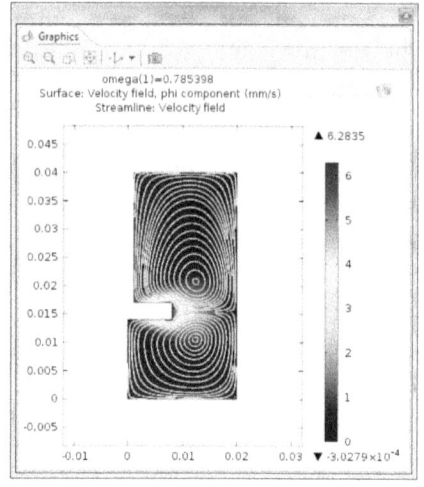

omega = 0.25π

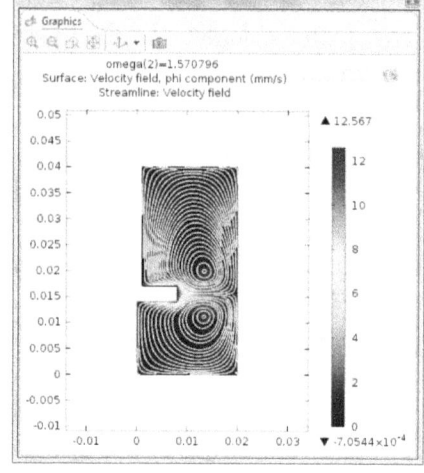

omega = 0.5π

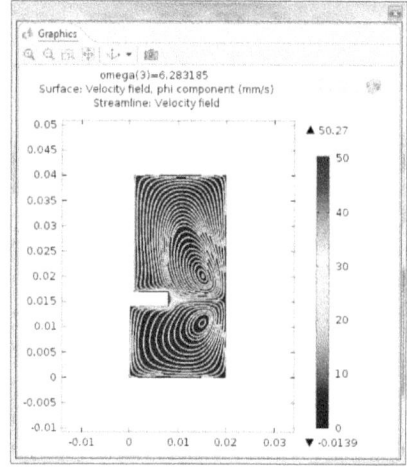

omega = 2π

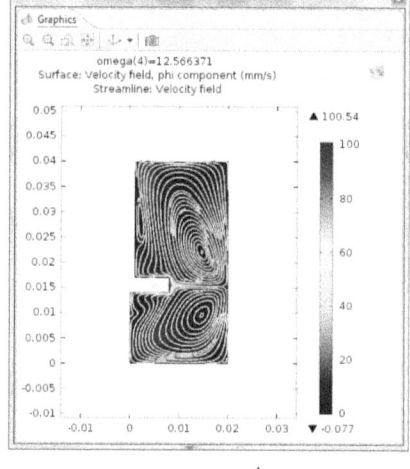

omega = 4π

FIGURE 4.22 Results for 2D surface velocity and streamlines for different values of omega.

13. To see the 3D axisymmetric results, click on the Velocity (*spf*) 1, rename it to Velocity 3D. From the ribbon toolbar click on Contour. In the Contour settings window enter p for Expression, deselect and deselect the Color legend box, and Uniform for Coloring and White for Color located under Coloring and Style. Zoom on the disk location to see the details. Results are shown in Figure 4.23.

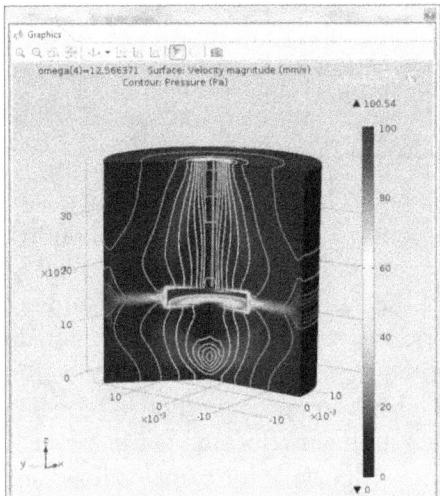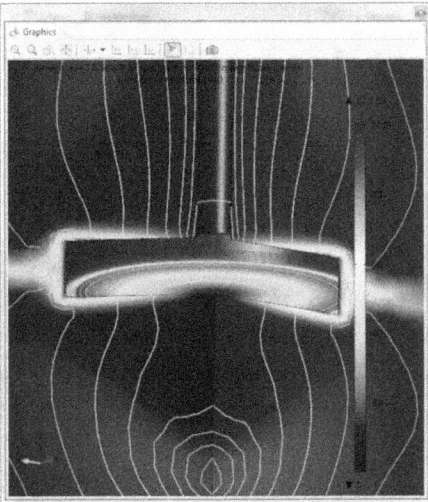

FIGURE 4.23 Results for 3D surface fluid velocity and pressure contour for a value of omega = 4π.

Example 4.3: Swirl flow around a rotating disk: Turbulent flow

This example is an extension of Example 4.2 with turbulent flow modeling. For high values of the angular velocity of the rotating disk, we may have to consider turbulence flow. The value of the Reynolds number will be calculated to determine the state of the flow regime. We define the Reynolds number as $R_e = \dfrac{\Omega r^2}{v}$ where Ω is disk angular velocity, r is disk radius, and v is fluid kinematic viscosity. COMSOL provides us with several turbulence models. For this example we use k-ε model, which is a RANS (Reynolds Averaged Navier Stokes equations) type model. The general approach in RANS is to assume decomposition of fluid velocity into a statistical average, such as the mean velocity, and a deviation from the mean. When governing N-S equations are averaged, the resulting equations contain additional stresses that are due to turbulence stress on the flow. These additional turbulence stresses are called Reynolds stress and require extra equations for their solutions. One approach is to calculate the turbulence kinetic energy k and its dissipation rate ε (hence k-ε model) as additional equations. Turbulence modeling is still an R&D subject and advanced modeling techniques are still being investigated and validated. Interested readers are referred to textbooks [18] and [19] and scientific journals for further study on this subject.

Solution:

1. Launch COMSOL and open the Example 4.2 file. Save it as Example 4.3.

2. Click on the Show icon in the Model Builder toolbar and select Advanced Physics Options, as shown in Figure 4.24.

3. Click on the Laminar Flow (*spf*) node. In the Laminar Flow settings window, in the Physical Model section locate Turbulence model type and select RANS. From the list under Turbulence model select $k - \varepsilon$ (it is recommended to expand the Equation section and study the equations listed; see [18]). Expand the Advanced Settings section and select the box for *Use pseudo time stepping for stationary equation form*, as shown in Figure 4.25. This is a common method in modeling; setting a pseudo time stepping for stationary flow models helps stability and convergence of the solution, especially for complex flows such as this one.

4. To build a mesh for the geometry, click on the Mesh 1 node in the Model Builder window. The Mesh settings window will open. In this window, select Fine from the list under Element size and click Build All. The total of elements built is displayed in the Messages window. Use Zoom Box to see the boundary elements adjacent to solid walls, as shown in Figure 4.26.

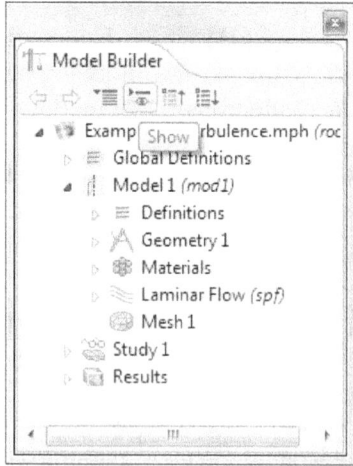

FIGURE 4.24 Model Builder window showing Advanced physics options setup.

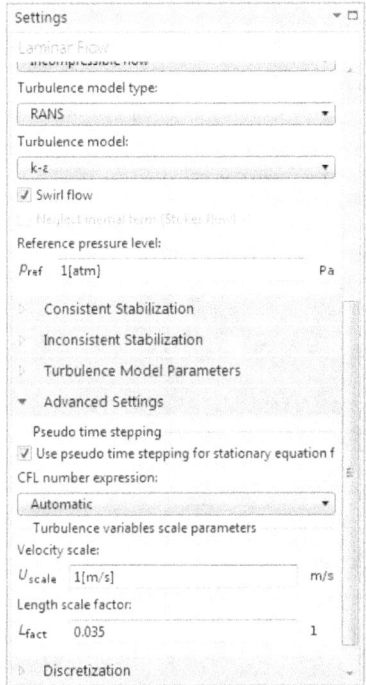

FIGURE 4.25 Laminar Flow settings window showing $k - \varepsilon$ turbulence model setup.

5. To set the values for omega, click on Step 1: Stationary under the Study 1 node in the Model Builder. The Stationary window will open. In this window, expand the Study Extensions section and in the table enter the following data for omega:

Continuation parameter	Parameter value list
omega (Disk angular velocity)	range(100*pi,200*pi,500*pi)

Omega values are then set to start from $100\,\pi$ to $500\,\pi$ with the step of $200\,\pi$.

6. To run the model, click on the Study tab and select Compute. Wait until the calculations are done. The default results should appear in the Graphics window, usually as 2D plots.

7. Reynolds stresses including turbulence viscous effects on the fluid are quantities of interest in a turbulent flow. To show this, right-click on Velocity (*spf*) and select Duplicate. The 2D Plot Group settings

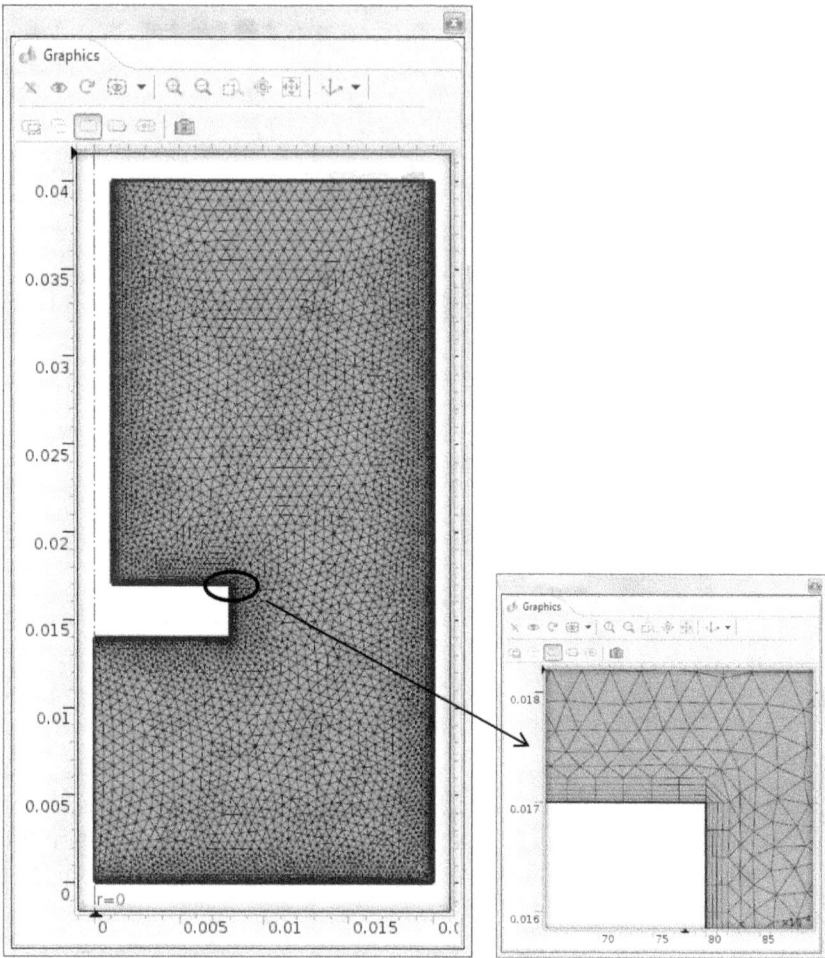

FIGURE 4.26 Graphics window showing details of finite elements mesh.

window will open. In this window, select 1570.8 (the value for $500\,\pi$) from the list for Parameter value (omega). Then expand the Velocity (*spf*) 1 node in the Model Builder window and click on Surface 1. In the Surface settings window, click the Replace Expression icon and type in turbulent in the search space, select (spf.muT- Turbulent dynamic viscosity) from the list. Also in this window, expand the Coloring and Style section and check the Color legend box. Finally, click on Plot. The results, as shown in Figure 4.27, display turbulent dynamic viscosity mapped with streamlines for disk angular velocity of $500\,\pi$. The value of turbulence dynamic viscosity (or eddy viscosity) ranges from

4.535e−3 to 0.3674 Pa.s. Comparing this to the fluid dynamic viscosity 1e−3 could be interpreted as the importance of turbulence viscous effects when flow becomes turbulent (see Table 4.1). Reynolds shear stress components are calculated using the mean velocity gradients multiplied by turbulent eddy viscosity, in COMSOL using Boussinesq's hypothesis. See Tabatabaian [18] for further details.

8. It is useful to calculate the Reynolds number for different values of disk angular velocity as well as the Cell Reynolds number. We defined

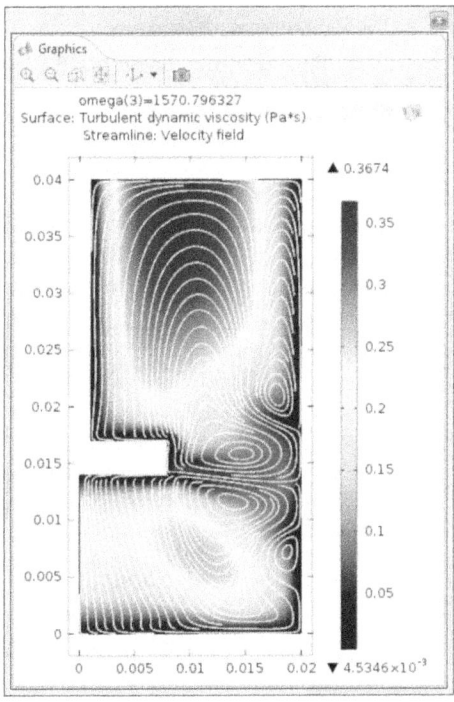

FIGURE 4.27 Results for turbulent surface velocity flow for omega = 500π.

Ω (Rad/s)	μ (Pa.s)	μ_{Tmax} (Pa.s)	Re	μ_{Tmax}/μ
100π	0.001	0.0765	2.0106E4	76.5
300π	0.001	0.223	6.0319E4	223
500π	0.001	0.3674	1.005E5	367.4

TABLE 4.1 Turbulent Dynamic Viscosity and Flow Reynolds Number

the Reynolds number $R_e = \dfrac{\Omega r^2}{v}$ as where is fluid kinematic viscosity (1E–6) and r is 0.008 m. Table 4.1 summarizes these results.

Example 4.4: Flow in a U-shape pipe with square cross-sectional area: Laminar flow

In this example, we model the flow in a pipe or duct with square cross-sectional area. Calculations for flow, mainly velocity and pressure, in pipes are needed in many engineering applications like ventilation and air-conditioning ducts. Semi-empirical formulae are available and commonly used for these types of calculations. For noncircular pipes, engineers usually calculate the equivalent hydraulic diameter and use the circular pipe formulations. For reference, hydraulic diameter is defined as $D_h = \dfrac{4A}{P}$, where A is the area of pipe cross section and P is the (wetted) perimeter.

COMSOL (since 4.3 version) has a new module for pipe network calculations and modeling that assumes 1D flow in the pipes and is a useful tool for modeling pipe networks. COMSOL 5 has a new tool (Pipe Connection) that could be used to connect 1D pipe networks to 3D flow domains. However, sometimes detailed 3D flow modeling is needed for engineering and industrial applications.

Solution:

1. Launch COMSOL and click on Model Wizard in the New window. Save the file as Example 4.4.

2. Select 3D from the list in the Select Space Dimension window. In the Select Physics window expand the list for Fluid Flow >Single-Phase Flow and select Laminar Flow (*spf*). Click on Add and then on the Study arrow icon. In the Select Physics window select Stationary from the list and click on Done. Make sure the file is saved.

3. To fully parameterize the geometry of the duct and inlet velocity, click on Model from the toolbar and select Parameters. Enter the data given in Figure 4.28, note that the parameter manes are case sensitive.

4. Now we draw the geometry of the pipe in COMSOL. You can also draw it using a CAD package and import it into this model. In the Geometry window, change the Length unit to mm by selecting it from the list. Right-click on the Geometry node in the Model Builder

window and select Work Plane from the list. Click on the Plane Geometry node, which is created under Work Plane 1 (*wp1*) node in the Model Builder window. See Figure 4.29.

5. In the Graphics window, click on the Square icon in the main toolbar and draw a square. In the Model Builder, expand the list under Plane Geometry and click on the Square 1 (*sq1*) node. The Square window will open. In this window, change the data as shown in Figure 4.30, and then click the Build Selected icon. The square with the given size and corner coordinates will appear in the Graphics window. In this window, click the Zoom Extents icon to adjust.

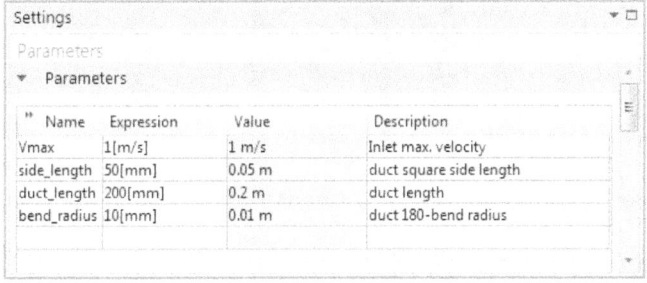

FIGURE 4.28 Entries for the Parameters for Example 4.4.

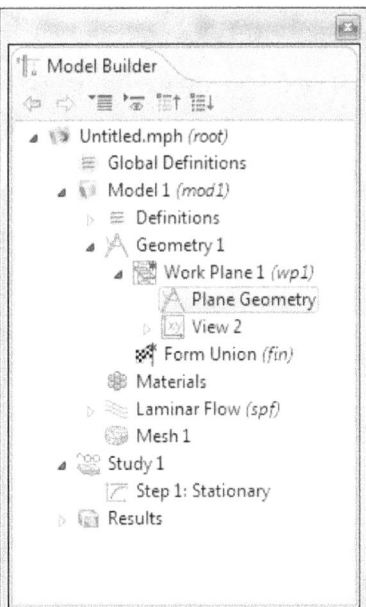

FIGURE 4.29 Model Builder window showing work plane setup.

6. Now we extrude this square to build the pipe geometry. Right-click on the Work Plane 1 (*wp1*) node in the Model Builder Window and select Extrude from the list. The Extrude window will open. Locate Distance from the Plane and type duct_length in the space under Distance. This will set the length of the pipe. Click the Build Selected icon. Click the Zoom Extents icon to adjust. See Figure 4.31.

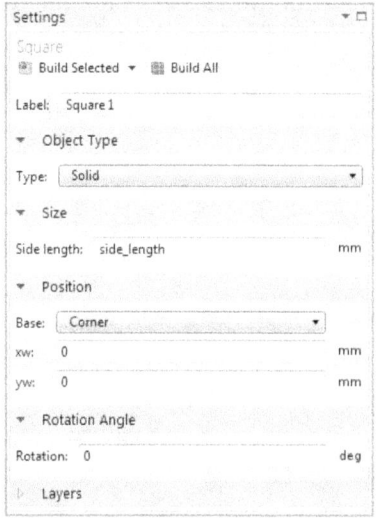

FIGURE 4.30 Square geometry setup.

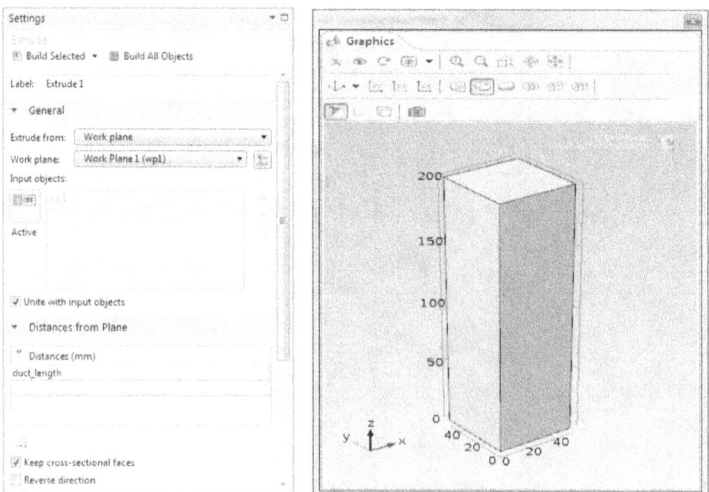

FIGURE 4.31 Windows showing Extrude setup and resulting geometry.

7. Create another work plane on the top surface of the pipe. Right-click on the Geometry node in the Model Builder window and select Work Plane from the list. Click on the Work Plane 2 (*wp2*) node in the Model Builder window. In the Work Plane window, change the Plane type and Origin as shown in Figure 4.32, and then click the Build Selected icon.

8. Click on the Plane Geometry node, which is created under Work Plane 1 (*wp1*) node in the Model Builder window. Draw a square in the Graphics window by following steps similar to those described in Step 5 above.

9. To build the bend, right-click on Work Plane 2 (*wp2*) and select Revolve from the list. The Revolve window will open. In this window, locate the Revolution Angles and Revolution Axis sections, input the data as shown in Figure 4.33, and then click the Build Selected icon.

10. To build the last part of the pipe, right-click on the Geometry 1 node in the Model Builder window and select Transforms > Copy. The Copy window will open. In this window, add the straight section of the pipe (which was built previously) to the Input objects list by clicking on it in the Graphics window and then right-click. In the Displacement, enter

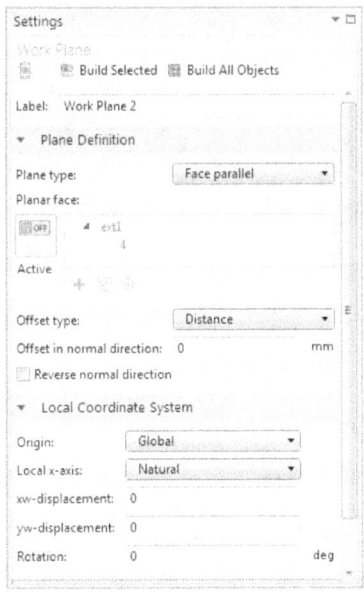

FIGURE 4.32 Work Plane 2 setup.

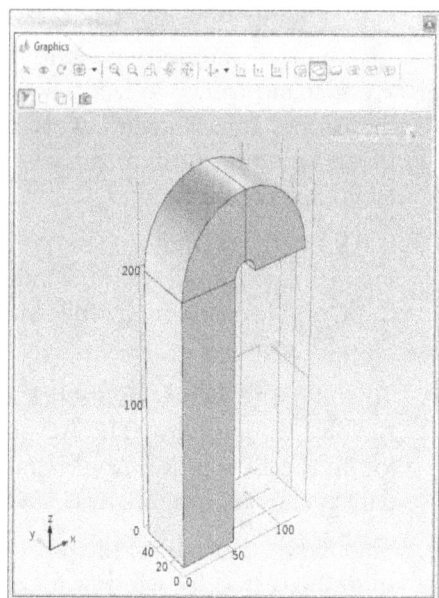

FIGURE 4.33 Windows showing work plane setup for the pipe bend.

side_length + 2*bend_radius for x and click on the Build Selected icon. See Figure 4.34.

11. We now add materials to the pipe domain. Click on the Materials tab from the toolbar and then click on Add Material. In the Add Material window select locate and select Liquids and Gases > Gases > Air. Click on Add to Component. Close Add Material window.

12. For the air flow we define a parameter as Vmax to be the maximum velocity at the entrance to the pipe. Right-click on the Global > Definitions node in the Model Builder and select Parameters. In the corresponding settings window, type the data Name: Vmax, Expression: 1 [m/s], and Description: Inlet max.velocity.

13. We now define the boundary conditions at the pipe inlet and outlet. The no-slip boundary condition is set by default for other pipe walls.

   13.1. For defining the boundary condition at the inlet, click on the Physics tab from the toolbar and select Boundaries > Inlet from the list. In the Inlet settings window, add one of the end surfaces (3) of the pipe to the Selection by clicking on the surface in the Graphics window. It is optional but useful to expand the Equation

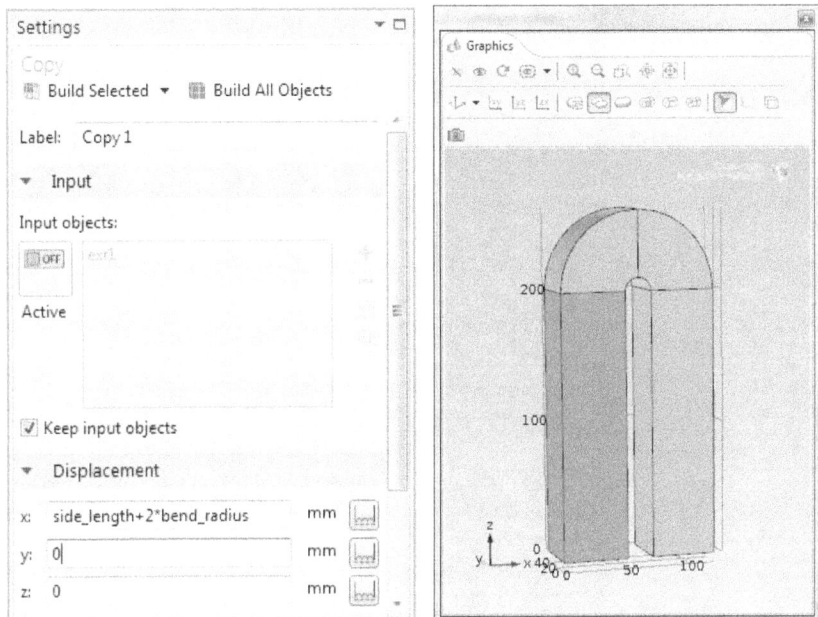

FIGURE 4.34 Windows showing Copy geometry setup for the pipe.

section, which shows the governing equations for the type of boundary conditions selected. Select Laminar inflow from the list in the Boundary Condition section. Select Average velocity and enter 0.5*Vmax. For the Entrance Length, enter 20 for L_{entr}. Also, check the box for Constrain outer edges to zero. This will force the laminar velocity profile to go to zero at the edges of the pipe inlet. A technical point should be mentioned here. COMSOL uses the value of L_{entr} to calculate a fully developed velocity outside the domain of the model, as if you have added an extension to the entrance. This is a very useful feature in COMSOL. The value of L_{entr} should be larger than *0.06ReD* (see *COMSOL Manual*), where *Re* is the Reynolds number and *D* is the pipe hydraulic diameter. For our model, we have $(0.06)(10^4/3)(0.05) = 10$ m, when Vmax = 1 m/s, air kinematic viscosity is 1.5×10^{-5}, and D = 0.05 m. For fully developed flow it can be shown that maximum velocity is 1.5 times the average velocity. See Figure 4.35.

13.2. For defining outlet condition, select Boundaries > Outlet. In the corresponding settings window, add the outlet surface (17) of the pipe to the Selection list.

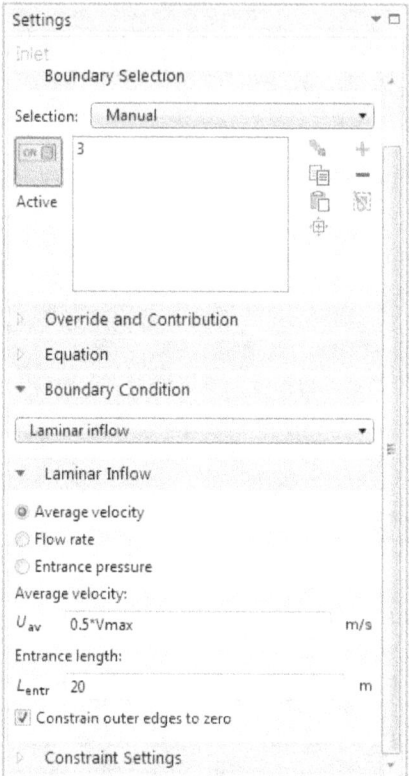

FIGURE 4.35 Inlet boundary condition setup.

14. Build a mesh by clicking on the Mesh node in the Model Builder window. In the Mesh settings window select Coarse for the Element size. Make sure that Sequence type is set to Physics-controlled mesh. Click the Build All icon. Results are shown in Figure 4.36.

Notice the hybrid mesh types, which include boundary-layer hexagonal and tetragonal elements for the internal flow region.

15. Next, we add a parametric study based on Vmax. Click on the Study tab from the toolbar and select Parametric Sweep. In the corresponding settings window, locate the Study Settings section and click on the plus sign (+) to add the existing parameter Vmax to the list in the table. Enter 0.5, 1, 1.5 for the Parameter value list, as shown in Figure 4.37. Notice that these values should satisfy the criteria $L_{entr} > 0.06ReD$.

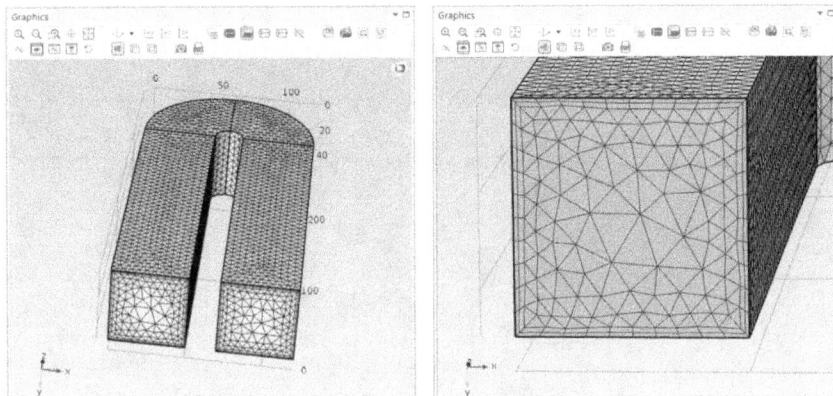

FIGURE 4.36 Windows showing finite elements mesh.

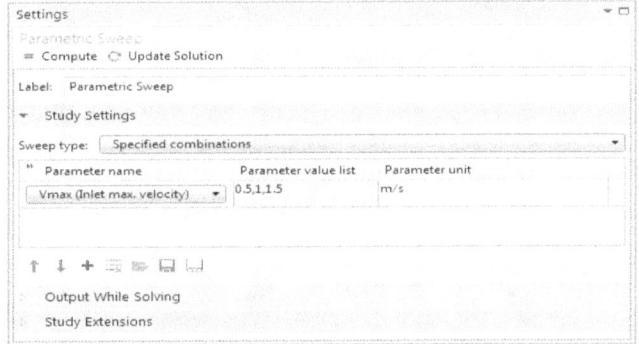

FIGURE 4.37 Parametric values for inlet velocity.

16. Run the model by clicking on the Compute button. Wait until computations are finished.

17. To manipulate the default results, expand the Velocity (*spf*) node and click on Slice 1. In the corresponding settings window, locate Plane Data section and select xy-planes from the list; enter 5 for Planes. Click the Plot icon. The results will show the air velocity magnitudes for the Vmax = 1.5 m/s. Notice the effect of the bend on the flow, especially through the downstream section of the pipe, as shown in Figure 4.38.

To study the results, it would be useful to use 1 plane in the Interactive mode. In the Slice settings window, enter 1 for Planes and check the box for Interactive. By sliding the slider nub, the plane will move

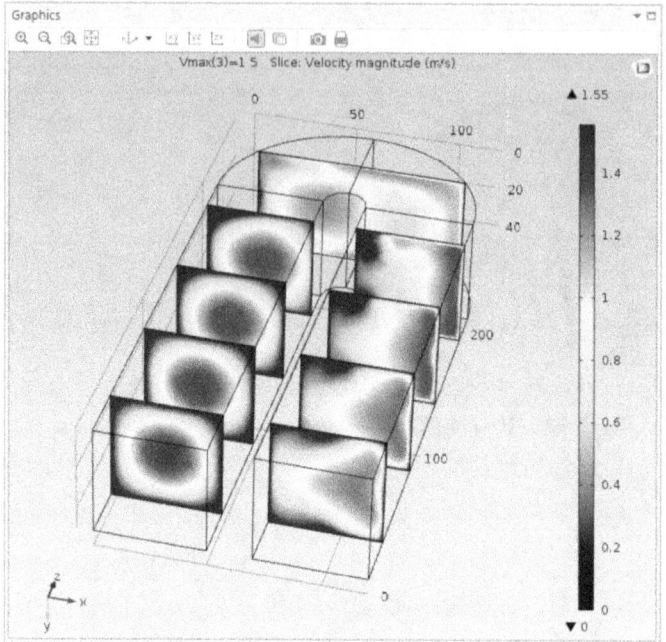

FIGURE 4.38 Flow velocity along the pipe cross sections.

through the pipe domain while showing the modeling results for velocity magnitude.

18. To show the results through the bend we create a Selection out of the domain and plot the results for this sub-domain, which is limited to the volume inside the pipe bend. Click on the Results tab from the toolbar and select Selection, from the ribbon. In the corresponding settings window, select Domain from the list for Geometric entity level. Click on the domain that represents the bend (i.e. domain 2) in the Graphics window to add it to the Selection list. See Figure 4.39.

Now click on the Slice 1 under Velocity (*spf*) node and in the corresponding settings window under Plane Data section change the Plane to yz-planes and un-check Interactive box. Click on Plot. The velocity magnitude will be displayed in the bend domain only, as shown in Figure 4.40.

19. To show the results for other values of Vmax, click on Velocity (*spf*) node in the Model Builder. In the corresponding window, locate the

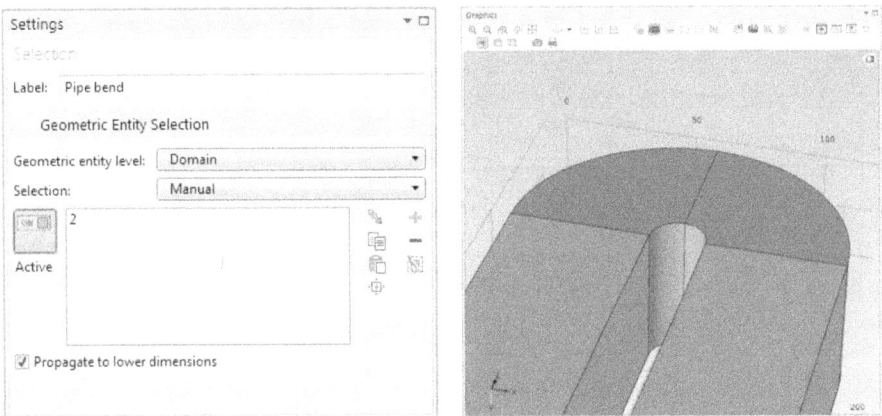

FIGURE 4.39 Windows showing Domain selection as the pipe bend.

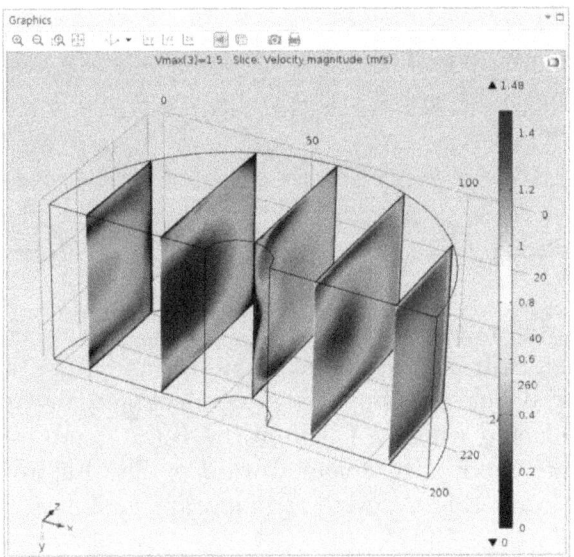

FIGURE 4.40 Flow velocity along the pipe bend cross sections.

Data section and select the desired Parameter value (*Vmax*). These values are the same as those set for these parameters.

20. To build an app for this model, launch COMSOL if not already running. From File menu select New. In the New window select Applications Wizard. In the Select Model for Application window click on Browse; locate and open Example 4.4.mph. The New

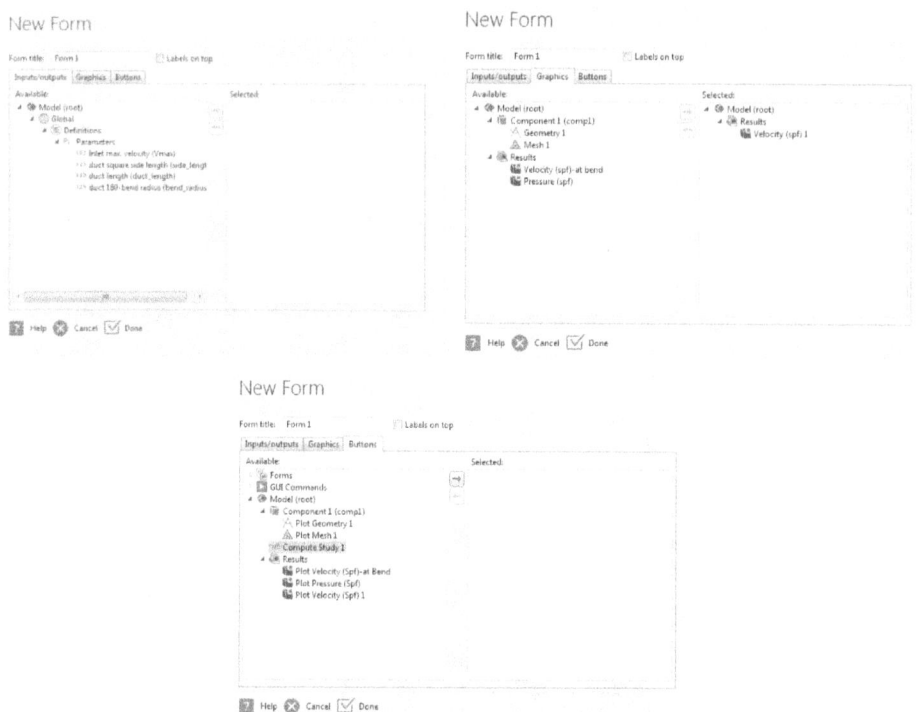

FIGURE 4.41 Form entries for Application for Example 4.4.

Form window will appear. In this window, select all entries under Parameters, from the Inputs/outputs tab and move them to Selected window. Click on Graphics tab; select and move Velocity (*spf*) 1 to Selected window. Click on the Buttons tab; select and move Compute Study 1 to Selected window. See Figure 4.41. Click on Done. Form Editor desktop window appears. Save the file as App_ Example 4.4.

21. Several editing tools are available in the Form Editor window. We use some of these tools in order to lay out the App interface. Click on Grid from the ribbon bar to relocate the objects in the Form 1. Use the Settings for each object to modify the size, affiliate a picture, and specify the function. Final results for the App interface are shown in Figure 4.42.

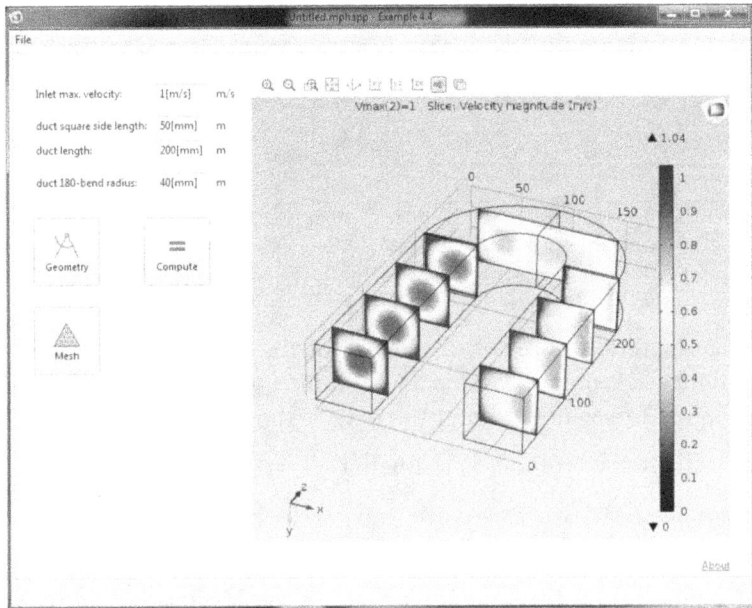

FIGURE 4.42 App interface for Example 4.4.

Example 4.5: Double-driven cavity flow: Moving boundary conditions

Driven cavity is a classic example of flow in a square shape geometry (2D) in which one of the square's side is a moving wall with a constant speed. Double-driven cavity is a version of this problem with two partially merged domains or squares and two walls moving in opposite directions [20]. In this example, we use COMSOL to find the solution for double-driven cavity. Cavity flow situations can happen in nature, such as in estuaries, and in industrial flows, such as mixing tanks.

Solution:

1. Launch COMSOL and in the New window click on Model Wizard button. Save the file as Example 4.5.

2. Select 2D in the Select Space Dimension window.

3. In the Select Physics window, select Fluid Flow > Single-Phase Flow > Laminar Flow (*spf*) and click on Add. Click the Study arrow button.

4. Select Stationary from the list in Study window and click on Done.

5. We will draw the geometry of the double-lid driven cavity. In the Geometry window, change the Length unit to cm. Next, we define

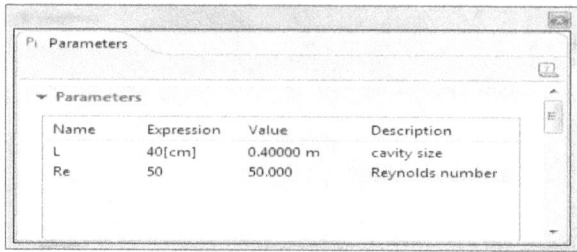

FIGURE 4.43 Parameters setting.

parameters for size of the length of the cavity and speed of its lids. Right-click on the Global Definitions and select Parameters. In the Parameters window, enter the data as shown in Figure 4.43.

The Reynolds number is defined as $Re = \dfrac{V_{wall} \cdot L}{v}$, where V_{wall} is the lid absolute speed, L is the cavity size, and v is the kinematic viscosity of the fluid inside the cavity.

6. Right-click on the Geometry node in the Model Builder window and Select Square from the list. In the Square window, enter L for the Side length under Size section, and then click the Build Selected icon. A square with size L will appear in the Graphics window. Draw another square by right-clicking again on the Geometry node in the Model Builder and selecting Square. In the corresponding settings window, enter L for the Side length and 0.4*L for both x and y under Position. Click Plot. To adjust the window, click on Zoom Extents, located in the toolbar of the Graphics window. For meshing purposes we would like to keep the edges/boundaries inside the flow domain. Click on Geometry tab from the toolbar and then select Virtual Operations > Mesh Control Edges. In the corresponding settings window select the edges inside the flow domain (4, 5, 7, 10) and add them to the selection under Edges to include. Results are shown in Figure 4.44.

7. To define material properties for the fluid, click on Materials tab from the toolbar and select Add Material. In the Add Material window click and select Liquid and Gases > Liquids > Water. Click on Add to Component. Water will be assigned to the whole domain of cavity space automatically. Notice the check marks that appear in the Material window under Material Contents for Dynamic Viscosity and Density properties. Also notice that these quantities will be calculated for a given temperature from the database. See Figure 4.45.

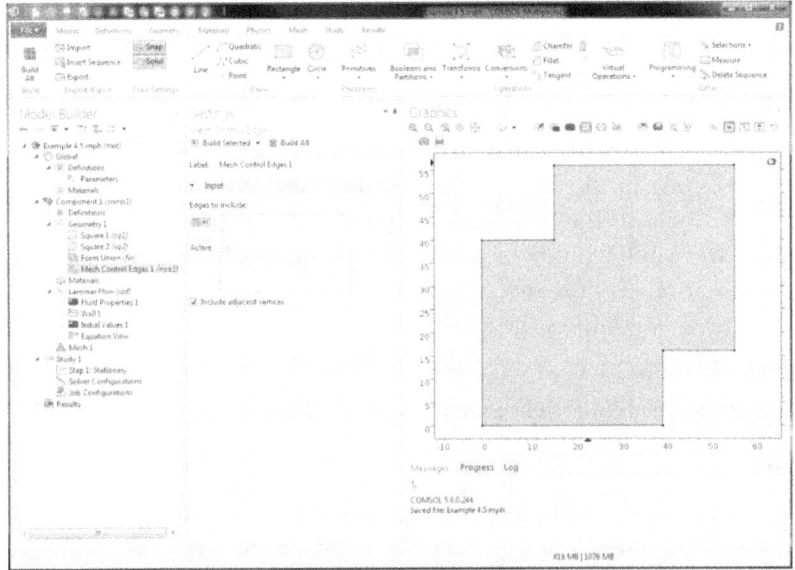

FIGURE 4.44 Double-cavity geometry setup and result.

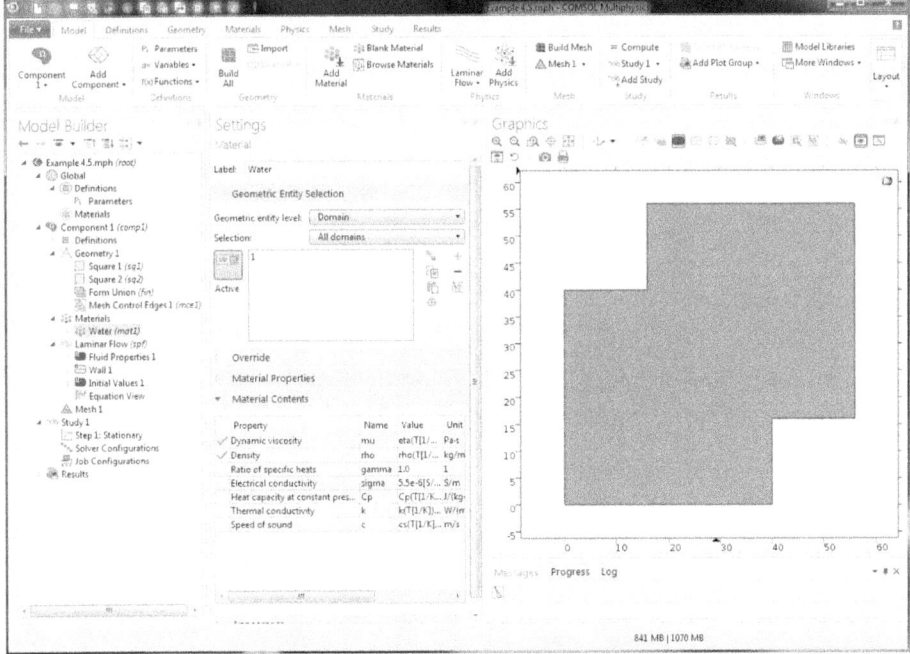

FIGURE 4.45 Adding water properties to the model.

8. We define the boundary conditions. Click on the Physics tab from the toolbar and select Wall from the list under Boundaries. A Wall 2 node will appear in the model tree. Right-click Wall 2, select Rename from the list, and change the name to Upper Lid. Repeat this last operation to create another wall and rename it to Lower Lid. Click on the Upper Lid node in the Model Builder, and then click on the upper edge of the cavity in the Graphics window to add this edge to the Selection. In the same window, select Moving wall from the list under Boundary condition section and enter $(Re * spf \cdot mu)/(L * spf \cdot rho)$ for x. This expression defines the speed of the upper lid, as shown in Figure 4.46.

Similarly, for Lower Lid define the speed as $-(Re * spf \cdot mu)/(L * spf \cdot rho)$. The reaming walls are defined as no-slip by default; no operations are needed to define them. We should define a point for pressure reference since the flow inside the cavity is a closed one. Since we have an internal flow, without inlet/outlet, we should define the reference pressure. Right-click on Laminar Flow (*spf*) and select Points > Pressure Point Constraint. In the corresponding settings window, click on point 6 (i.e. vertex at the lower corner) in the cavity geometry and add it to the list of Selections. See Figure 4.47.

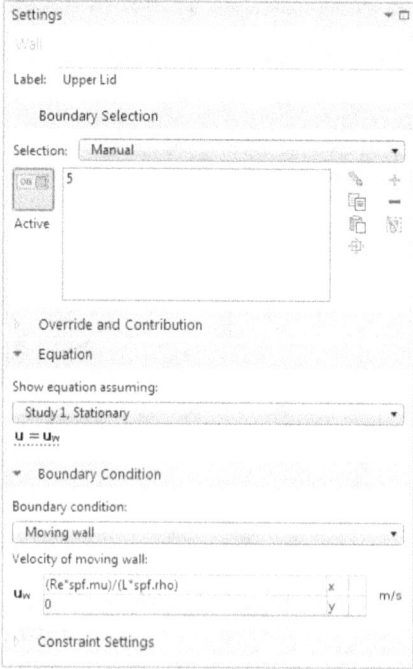

FIGURE 4.46 Moving Wall velocity boundary condition setup.

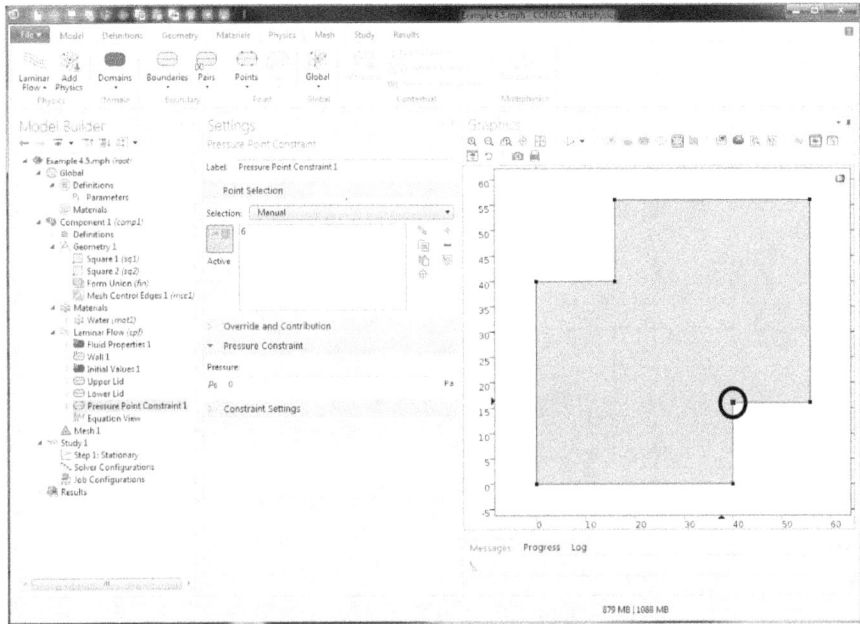

FIGURE 4.47 Pressure reference point boundary condition setup.

9. For complex flow models like this example, it is recommended to define a few other features to enhance the convergence of the solution. To show the advanced physics options, click on the Show icon from the toolbar of the Model Builder window and select the Advanced Physics Options. Then click on the Laminar Flow (*spf*) node and in the corresponding window locate Advanced Settings section and expand it. Check the box for *Use pseudo time stepping for stationary equation form*. Also in this window, select Incompressible flow from the list under Compressibility in the Physical Model section. See Figure 4.48.

10. To build a mesh, click on the Mesh in the Model Builder window. In the Mesh window, change the Element size to Fine and click on Build All icon. To obtain the mesh statistics and quality, right-click on Mesh 1 and select Statistics. See Figure 4.49.

11. To run the model, click on the Study tab from the toolbar and Compute button. Wait for the computations to finish. The default results will appear in the Graphics window, as shown in Figure 4.50.

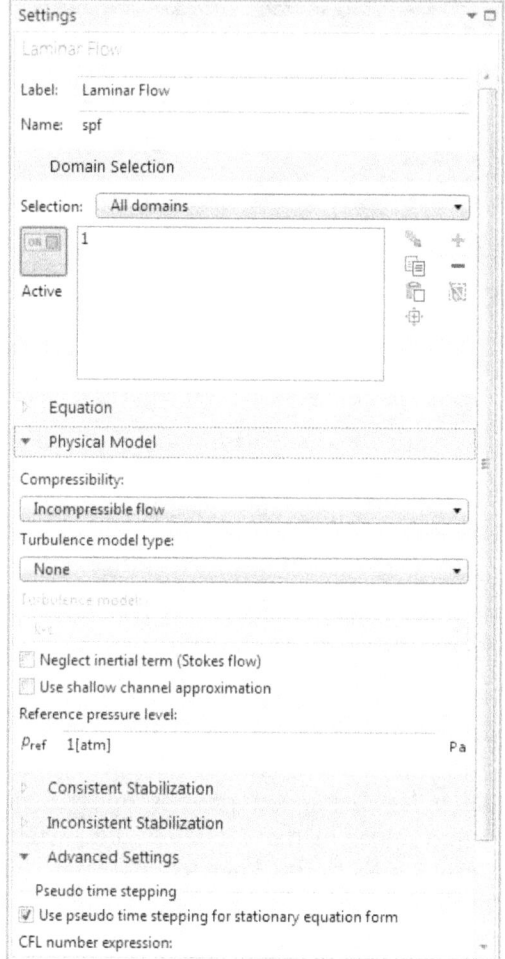

FIGURE 4.48 Laminar flow physics and pseudo time stepping setup.

12. To distinguish the solution, we rename it. Expand Data Sets, located under Results node, and rename Solution 1 to Solution Re 50, to indicate that these are the solutions for Re equal to 50.

13. To analyze the effect of mesh resolution on the results, we create another mesh with higher resolution. Rename the Mesh 1 to Mesh Fine. Right-click on the Mesh Fine node in the Model Builder window and select Duplicate from the list. A new Mesh node will be created in the model tree; rename it to Mesh Extra fine. In the corresponding Mesh settings window, select Extra fine for the Element size and click the Build All icon. A total of 21776 finite elements will

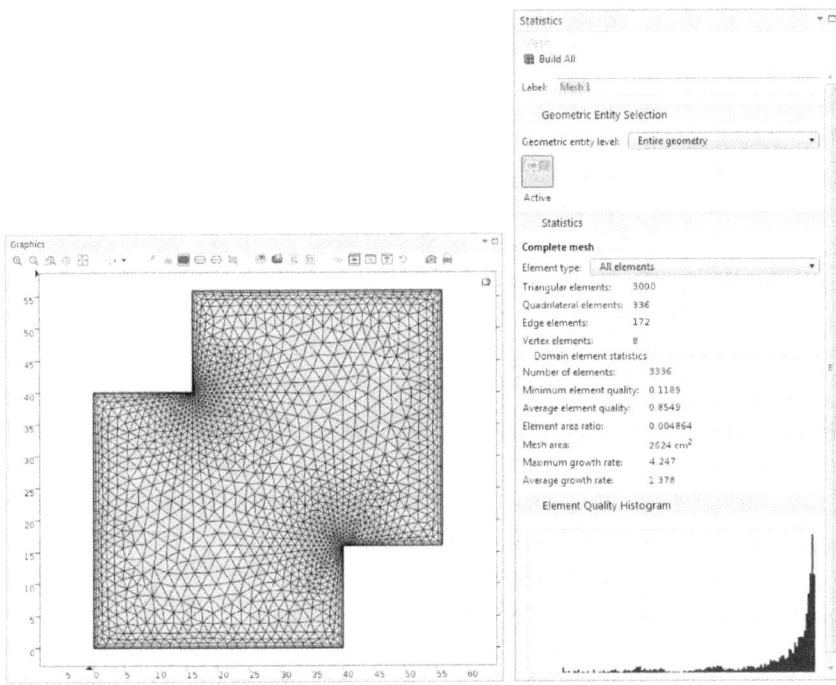

FIGURE 4.49 Finite elements mesh and statistics.

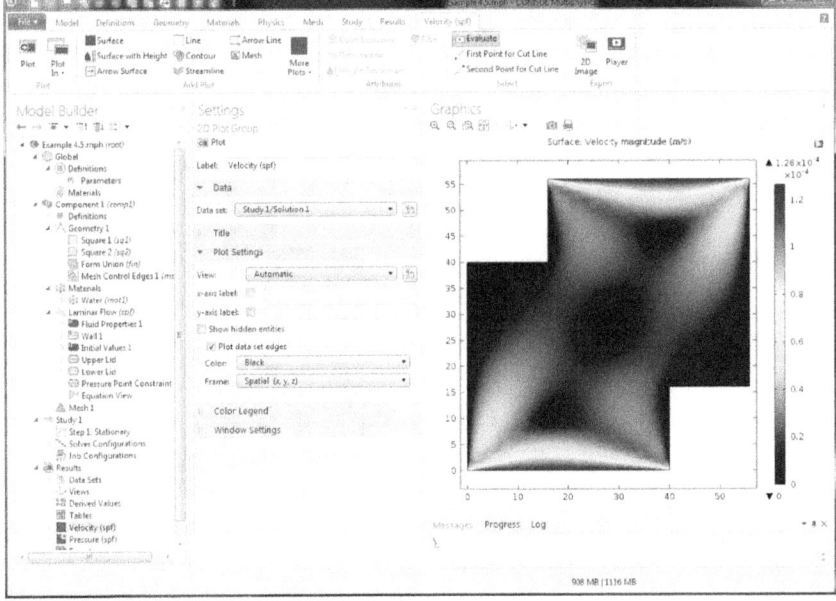

FIGURE 4.50 Default results for flow velocity in Double cavity domain.

be created. Now we have two mesh sets, which can be shown in the Graphics window when clicking on Mesh Fine or Mesh Extra fine nodes. See Figure 4.51.

14. Expand Results and right-click Data Sets > Solution to a new Solutions as data subset. Rename this node as Solution Mesh Fine. To run the model using the Extrafine mesh, click on the Study tab from the tool-bar and select Study Steps > Stationary. In the corresponding settings window, expand the Mesh Selection section and select Mesh Extra fine from the list under Mesh. Right-click on the Study 1 node and select Compute. Wait for computations to finish. Rename the solution to Solution Mesh Extra fine. Default results are shown in Figure 4.52.

15. Now we have solutions with normal mesh stored in Solution Mesh Normal and those for finer mesh in Solution Mesh Finer. We create another Data set that stores the difference between these two solu-tions. Right-click on Data Sets and select Join from the list. Click on the newly created node Join 1. In the corresponding settings window, select Solution Mesh Fine from the list for Data 1, select Solution Mesh Extra fine for Data 2, and then select Difference for Combination Method. See Figure 4.53.

16. We create a new plot group to graph the results stored in Join 1. Click on the Results tab from the toolbar and select 2D Plot Group. A new plot group will appear as 2D Plot Group 3. Click on Surface from the ribbon bar. In the Surface settings window locate Data section; select

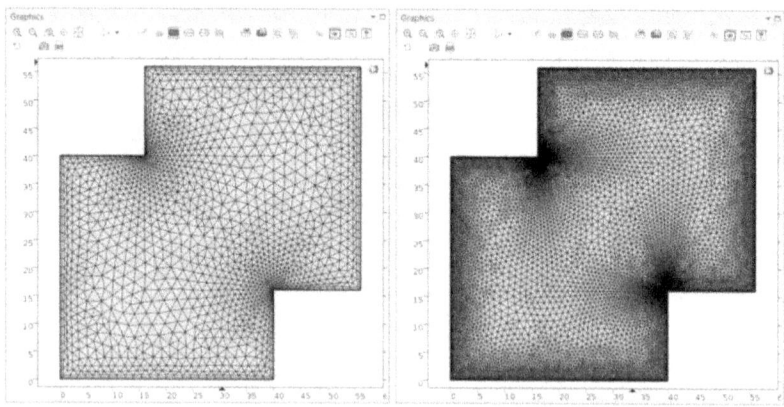

FIGURE 4.51 Graphics windows showing two mesh resolutions.

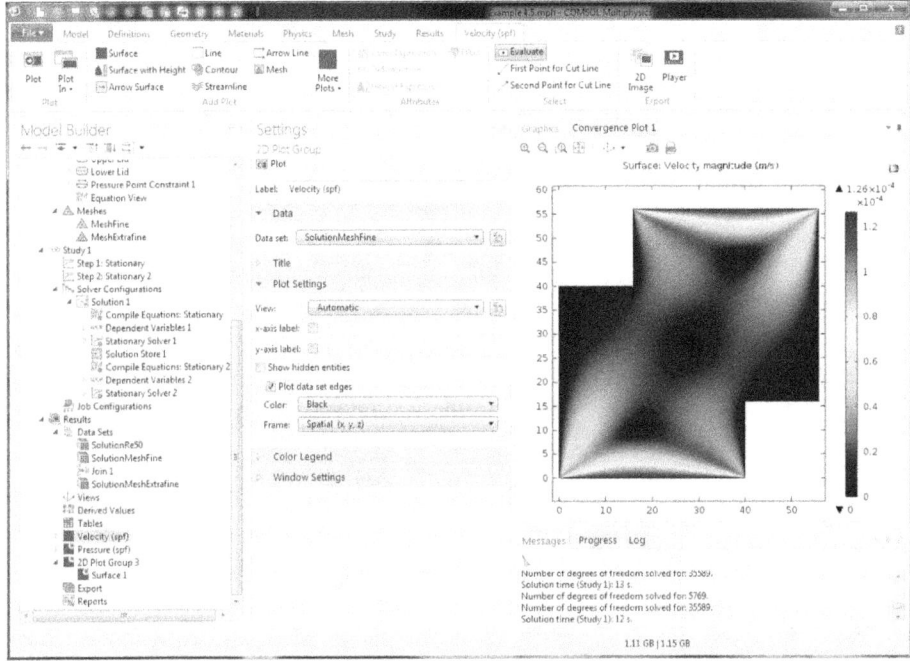

FIGURE 4.52 Results for Extrafine mesh option.

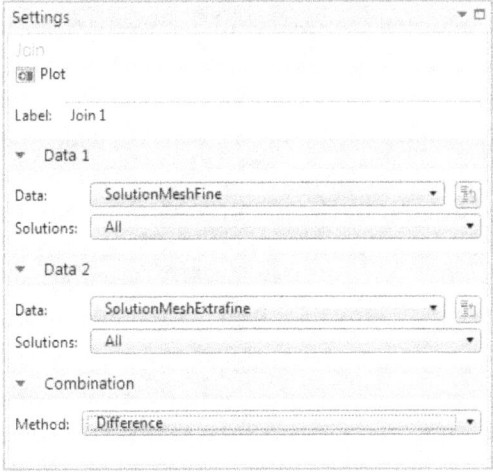

FIGURE 4.53 Combination option Difference setup for two solutions.

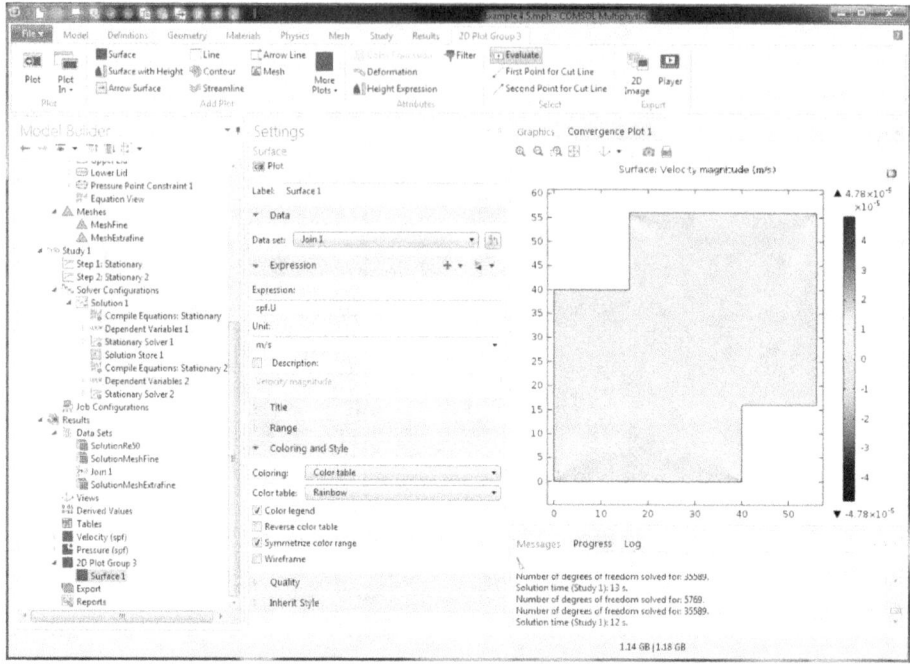

FIGURE 4.54 Join 1 setup and solution for resulting flow velocity.

Join 1 for Data set. Also expand Coloring and Style and change the options as shown in Figure 4.54. Click on Plot to plot the results. The difference between the two solutions is very small and negligible.

17. Next, we run the model for several values of Reynolds number. We use Mesh Fine for these calculations. Disable Solution Mesh Finer, Step 2: Stationary 2, and Join 1 by right-clicking on them and selecting Disable from the list. Click on the Study tab from the toolbar and select Parametric Sweep. In the Settings window click on the plus icon (+) and from the list of parameters select Re (Reynolds number). For its values enter 50, 100, 400, 1000 in the space under the Parameter value list, as shown in Figure 4.55. Run the model by clicking on Compute.

18. From the ribbon list, for Velocity (*spf*) 1, select Contour. In the Contour settings window, type u in the space available under Expression. Click on Plot. To draw the contours for a different Re, click on Velocity (*spf*) 1 and select a value for the Parameter value (Re). The results for Re = 50, 100, 400, 1000 are shown in Figure 4.56.

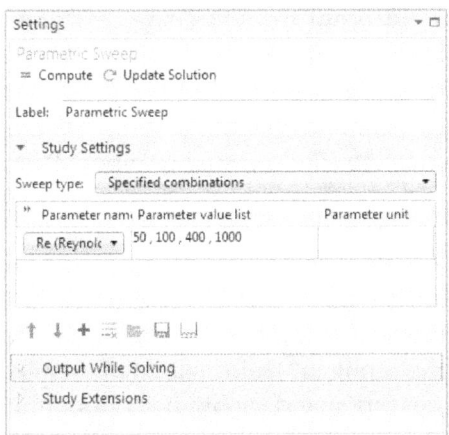

FIGURE 4.55 Parametric Sweep settings for Reynolds number values.

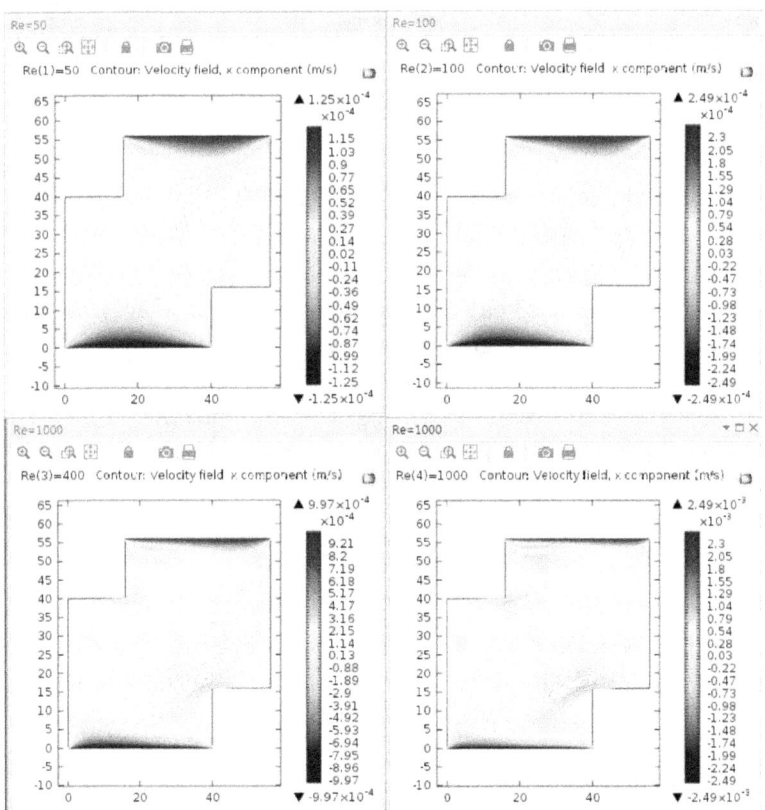

FIGURE 4.56 Windows showing results for flow x-component of velocity for several Reynolds numbers.

Example 4.6: Water Hammer model: Transient flow analysis

Water hammer is a phenomenon that occurs as a result of a sudden pressure change or pressure pulse, usually in pipelines that conduct water, but it can also occur in steam or multiphase fluids conduits. For this example, we consider water as the fluid moving in pipes. The pressure pulse could be the result of sudden closure of a valve or some other changes in the fluids that result in a pressure wave propagation. The pressure pulse creates a pressure wave that propagates through water and the pipe material. For many applications, water can be considered an incompressible fluid; for water hammer, water compressibility should be considered for a more exact solution to the problem. Deformation of the pipe material affects the pressure wave propagation and should be considered, as well. Water hammer governing equations can be derived from Navier-Stokes equations for compressible fluids [21]. The speed of pressure wave propagation is much larger than average water velocity and a 1D mathematical model is usually used for modeling:

$$\frac{g}{c^2}\frac{\partial H}{\partial t} + \frac{\partial V}{\partial x} = 0$$

$$\frac{\partial V}{\partial t} + g\frac{\partial H}{\partial x} + \frac{\tau_w \pi D}{\rho A} = 0$$

where H is piezometric head, c pressure wave speed, V average fluid velocity, τ_w wall shear stress, g gravitational acceleration, A pipe cross-sectional area, and D pipe diameter. In early cycles of water hammer and for many cases in practice the wall shear stress can be neglected, which simplifies the above equations, and pressure wave speed c is given by the following equation:

$$c = \left(\frac{d\rho}{dP} + \frac{\rho}{A}\frac{dA}{dP}\right)^{-0.5}$$

The first term in the bracket is the square of the inverse of speed of sound in water $c_0 = \sqrt{dP/d\rho}$. This term represents the compressibility of the liquid water. The second term in the bracket represents the effect of pipe material flexibility. Juokowsky's relation (see [21]) for water hammer could be derived from the 1D mathematical model equations as:

$$\Delta P = \pm \rho c V$$

where ΔP is the change in pressure. COMSOL (since version 4.3) has a water hammer module that employs these equations. In this example, we use this module to model and analyze the results of a water hammer in a pipe network as shown in Figure 4.57, which has two measurement points designated as B-gauge and E-gauge.

Solution:

1. Launch COMSOL and in the New window click on Model Wizard. Save the file as Example 4.6.

2. In the Select Space Dimension window select 3D. In the Select Physics window, select Fluid Flow > Single-Phase Flow > Water Hammer *(whtd)* and click on Add. Click on the Study arrow button.

3. In the Select Study, select Time Dependent and click Done. Make sure file is saved.

4. To define the parameters, click on the Model tab from the toolbar and select Parameters. In the Parameters settings window, enter the following data, as shown in Figure 4.58.

   The Expression for parameter time step dt should be explained here. Since the pressure surge is a sudden jump in pressure at a point, the mesh size and the time step should be carefully selected to have a stable and converged solution for the model. The pressure wave travels with velocity c and it takes $dt = dx/c$ seconds to travel through a distance dx, which is the mesh size. For a given typical pipe length L with number of elements N, we

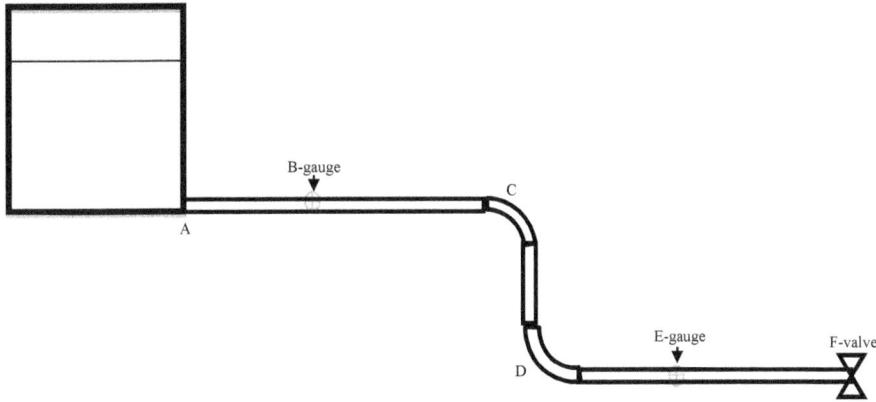

FIGURE 4.57 Geometry and components for a water network and storage.

Name	Expression	Value	Description
Lac	10[m]	10.000 m	Pipe length
Lcd	3[m]	3.0000 m	Pipe length
Ldf	10[m]	10.000 m	Pipe length
GaugeB	5[m]	5.0000 m	Measurement point
GaugeE	12[m]	12.000 m	Measurement point
R	8[in]	0.20320 m	Pipe inner radius
w	8[mm]	0.0080000 m	Pipe wall thickness
E	210[GPa]	2.1000E11 Pa	Pipe Youngs modulus
Q0	0.5[m^3/s]	0.50000 m³/s	Initial flow rate
u0	Q0/(R^2*pi)	3.8545 m/s	Initial velocity
p0	1[atm]	1.0133E5 Pa	Initial pressure
L	10[m]	10.000 m	max (Lac,Lcd,Ldf)
N	400	400.00	Number of elements/m...
dt	0.2*L/N/1200[m/s]	4.1667E-6 s	Time step
dx	L/N	0.025000 m	

FIGURE 4.58 Parameters for water hammer quantities and their values.

have $dx = L/N$. Thus dt is equal to L/cN. However, we would like to have a much smaller time step so the wave can be captured within a mesh size, say:

$$dt = \frac{0.2L}{cN}$$

Speed of wave should be estimated, which is equal to 1200 m/s here.

5. To draw the pipe network laying in the x-y plane, right-click on the Geometry 1 node in the Model Builder window and select More Primitives > Polygon. In the Polygon settings window, enter the following data in the Coordinates section and click Build All Objects. The pipeline geometry will appear in the Graphics window. Nodes at x = 5 and 12 are measurement/gauge points. Results are shown in Figure 4.59.

6. To add material to the model, click on the Materials tab from the toolbar and select Open Add Material. In the Add Material select Liquid and Gases > Liquids > Water and add it to the model by clicking on Water and selecting Add to Component. Close Add Material window.

   Now we set up the boundary conditions and pipe shape and dimensions.

7. Click on the Pipe Properties 1 node in the Model Builder, located under Water Hammer (*whtd*). In the corresponding settings window,

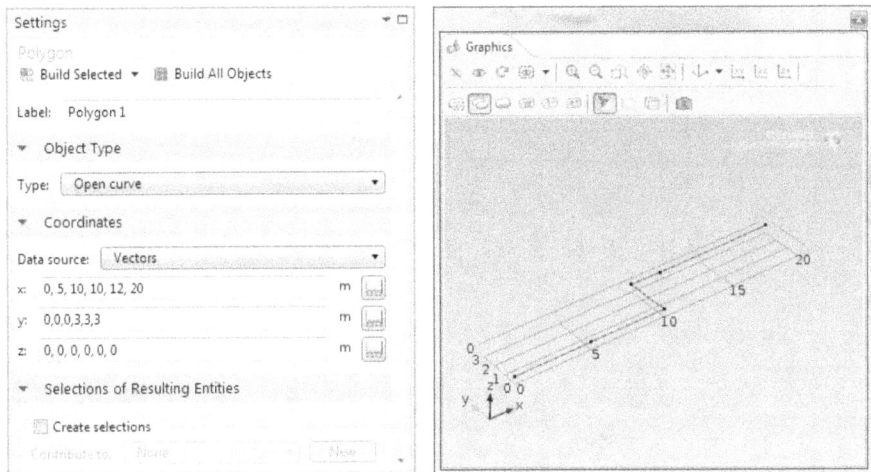

FIGURE 4.59 Values for building the pipeline geometry and results.

locate Pipe Shape section and select Round from the list. Enter 2*R for Inner diameter. In the Pipe Model section, select User defined for Young's modulus and enter E. Similarly, enter w (lowercase font) for Wall thickness. In the Flow Resistance section, select User defined for Friction model. Leave Darcy friction factor as zero. We accept this since the pipe's wall friction has minimal effects on the water hammer phenomenon, at least for most practical cases and during early cycles of oscillations. See Figure 4.60.

8. Click on the Physics tab from the toolbar and select Points >Pressure from the list. In the Pressure settings window add point 1 to the Selection list by clicking on the point at the start of the pipe network. Enter p0 for Pressure and check the box for Use weak constraints. This will help convergence of the solution for friction locations. Similarly, add Points > Local Friction Loss. In the corresponding settings window enter 0.9 for Loss coefficient (K_f for 90° bend). Click on the Initial Values 1 node in the Model Builder window and in the corresponding settings window, under Initial Values section, enter 87958 Pa for Pressure and 3.85 m/s for Tangential velocity. The numerical values can be calculated for the pipe flow with set pressure p0 at the entrance (point A) and velocity u0 at the exit (point F). Alternatively, we could add a pipe flow to this model to calculate the pressure and velocity.

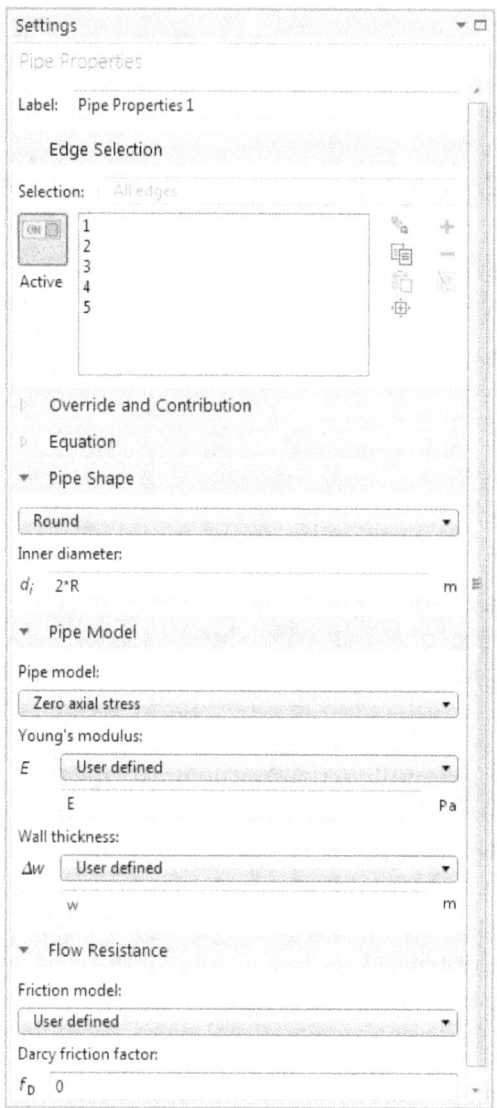

FIGURE 4.60 Pipe properties and shape settings.

9. To build the mesh, click on the Mesh 1 node in the Model Builder window. In the Mesh settings window, select User-controlled mesh from the list under Sequence type. Click on the Size node under Mesh 1 and in the corresponding settings window select Custom, then enter

L/N for Maximum element size and 1 [mm] for Minimum element size. Click the Build All icon to build the mesh. Refer to Figure 4.61 for mesh parameters.

10. To set up the study, click on the Step 1: Time Dependent node in the Model Builder window. In the corresponding window, enter range (0,1e−3,0.25) for the Times. This will set a range starting from 0 seconds up to 0.25 seconds, with a time step of 0.001 seconds for capturing and storing the solutions into the solution database. The maximum run time can easily be estimated by multiples of the time required for the pressure wave to travel once over the length of the pipe line (about 25/1200 sec.).

11. For the solver time step, we use dt. Right-click on the Study 1 node in the Model Builder and select Show Default Solver from the list. A new Solution 1 node will appear in the model tree. Expand the Solution 1 node and click on Time-Dependent Solver 1. In the corresponding settings window, locate the Time Stepping section and check Maximum step to enter dt. For initial step, enter 0.5*dt. See Figure 4.62.

12. Click on the Study tab and select Compute. Wait for computations to finish.

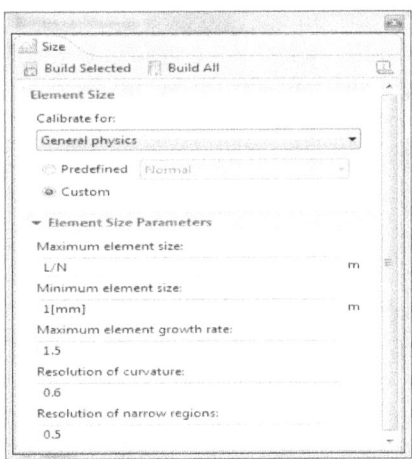

FIGURE 4.61 Custom mesh size and parameters.

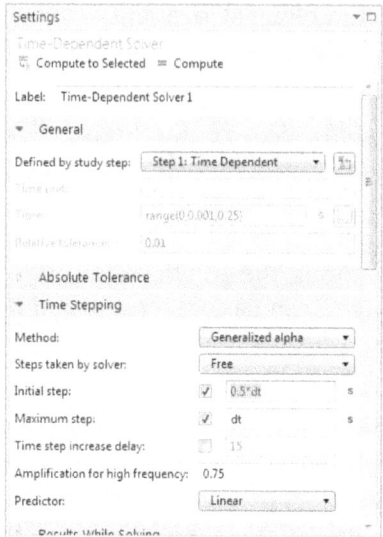

FIGURE 4.62 Transient solver parameters.

13. For visualization of the results, we use line graphs. Click on the Results tab from the toolbar and select 1D Plot Group. Rename the newly generated line group to Pressure Line Plot. In the corresponding settings window, choose From list, for Time selection options, and then click to highlight 0.244 from the values listed under Times. Select Line Graph from the list under Pressure Line Plot tab. In the corresponding settings window select All edges under Selection section, expand the y-Axis Data section, and enter p for Expression. Click Plot. The results for the water wave pressure at time 0.244 seconds appears in the Graphics window. See Figure 4.63.

14. Another useful graph is to draw the pressures at the gauges (measurement points 2 and 5). Click on the Results tab and create another 1D Plot Group; rename it to Pressure point Plot. Right-click on the Pressure point Plot node and select Point Graph. In the corresponding settings window, select and add points 2 and 5 to the Selection list. In the y-Axis Data, type in p. Click Plot. The results appear in the Graphics window. Pressure variations for points 2 and 5 versus time clearly show the water wave moving back and forth through the pipe and measured at these points. Results are shown in Figure 4.64.

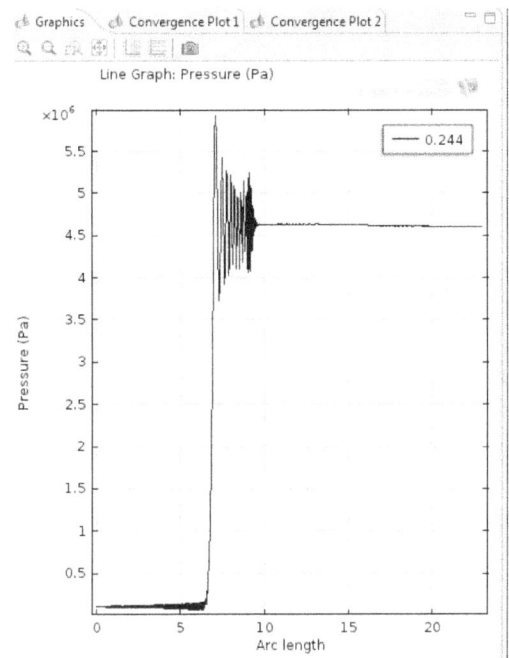

FIGURE 4.63 Pressure pulse for water hammer phenomenon at t = 0.244 s.

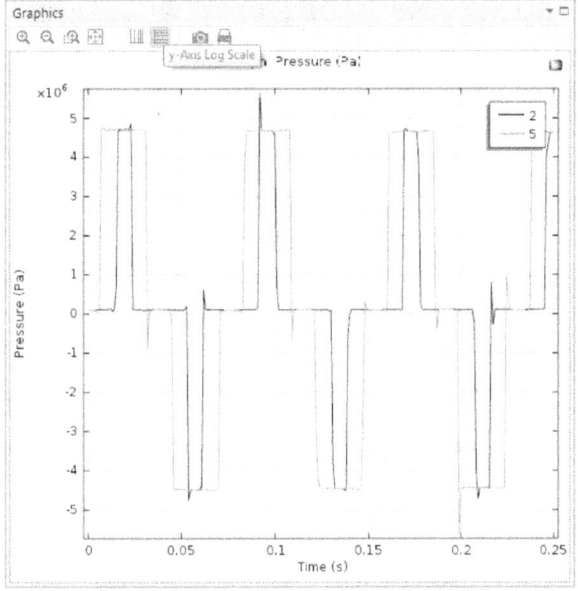

FIGURE 4.64 Variations of pressure due to water hammer phenomenon at gauge points 2 and 5.

Example 4.7: Static fluid mixer model

In this example, we use flow analysis tools (Computational Fluid Dynamics, or CFD) in COMSOL for analyzing the flow in a static mixer. The mixer is a 2D box mixer with dimensions 60 cm by 40 cm and has two inlets and one outlet. Three internal plates separate the internal space of the mixer to guide the flow for mixing.

Solution:

1. Launch COMSOL, and in the New window click on Model Wizard. Save the file as Example 4.7.

2. Select 2D from the list in Select Space Dimension window. In the Select Physics window, expand the list for Fluid Flow and select Single-Phase Flow > Laminar Flow (*spf*). Click on Add. Then click on the Study arrow button to go to the next window. In the Selected Study window select Stationary and click on Done.

3. To build the geometry of the mixer:

   3.1. Click on the Geometry 1 node in the Model Builder window (if not already highlighted). In the Geometry settings window under Units, change the Length unit to cm from the list (open the list to see unit options). Draw a box by right-clicking on the Geometry 1 node and select Rectangle. The Rectangle window will open. In this window, enter 80 for Width and 40 for Height in the Size section, and 0 for x and 0 for y under Position. Make sure that Corner is selected for Base. Click on the Build Selected icon.

   3.2. Similarly, draw three more boxes for inlets and out manifolds. Right-click on the Geometry 1 node in the Model Builder window and select Rectangle. In the Rectangle settings window, enter 4 for Width and 6 for Height in the Size section, and 6 for x and 40 for y under Position. Make sure that Corner is selected for Base. Click on the Build Selected icon. Again, right-click on the Geometry 1 node in the Model Builder window and select Rectangle. In the Rectangle settings window, enter 6 for Width and 8 for Height in the Size section, enter 80 for x and 4 for y under Position, and make sure that Corner is selected for Base. Finally, right-click on the Geometry 1 node in the Model Builder window and select Rectangle. In the Rectangle settings window, enter 6 for Width and 4 for Height in the Size section, enter −6

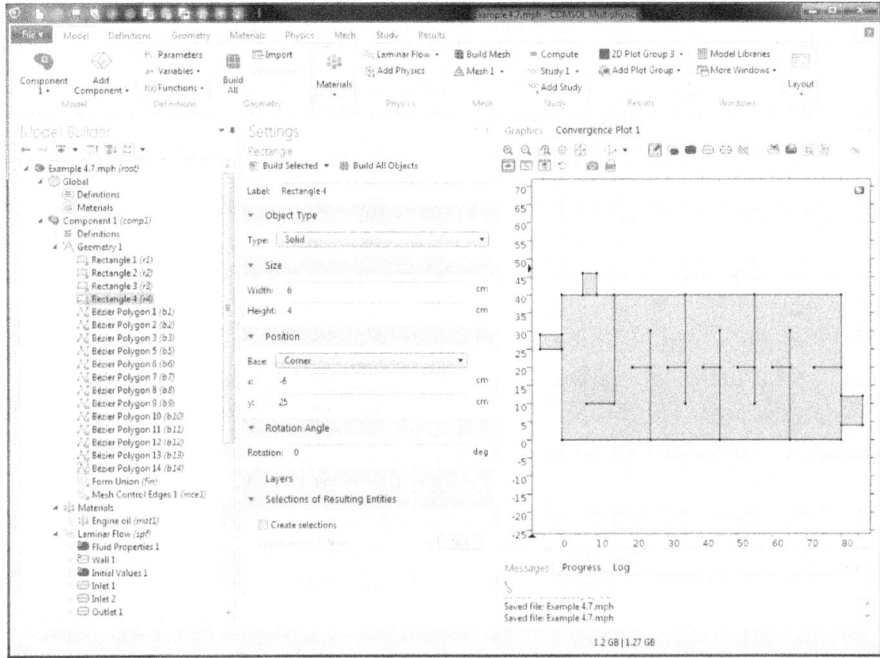

FIGURE 4.65 Mixer geometry with internal baffles.

for x and 25 for y under Position, and select Corner for Base. Click on Build All Objects. Results are shown in Figure 4.65.

3.3. Next, draw the baffles. Click on the Geometry tab from the toolbar. Click on the Draw Line icon in the toolbar and draw a vertical line inside the mixer box (click on a point inside the box; then move the mouse and click again—to release, right-click). In the Bezier Polygon settings window, click on Segment 1 (linear) located in the area under the Added segments. In the Control points section, enter the coordinates of the line points as (15, 40) and (15, 10) and then click the Build Selected icon. Similarly, draw 12 more lines with coordinates as {(25, 0), (25, 30)}, {(35, 40), (35, 10)}, {(45, 0), (45, 30)}, {(55, 40), (55, 10)}, {(65, 0), (65, 30)}, {(80, 20), (72, 20)}, {(15, 10), (7, 10)}, {(65, 20), (60, 20)}, {(55, 20), (50, 20)}, {(45, 20), (40, 20)}, {(35, 20), (30, 20)}, and {(25, 20), (20, 20)}. The final Geometry should be as shown in Figure 4.65.

3.4. For mesh control purposes, click on the Geometry tab in the toolbar. Select Virtual Operations > Mesh Control Edges. In the corresponding settings window, assign the edges/lines located at the end of inlets and outlet domains to the list. Click Build All.

4. To assign the fluid properties, Click on the Materials tab from the toolbar. In the Add Material window, expand the Liquid and Gases > Liquids node and click on Engine Oil. Click on Add to Component. The Material Settings window will open where flow domain will be listed in the Selection area. Note that the Dynamic viscosity and density of the fluid is checked. In the Value column under Material Contents section, these properties are given as a function of T, temperature. Therefore, we should define the value of temperature for which we would like to calculate these properties. To do this, right-click on the Global > Definitions node in the Model Builder window and select Parameters. The Parameters window will open. In this window, type T under Name and 40 °C under Expression. Also add two parameters for inlet velocities of the mixer, as Vin 1=1.5 [m/s] and Vin 2=1 [m/s].

5. To assign the boundary conditions:

5.1. For inlets, click on the Physics tab from the toolbar and select Boundaries > Inlet. The Inlet settings window will open. In this window, add the entrance edge of the inlet located on the top of the mixer to the Boundary Selection list by clicking on the corresponding edge of the mixer inlets in the Graphics window. In the Velocity section, choose Normal inflow velocity and enter Vin 1. Similarly, assign the boundary condition for the inlet on the left side of the inlet, set the velocity as Vin 2.

5.2. Similarly, assign the boundary conditions for the outlet. Click on Physics tab from the toolbar and select Boundaries > Outlet. The Outlet settings window will open. In this window, add the edge of the outlet to the Boundary Selection list by clicking on the corresponding edge of the mixer outlet in the Graphics window. In the Boundary Condition section, choose Pressure and enter 0 for value of p_0. Assuming zero pressure for the atmospheric pressure gives the pressure field inside the mixer as gauge pressure.

5.3. We should define the baffles as no-slip internal boundaries. Click on Physics tab and select Boundaries > Interior Wall. The

Interior Wall settings window will open. In this window, add the lines/baffles, by clicking on them in the Graphics window, to the Selection list. Make sure No slip option is selected in the Boundary condition section. See Figure 4.66.

6. For creating a mesh, click on the Mesh 1 node in the Model Builder, and in the Mesh settings window select Normal for Element size under Mesh Settings. Make sure the default sequence type Physics-controlled mesh is selected. Click the Build All icon to create the mesh. The mesh will appear in the Graphics window. Note that the mesh is a hybrid one that is a combination of quadrilateral boundary layer and triangular elements, as shown in Figure 4.67.

7. To run the model, click on the Study tab and Compute. The default velocity magnitude will appear in the Graphics window.

8. To manipulate the results, click on the Contour 1 under Pressure (*spf*) to show the pressure contours in the Graphics window. Then right-click on the Pressure (*spf*) and select Surface. A Surface 1 node

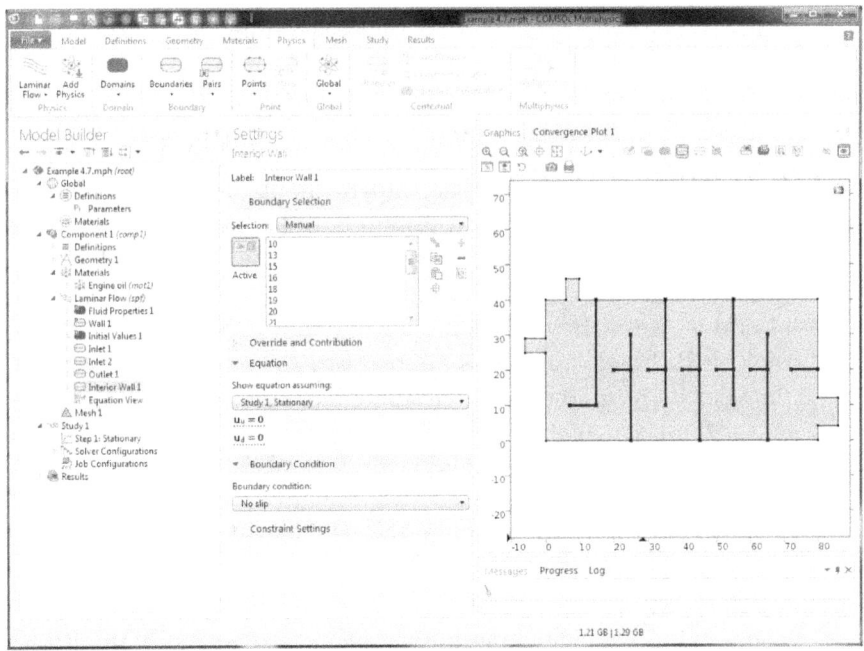

FIGURE 4.66 Boundary conditions setup for inlets, outlet, and baffles.

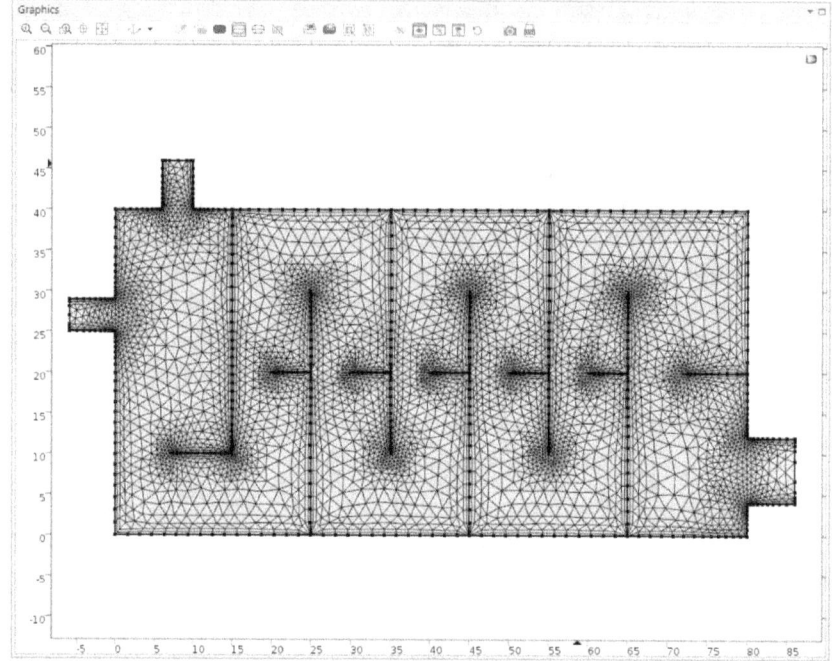

FIGURE 4.67 Finite elements mesh for the mixer.

will appear under Contour 1. Click Plot to have velocity magnitude mapped on the pressure contour, as shown in Figure 4.68.

9. It would be useful to calculate the Reynolds number for fluid flow, which helps to confirm the assumption of laminar flow made for this analysis. For the given fluid (kinematic viscosity is about 2.38E–4 Pa.s) and maximum velocity magnitude (3.65 m/s), based on inlet dimensions (4 cm) we get a Reynolds number of about 615. Also, in COMSOL, the cell Reynolds number is defined as $Rec = \rho|u|h/(2\mu)$, where $|u|$ is the velocity magnitude, h is the finite element length, ρ fluid density, and μ fluid dynamic viscosity. Click on the Surface 1 under Velocity (*spf*) node; in the Surface window click Replace Expression, and click on Laminar Flow > Cell Reynolds number (spf.cellRe) and click Plot. The values of the Reynolds number will appear in the Graphics window. The maximum value is about 31. Note that this is a cell Reynolds number and its value depends on the resolution of the mesh used. Therefore, if we build a mesh with smaller element size, the Reynolds number will be smaller, as well. This is typical for Reynolds number

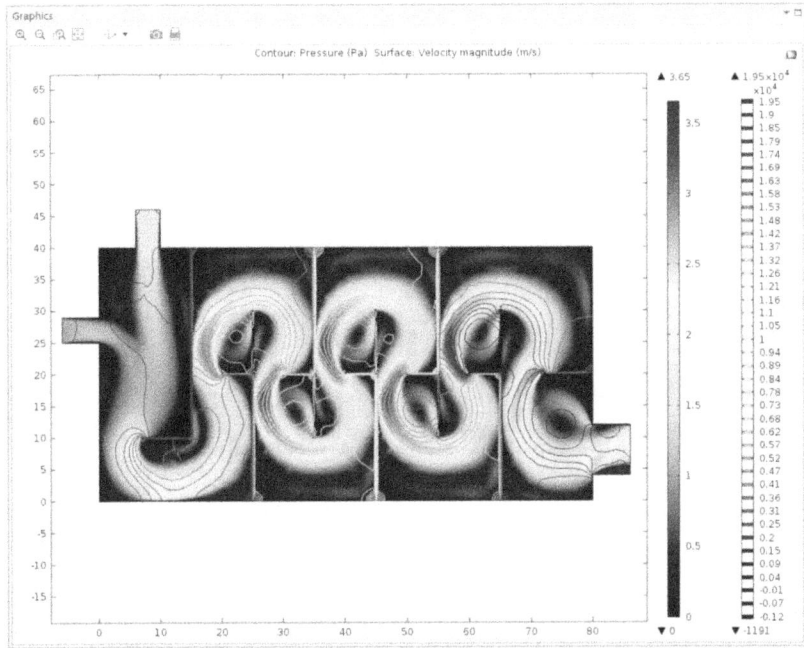

FIGURE 4.68 Results for flow velocity magnitude and pressure contours.

calculation and for any specific problem we need to identify the characteristic length scale based on which Reynolds number is defined. For this example, a Normal mesh has a maximum element size of about 3 cm, which is close to the zone that the maximum velocity is. Therefore, we can accept a cell Reynolds number of about 31. See Figure 4.69. Save the model.

10. To build an app for this model, launch COMSOL if not already running. From File menu select New. In the New window select Applications Wizard. In the Select Model for Application window click on Browse; locate and open Example 4.7.mph. The New Form window will appear. In this window, select entries Vin 1 and Vin 2 under Parameters, from the Inputs/outputs tab, and move them to Selected window. Click on Graphics tab, select and move Velocity (spf) to Selected window. Click on the Buttons tab; select and move Compute Study 1 to Selected window. See Figure 4.70. Click on Done. Form Editor desktop window appears. Save the file as App_ Example 4.7.

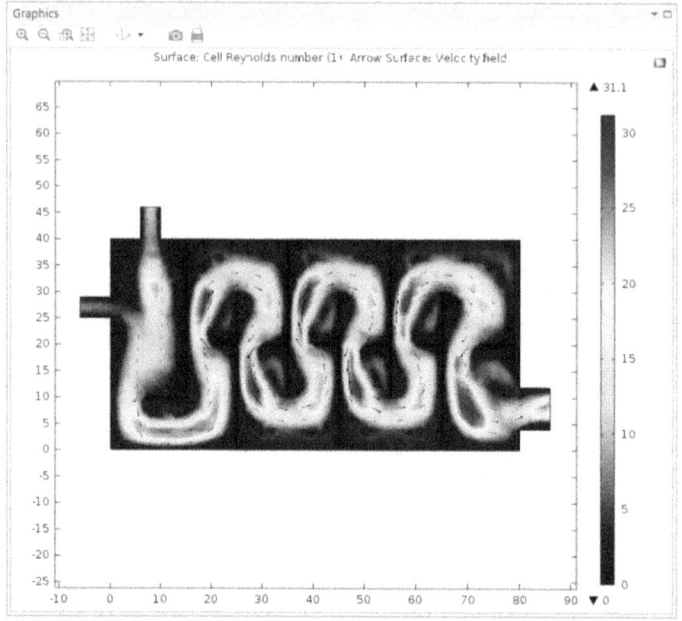

FIGURE 4.69 Cell Reynolds number distribution.

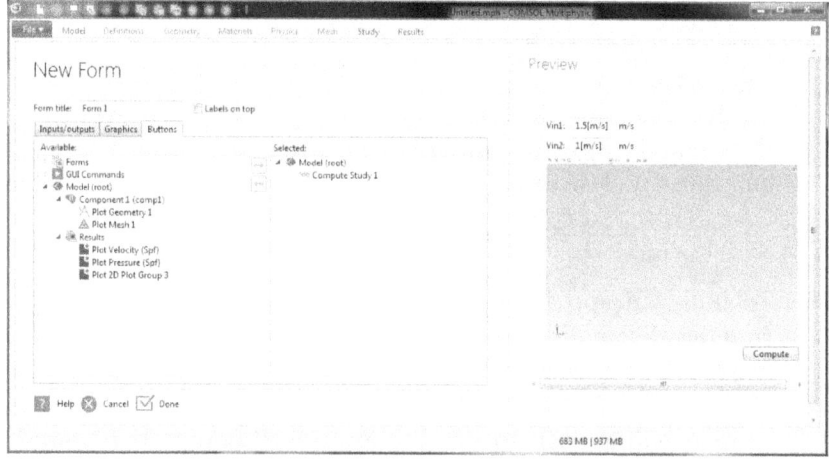

FIGURE 4.70 Application settings for Example 4.7.

11. Several editing tools are available in the Form Editor window. We use some of these tools in order to layout the App interface. Click on Grid from the ribbon bar to relocate the objects in the Form 1. Use the Settings for each object to modify the size, affiliate a picture, and specify the function. Refer to detailed instructions given for Example 2.1, for a guide. Final results for the App interface are shown in Figure 4.71.

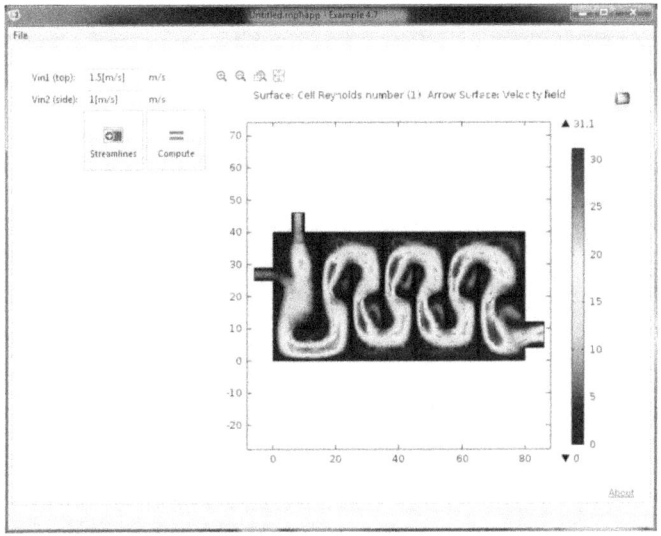

FIGURE 4.71 Application GUI for Example 4.7.

CHAPTER 5

MODEL EXAMPLES FOR HEAT TRANSFER IN MEDIA
Steady and Transient

In this chapter, we use COMSOL modules to model some examples in heat transfer. Models include steady, transient conduction, and convection in two- and three-dimensional media. The main objective is to provide, for users and readers, some solved examples that can be used directly or lead to further solutions of similar or more complex structures using COMSOL. It is assumed that readers are familiar with relevant engineering principles and governing equations related to heat transfer [22]. In each example, we provide brief explanations of physics involved along with governing equations and phenomena, as applicable. It is recommended that readers cover Chapters 1 and 2 before attempting examples in this chapter.

Example 5.1: Heat transfer in a multilayer sphere

In this example, we model a steady state heat transfer in a hollow multilayer sphere. The governing equation for steady state heat conduction is:

$$\nabla \cdot (k \nabla T) + Q = 0$$

where k is material thermal conductivity, T temperature, and Q heat source/sink per unit volume. The quantity $|k\nabla T|$ is the heat flux or thermal energy per unit area per unit time, according to Fourier's law. For the domain geometry, we assume a composite multilayer hollow sphere. We use the axisymmetric feature of the sphere for creating the geometry of the model and therefore solve a 3D problem as a 2D axisymmetric one. This approach saves on computation time and computer memory and resources.

Solution:

1. Launch COMSOL and in New window select Model Wizard. Save the file as Example 5.1.

2. In the Select Space Dimension window click on 2D axisymmetric.

3. From the Select Physics list, expand Heat Transfer and select Heat Transfer in Solids (ht) and click on Add. Click on the Study arrow.

4. From the Select Study, click on Stationary and then click on Done. Make sure file is saved.

To build the geometry, we create several half circles using the parametric curves tool available in COMSOL.

5. Click on the Geometry node in the Model Builder window and select cm from the list for Length unit for Units. Click on Geometry tab from the toolbar and select Primitives > Parametric Curve. In the corresponding settings window, enter the following data and click Build All, to create a half circle of radius 2 cm. See Figure 5.1.

6. Right-click on Parametric Curve 1 (*pc1*) and select Duplicate. A new node will appear. In the corresponding settings window, change the data under Expressions for r: to 3*cos(s) and z: to 3*sin(s), only. Periodically click Zoom Extents icon in the Graphics window to see the entire geometry. Click Build All Objects.

7. Repeat actions described in Step 6 to create two more half circles with radii 5 cm and 8 cm. See Figure 5.2.

8. To finish building the model geometry, we should create solids or domains by drawing several lines that connect the end points of the circles located on the r = 0 (or z) axis. From the ribbon, under the Geometry tab, select Draw Line. Click on an end point and then the next node to draw a line; to release, right-click. Make sure that the two ends of the central half circle are not connected, since we would like to have a hollow space at the center of the sphere. Finally, click on the Conversions > Convert to Solid, located in the ribbon toolbar. In the corresponding settings window select all the curves and add them to the Selection list. Click on Build All Objects. Result is shown in Figure 5.3.

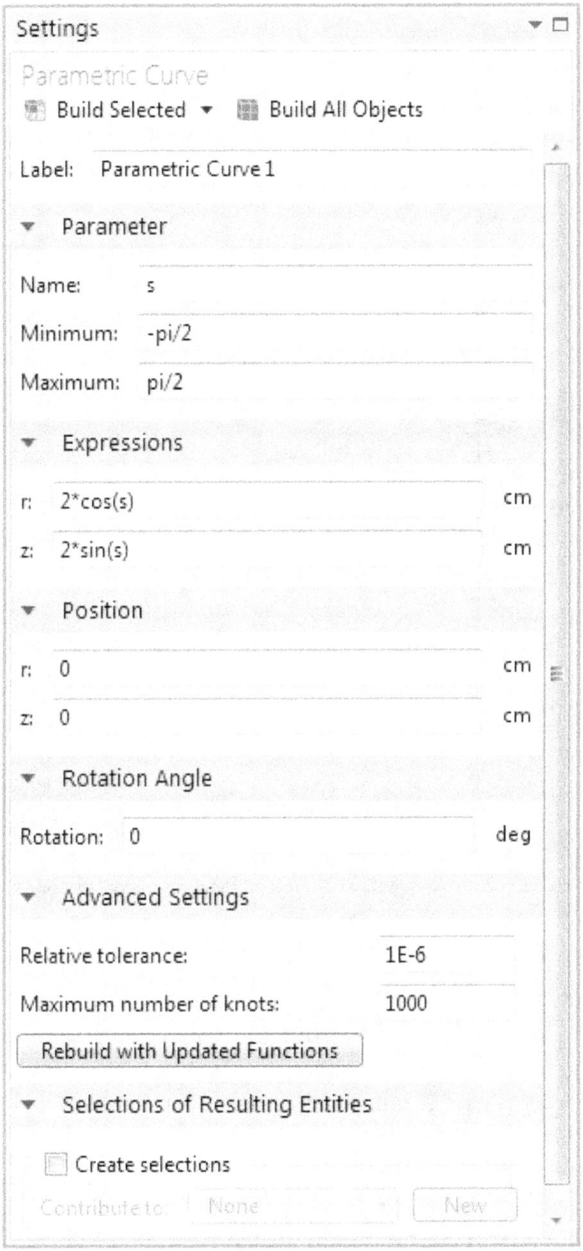

Settings

Parametric Curve

🔲 Build Selected ▼   🔲 Build All Objects

Label: Parametric Curve 1

▼   Parameter

Name: s

Minimum: -pi/2

Maximum: pi/2

▼   Expressions

r: 2*cos(s)   cm

z: 2*sin(s)   cm

▼   Position

r: 0   cm

z: 0   cm

▼   Rotation Angle

Rotation: 0   deg

▼   Advanced Settings

Relative tolerance: 1E-6

Maximum number of knots: 1000

[Rebuild with Updated Functions]

▼   Selections of Resulting Entities

☐ Create selections

Contribute to: None   ▼   [New]

FIGURE 5.1 Parametric curve data for a half circle or radius 2 cm.

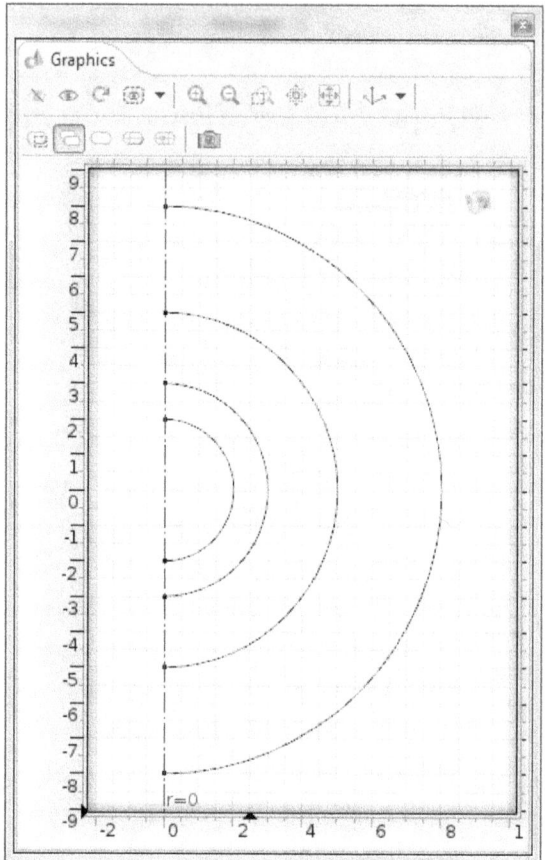

FIGURE 5.2 Resulting 2D axisymmetric geometry curves for spheres.

9. To add materials to the model, click on the Materials tab from the toolbar and select Add Material. In the Add Material window, expand Built-in, select Cast iron from the list by and then click on Add to Component. In the corresponding settings window, add/choose the outer layer (layer 1, only) to the Selection list by clicking on this layer in the Graphics window.

10. Repeat actions described for Step 9 to add material Aluminum 6063-T83 to the middle layer (layer 2) and Silicon to the inner layer (layer 3). We assume zero-thermal resistance interface for all layers.

11. To define the boundary conditions, we set a temperature at the inner surface and a convective cooling one at the outer surface. Click on the Physics tab from the toolbar and select Heat Transfer in Solids

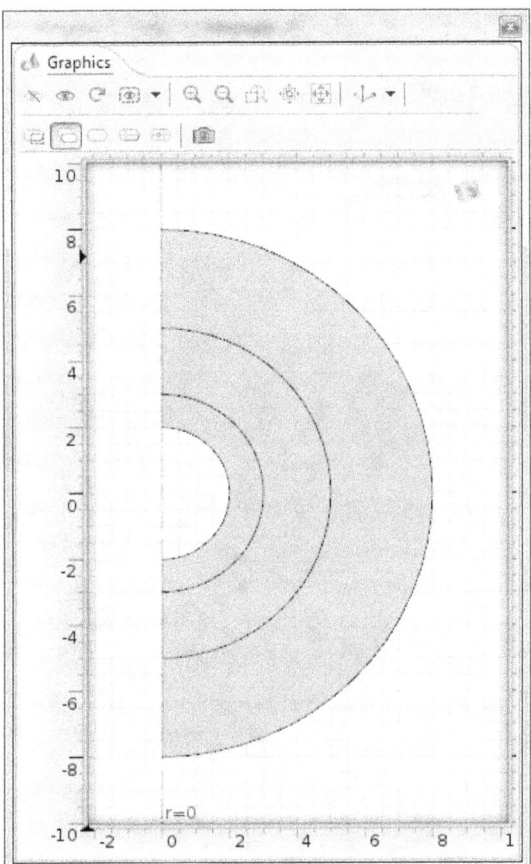

FIGURE 5.3 Resulting 2D axisymmetric domains.

(*ht*) node in the Model Builder window and select Boundaries > Temperature from the list. In the corresponding settings window, add the inner boundary (number 10) to the Selection list in the Boundary Selection section. Enter 450°C for Temperature. Similarly, select Boundaries > Heat Flux. In the corresponding settings window, assign the outer boundary (number 7) to the Selection list. In this window in the Heat Flux section, select Convective heat flux and enter 5 for heat transfer coefficient and 15°C for External temperature. See Figure 5.4.

12. To build a mesh, click on the Mesh 1 node. In the Mesh window, select Fine from the list for Element size and click Build All. A total of 464 finite elements will be created, as shown in Figure 5.5.

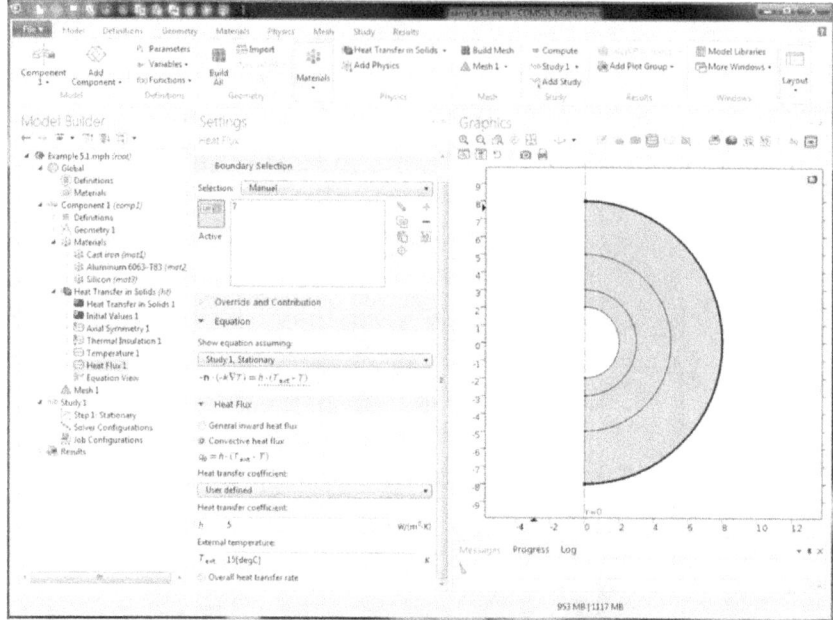

FIGURE 5.4 Boundary conditions assigned to spherical layers.

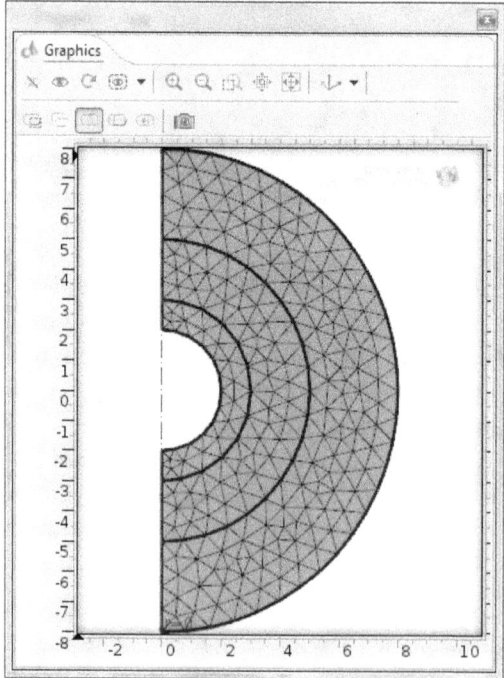

FIGURE 5.5 Finite elements mesh for spherical layers.

13. To run the model, click on the Study tab and select Compute. Wait for the computations to finish. The default results will appear in the Graphics window for the temperature values. Expand the Temperature, 3D (ht) node, and click on Surface 1. In the corresponding settings window, change Unit to degC in the Expression section and Color table to Rainbow in the Coloring and Style section. See Figure 5.6.

14. To draw the temperature on a graph, click on the Results tab from the toolbar and select 1D Plot Group. A new tab appears, as 1D plot Group 3, click on Line Graph. In the corresponding settings window for Line Graph, we can draw a line to plot the value of a quantity from

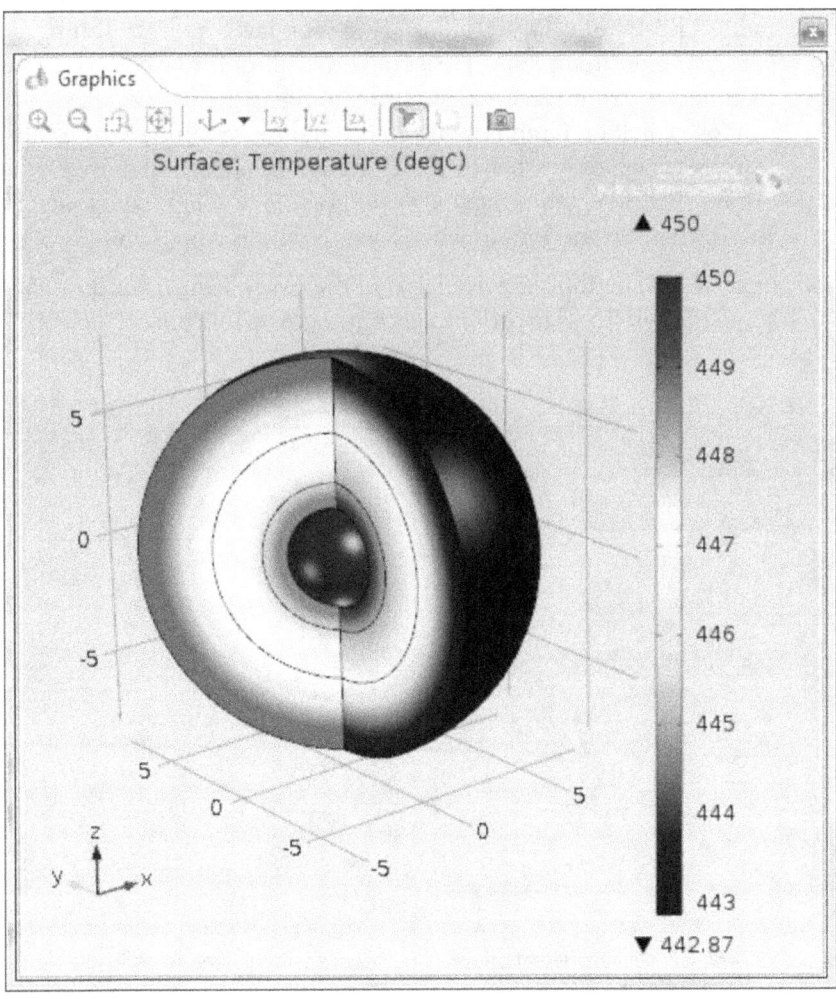

FIGURE 5.6 Results for temperature distribution.

the model. In this example, any radial line (due to symmetry) can be selected for this purpose. For example, select all boundaries along the upper part of the axis of symmetry (boundaries 4, 5, 6) from the Graphics window and add them to the Selection list (click on each boundary). In the y-axis Data section, select degC from the list for Unit. Click the Plot icon. See Figure 5.7.

15. To build an app for this model, launch COMSOL if not already running. From File menu select New. In the New window select Applications Wizard. In the Select Model for Application window click on Browse; locate and open Example 5.1.mph. The New Form window will appear. In this window, select entries R1-cavity core radius, R2-inner layer radius, R3-middle layer radius, R4-outer layer radius, Temp. at the core, Ambient temp, and Heat trans. Coeff.-outer layer surface under Parameters, from the Inputs/outputs tab and move them to Selected window. Click on Graphics tab, select and move Temperature, 3D (ht) to Selected window. Click on the Buttons tab; select and move Compute Study 1 to Selected window. See Figure 5.8. Click on Done. Form Editor desktop window appears. Save the file as App_Example 5.1.

16. Several editing tools are available in the Form Editor window. We use some of these tools in order to lay out the App interface. Click on Grid,

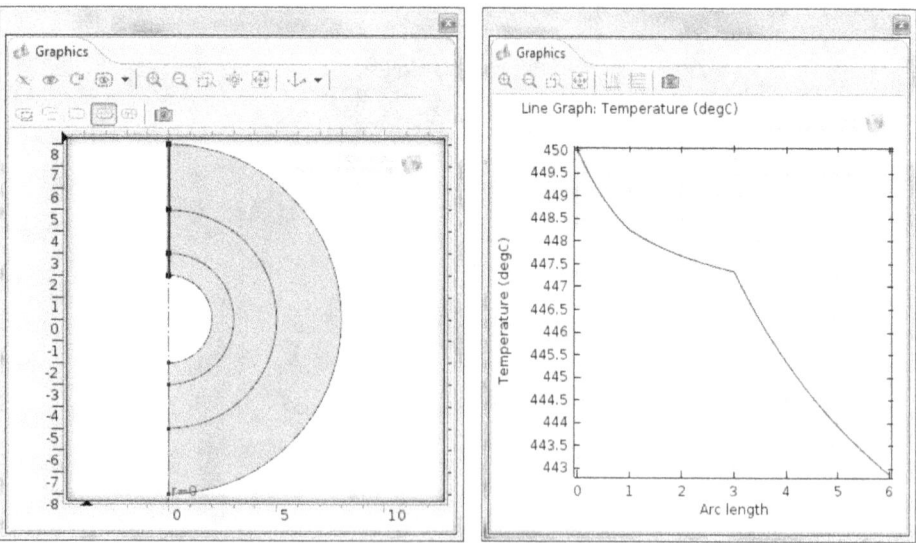

FIGURE 5.7 Windows showing results for temperature distribution across layers.

from the ribbon bar to relocate the objects in the Form 1. Use the Settings for each object to modify the size, affiliate a picture, and specify the function. Refer to detailed instructions given for Example 2.1, for a guide. Final results for the App interface are shown in Figure 5.9.

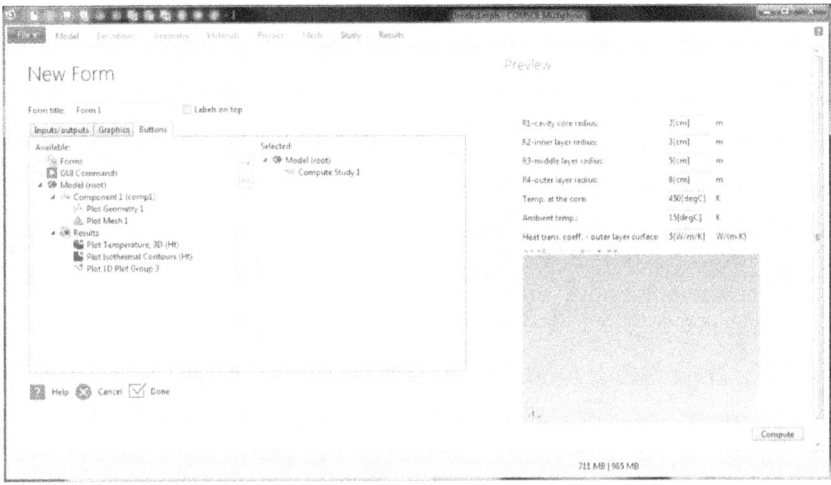

FIGURE 5.8 Application settings for Example 5.1.

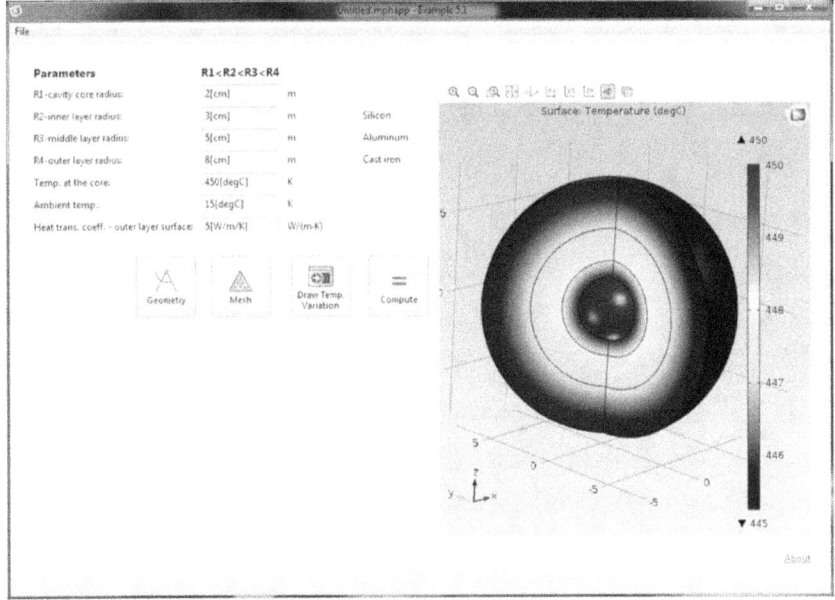

FIGURE 5.9 Application GUI for Example 5.1.

Example 5.2: Heat transfer in a hexagonal fin

Temperature variation and heat transfer calculations in fins are needed for design of many heat transfer systems, such as heat exchangers. In this example, steady state heat transfer equations are solved for a hexagonal shape fin, as shown in Figure 5.10, including convective cooling boundary conditions.

Solution:

1. Launch COMSOL and in the New window click on Model Wizard. Save the file as Example 5.2.

2. In the Model Wizard window, click on 2D.

3. From the Select Physics list, expand Heat Transfer, select Heat Transfer in Solids (ht), and click on Add. Click on the Study arrow button.

4. From the Select Study, select Stationary and then click on Done.

5. In the Geometry settings window, select cm from the list for Length unit.

6. To draw a hexagonal, right-click on Geometry 1 in the Model Builder window and select Polygon from the list. In the Polygon settings

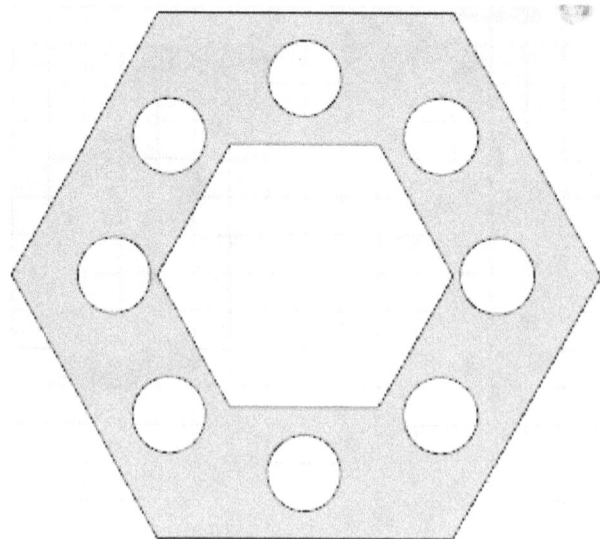

FIGURE 5.10 Geometry of the hexagonal fin.

window, enter the data as shown in Figure 5.11 for Coordinates of polygon vertices.

7. Similarly, create another polygon and enter the data as shown in Figure 5.12 for its Coordinates.

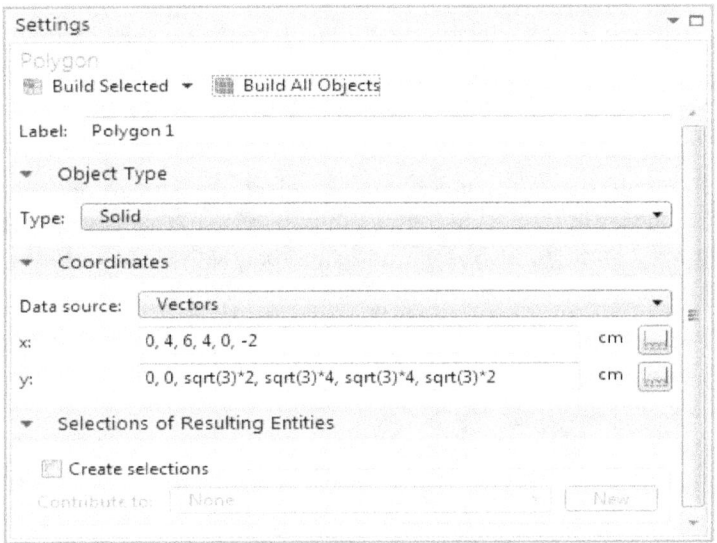

FIGURE 5.11 Polygon vertices coordinates for building outer hexagonal fin geometry.

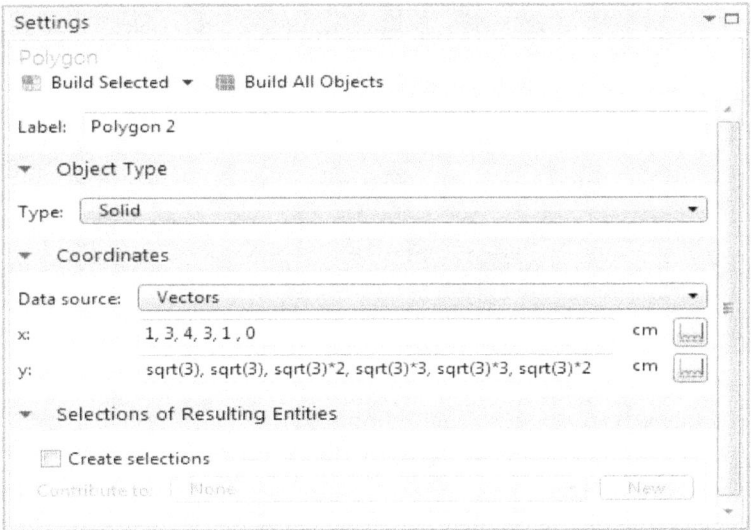

FIGURE 5.12 Polygon vertices coordinates for building inner hexagonal fin geometry.

8. To subtract the second polygon from the first one, select Booleans and Partitions > Difference, from the ribbon under Geometry tab. In the Difference settings window, specify the first polygon as the Object to add and the second polygon as the Object to subtract. Then click Build All Objects.

9. To draw a circle, click on the Geometry tab and select Circle. Draw an arbitrary circle in the Graphics window. In the corresponding settings window, enter the data as shown in Figure 5.13 for Radius and Center coordinates of the circle. Click Build All Objects.

10. To create the other seven circles, select Transforms > Rotate. In the Rotate settings window, enter range(0, 45, 315) for Rotation Angle and enter 2 and 2*sqrt(3) for x and y, respectively, for Center of Rotation coordinates. Click Build All Objects. See Figure 5.14.

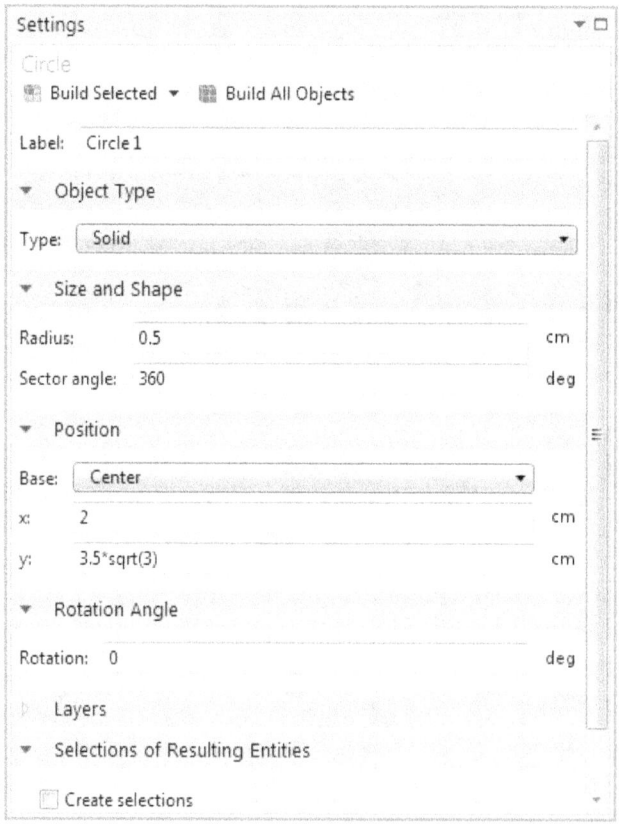

FIGURE 5.13 Parameters for building circular hole geometry.

11. Similar to Step 8 above, subtract the circles from the fin geometry. This time we use a different method. Click anywhere in the Graphics window and then press CTRL+A. All the geometry entities will change color to indicate they are selected. Then click the Difference icon located in the main toolbar under Booleans and Partitions. The final fin shape will appear in the Graphics window, as shown in Figure 5.15.

12. To add materials to the model, right-click on the Materials node in the Model Builder and select Add Material. In the Add Material window, expand Built-In and select Aluminum. Add it to the model by clicking on Add to Component. Close Add Material window.

13. Next, we add the boundary conditions to the model. Right-click on the Heat Transfer in Solids (*ht*) node in the model builder and select Heat Flux. In the corresponding settings window, add all the outer edges (1, 2, 3, 6, 11, 12) of the fin to the Selection list. To do this, click on each edge in the Graphics window. Expand the Heat Flux section and enter the Heat transfer coefficient 5 and External temperature 18°C, as shown in Figure 5.16.

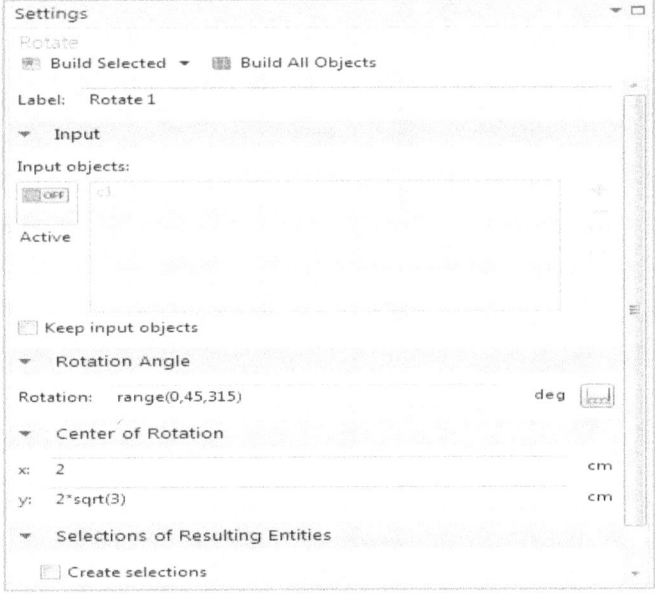

FIGURE 5.14 Parameters for building Rotate circular holes geometry.

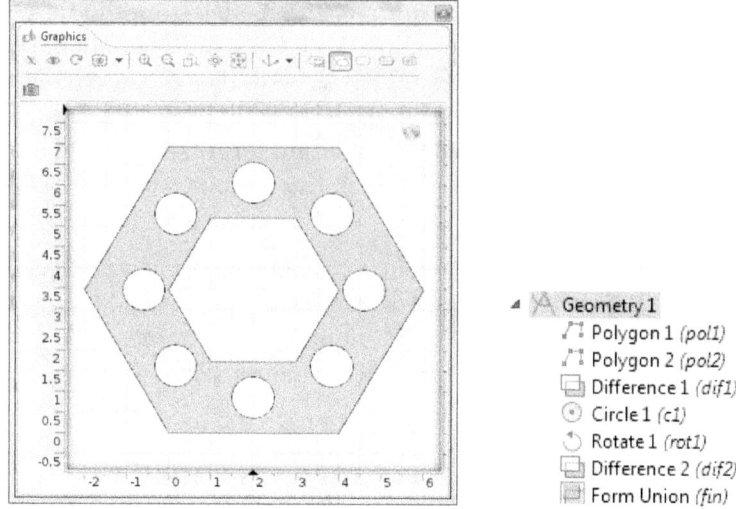

FIGURE 5.15 Resulting fin geometry and tree.

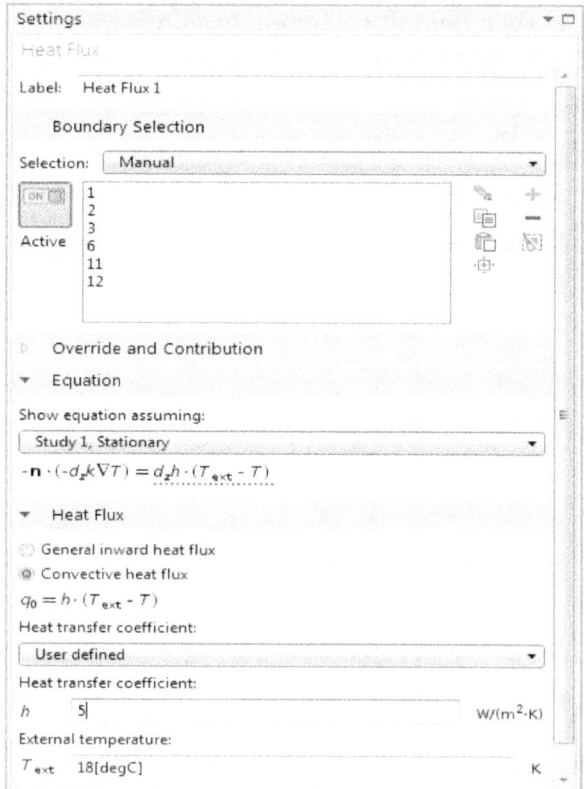

FIGURE 5.16 Convective boundary condition setup for outer edges.

14. Similarly, create another Convective Cooling node and assign the inner edges of the fin to the Selection list. Enter the Heat transfer coefficient 5 and External temperature 30°C.

15. We assume four pipes passing through the fin circular holes with surface temperature of 250° and another four pipes with surface temperature of 80°. To assign these boundary conditions, right-click on the Heat Transfer in Solids (*ht*) node in the Model Builder and select Temperature. In the corresponding settings window, add the boundary of the circular holes to the Selection list, as shown in the left window in Figure 5.17. For Temperature, enter 250 °C. Similarly, create another Temperature node, assign the remaining circular holes boundaries to it, and enter 80 °C for Temperature, as shown in the right window in Figure 5.17.

16. To build a mesh, click on the Mesh 1 node and select Fine for Element size in the Mesh window. Click Build All. Results are shown in Figure 5.18.

17. To run the model, click on Study tab and select Compute. The default results will appear in the Graphics window after computations finish, as shown in Figure 5.19.

18. To show the direction of heat flux vector, right-click on the Isothermal Contours (ht) node and select Arrow Surface. Click Plot. Results appear in Graphics window as shown in Figure 5.20.

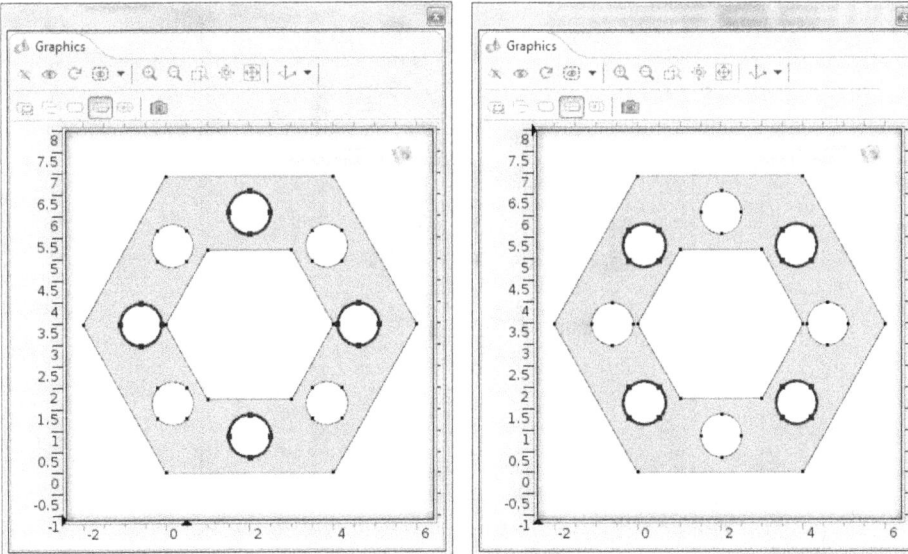

FIGURE 5.17 Windows showing temperature boundary condition setup for circular holes.

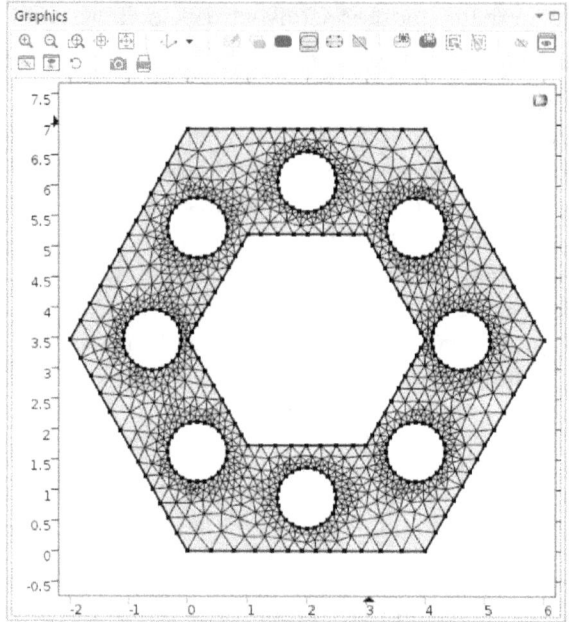

FIGURE 5.18 Finite elements mesh for the fin.

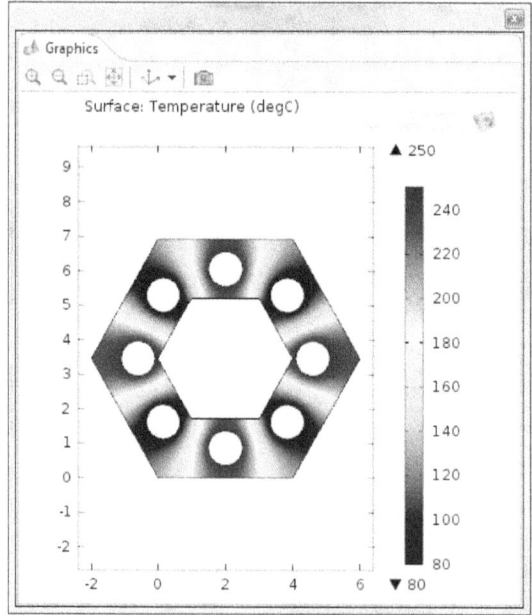

FIGURE 5.19 Results for temperature distribution for the fin.

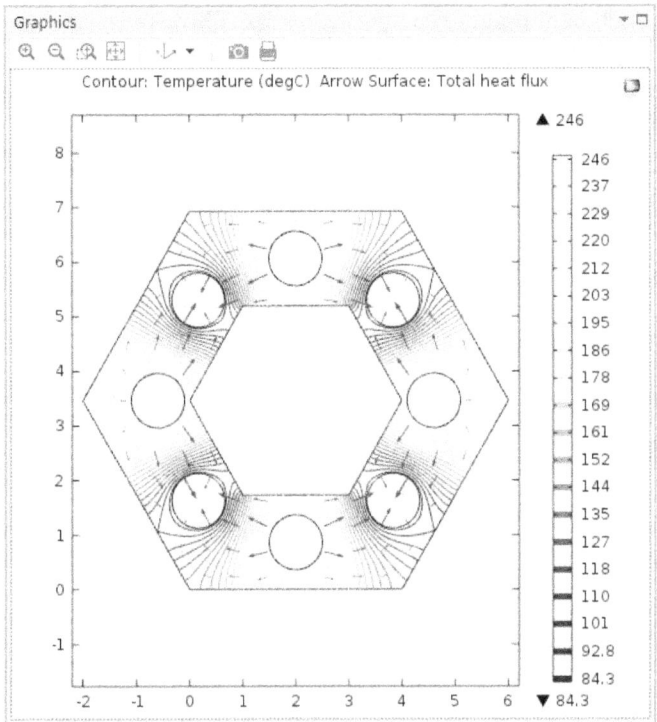

FIGURE 5.20 Results for temperature contours and heat flux vectors in the fin.

Example 5.3: Transient heat transfer through a nonprismatic fin with convective cooling

In this example, we model transient heat transfer and variation of temperature through the body of a nonprismatic fin. For heat transfer modes, we consider heat conduction inside the fin material and heat convection from its surface to the ambient. The governing PDE for transient heat transfer is given as:

$$\rho C \frac{\partial T}{\partial t} = \nabla \cdot (k\nabla T) + Q$$

where ρ is material density, C heat capacity, k thermal conductivity, T temperature, and Q heat source/sink per unit volume. Two dimensionless numbers—Biot number ($Bi = hL/k$) and Fourier number ($F_o = \alpha t/L^2$)—are used for scaling the unsteady heat transfer and nondimensionalizing the governing equation. In these equations α is thermal diffusivity ($k/\rho C$, ratio of how much heat a material conducts versus how much it stores), h is heat transfer coefficient, and L is a length scale (usually, the ratio of volume of the heat transfer domain over the surface where heat convection occurs). For example, for a sphere L is equal to one-third of its radius.

The physical meanings of Bi and Fo are useful when dealing with modeling transient heat transfer problems.

$$Bi = \frac{convection\, at\, the\, surface\, of\, the\, body}{conduction\, inside\, the\, body}$$

$$= \frac{conduction\, resistance\, inside\, the\, body}{convection\, resistance\, at\, the\, surface\, of\, body}$$

Therefore, when $Bi < 1$, conductive heat transfer is faster than convection and the spatial variation of temperature inside a solid body can be neglected, as if whole body temperature changes only with respect to time. When $Bi > 1$, conduction resistance inside the body is larger than that of convection at its surface. For the latter case, spatial variation of temperature is considerable as compared to its temporal variation inside the body, and both should be included in the model. The Fourier number is basically a dimensionless time scale defined as:

$$F_O = \frac{rate\, of\, conduction}{rate\, of\, heat\, storage}$$

For modeling an unsteady heat transfer, it is crucial to have the right time step/scale for numerical computation. If time scale is too large (versus characteristic time scale of the problem), then transient variation cannot be modeled/captured. Whereas if time step is too small, then unrealistic temperature fluctuation may show up in modeling results. When Bi is small (usually, smaller than 0.1), then we can neglect the spatial variation of temperature inside the body. In other words, for small Bi the body temperature changes mainly with respect to time. The product $Bi \times Fo$ can be used to estimate characteristic time.

$$Bi \times F_o = \left(\frac{hL}{k} \right)\left(\frac{\alpha \Delta t}{L^2} \right) = \left(\frac{h}{\rho C L} \right)\Delta t = b\Delta t$$

The inverse of b (or $\rho CL/h$) is the characteristic time. In other words, a large value of b indicates that temperature reaches the ambient temperature in a short amount of time (see [22]).

For this example, we consider a fin attached to a section of a wall, as shown in Figure 5.21.

Solution:

1. Launch COMSOL and in the New window select Model Wizard. Save the file as Example 5.3.

2. In the Model Wizard window, select 3D and click Next.

3. In the Select Space Dimension window, select 3D. In the Select Physics window, select Heat Transfer in Solids (ht) and click Add. Click on the Study arrow.

4. Under Select Study, click on Time Dependent and then click on Done. In the Geometry settings window, select mm from the Length unit list.

5. To build the geometry, click on Geometry tab from the toolbar and select Block from the list. Enter the data shown in Figure 5.22 for Width, Depth, and Height. Rename the Block node to Wall.

6. Again, from the ribbon bar select Cone. Enter the data shown in Figure 5.23, and then click Build All Objects. Rename the Cone node to Fin.

7. To add material to the model, right-click on Materials node in the Model Builder window and select Add Material. In the Add Material window, expand Built-In and select Aluminum; then click on Add to Component. Notice that Aluminum material is added to both the wall and fin (Domains 1 & 2). We use Cast iron for the wall, by adding this

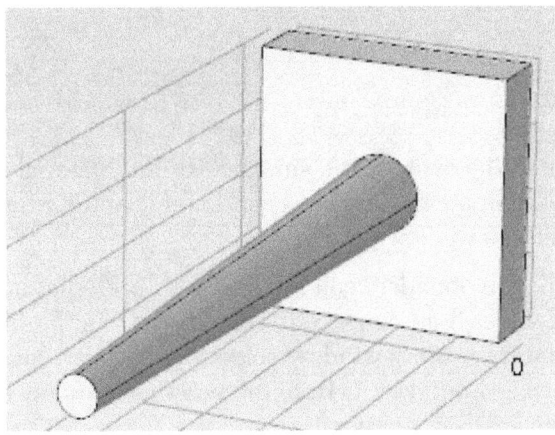

FIGURE 5.21 Fin geometry attached to a section of wall.

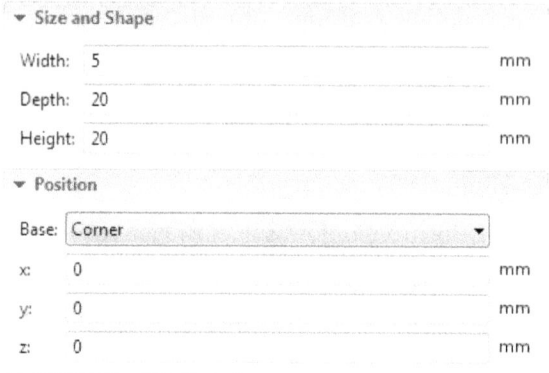

FIGURE 5.22 dimensions for wall block geometry.

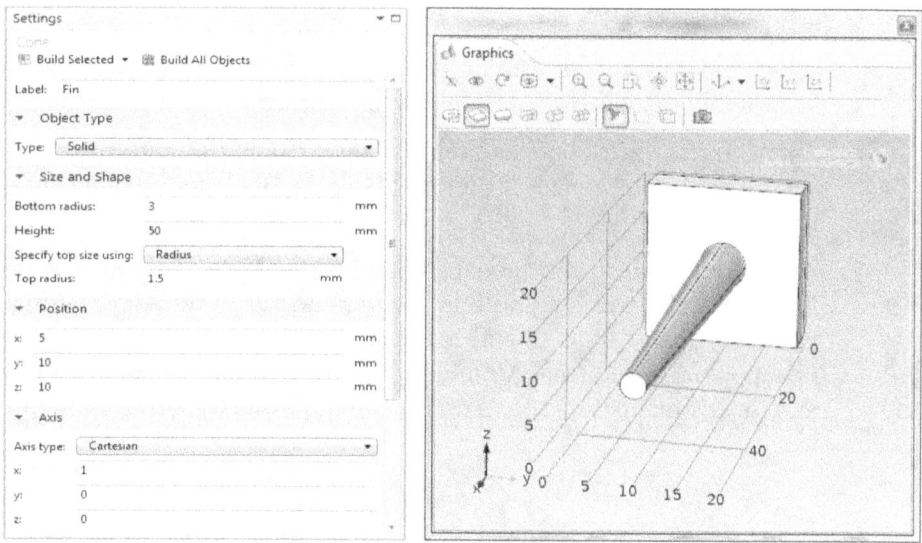

FIGURE 5.23 Fin cone parameters setup and final geometry of fin and wall.

material to the component, similarly. In the Material settings window select Wall domain (Domain 1) and add it to the Selection list. Close Add Material window.

8. To define the boundary conditions, click on Physics tab from the toolbar and select Boundaries > Temperature from the ribbon list. In the Temperature settings window, select the back face of the wall located at y-z plane (Boundary 1) from the Graphics window and add it to the Selection list. For Temperature, enter 450°C.

9. Similarly, select Boundaries > Heat Flux. In the corresponding settings window, select all surfaces of the fin and only the front face of the wall (Boundaries 6, 8, 9, 10, 11, 12). In the Heat Flux section, select Convective heat flux and User defined. Enter 50 for Heat transfer coefficient and 30°C for External temperature. See Figure 5.24.

10. To build a mesh, we use Free Tetrahedral tool. Click on Mesh tab from the toolbar and select Free Tetrahedral. In the corresponding settings window, select Domain from the list for Geometry entity level and add Wall (Domain 1) to the Selection list. Right click on Free Tetrahedral 1 node and select Size. In the Corresponding settings window select Extra fine. Similarly, create another Free Tetrahedral node and assign Fin domain to it with a Fine mesh resolution attribute. Click Build All. Final mesh is shown in Figure 5.25.

The value of b for this problem is; $\dfrac{50}{2700\times900\times1.155\times10^{-3}}=0.01781$

(using Fin material properties and dimensions), which corresponds to a time constant of about 56 seconds (inverse of b). Therefore, it takes about 56 seconds for temperature to reach its equilibrium (i.e. temperature would

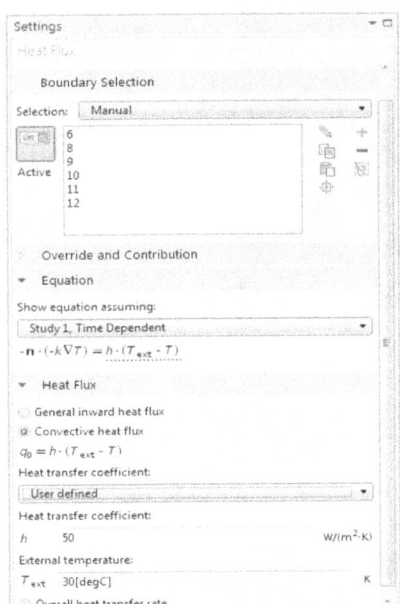
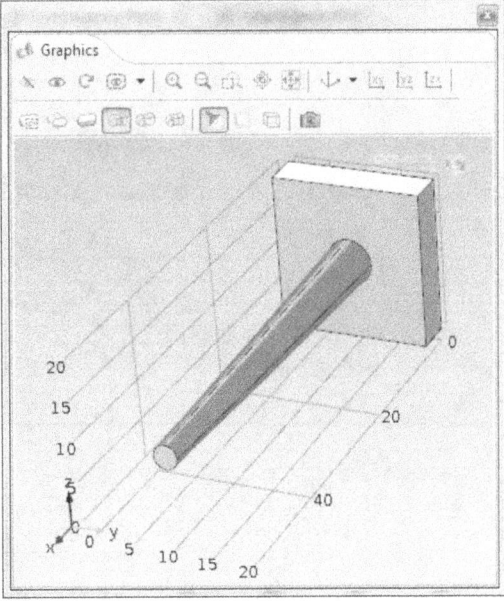

FIGURE 5.24 Convective boundary condition set up for fin.

not vary versus time at a given point). For the transient solver, we use a small time step of 0.5 second for the first 5 seconds and a 2-second time step for the remaining times, up to 60 seconds.

11. To run the model, expand the Study 1 node and click on Step 1: Time Dependent node. In the corresponding settings window, enter range(0, 0.2, 5), range(6, 2, 60) for Times. Then click on the Study tab from toolbar and select Compute. Default results for temperature will appear in the Graphics window, after calculations are complete. See Figure 5.26 for results at t = 30 sec., as an example.

12. To plot the temperature along the center-line of the fin, click on the Results tab from the toolbar and select Cut Line 3D. In the corresponding settings window locate Line data section and enter (0, 10, 10) for Point 1 and (70, 10, 10) for Point 2. Click Plot to see the line. To

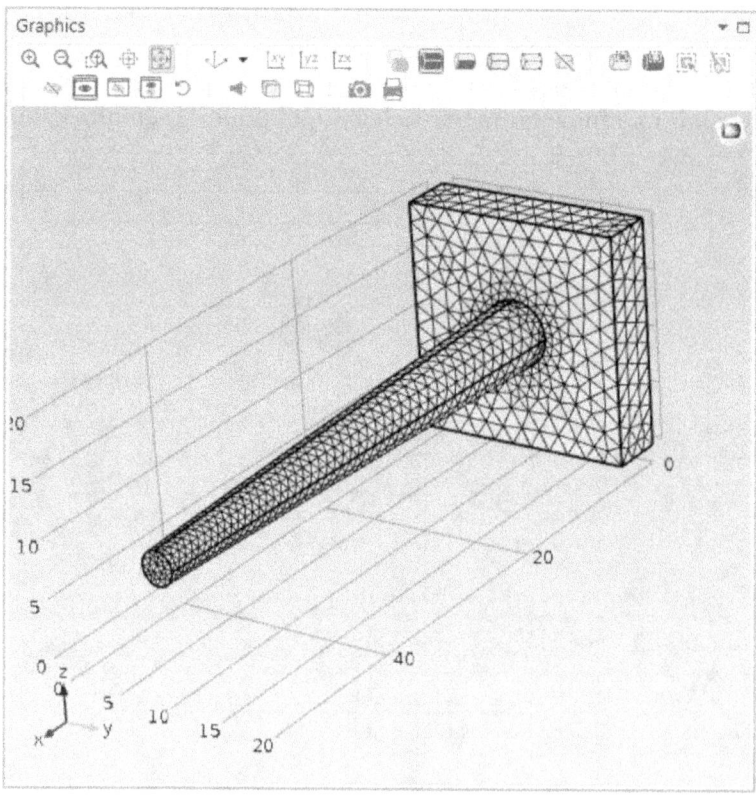

FIGURE 5.25 Finite elements mesh for fin and wall section.

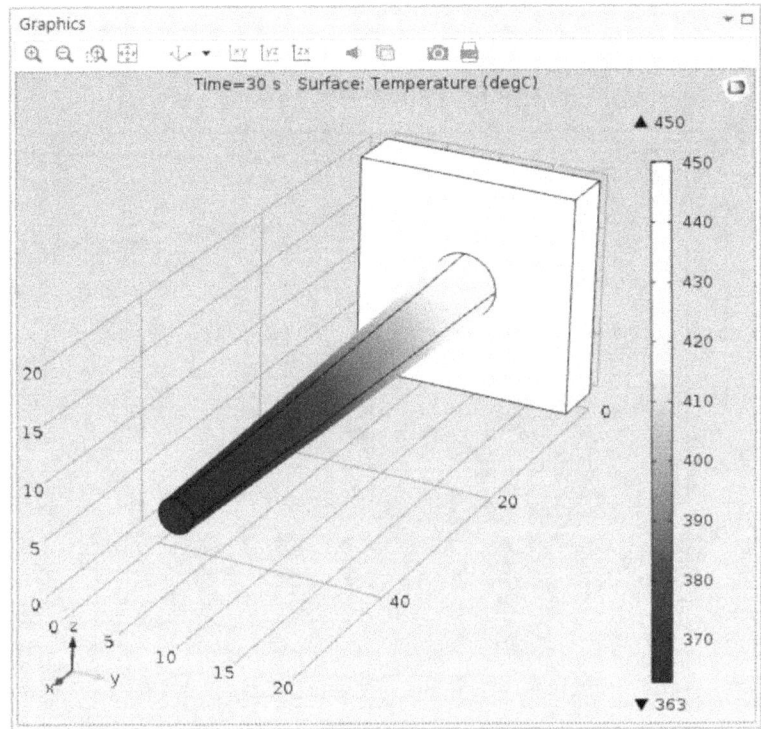

FIGURE 5.26 Temperature distribution for fin and wall section at time 30 sec.

draw the temperature along this line, click on 1D Plot Group, from the ribbon toolbar, and then select Line Graph. In the Line Graph settings window select Cut Line 3D 1 from Data set and change the Unit to degC. Click Plot to see the solution curves for temperature. The variation of temperature at different time steps (according to time step selection 0 to 60 seconds) for points along the selected line will appear in the Graphics window. As expected, at early times temperature variations are much larger than those at later times. Results are shown in Figure 5.27.

13. Another useful plot is the variation of temperature versus time at a fixed point. Right-click on the Results node in the Model Builder and select 1D Plot Group. A new node, 1D Plot Group 4 will appear in the model tree. Right-click on this node and select Point Graph. In the Point Graph window, change Unit to degC. Select two points from the Graphics window, one at the fin base (e.g. point 9) and another one at the fin tip (e.g. point 15) and add them to the Selection list. Expand

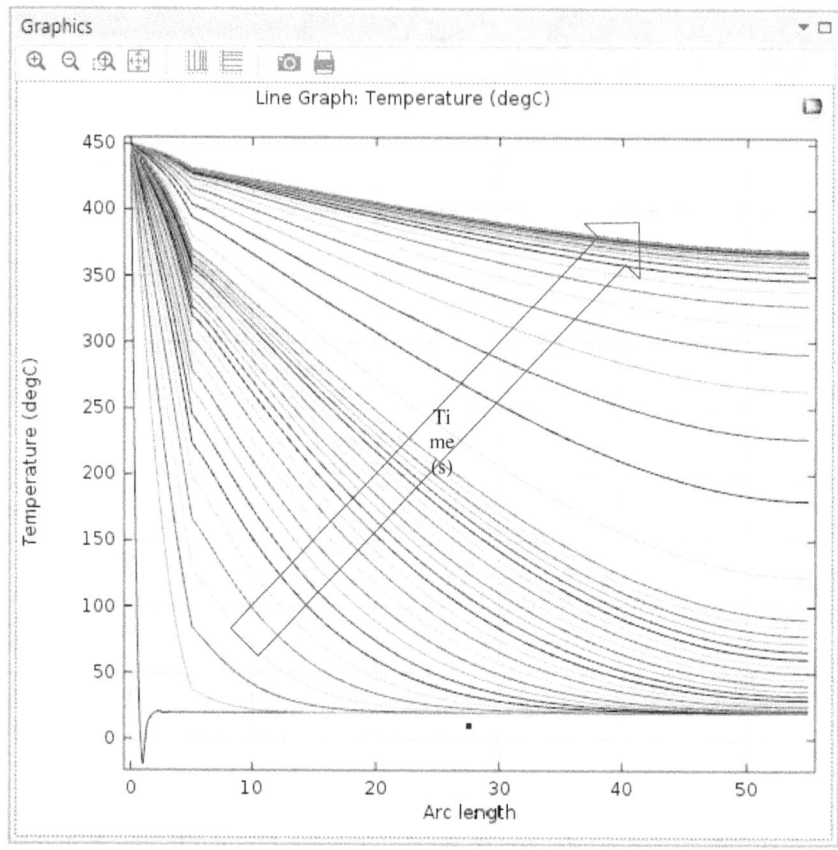

FIGURE 5.27 Temperature variation along the fins axis for different time steps.

Legends and check the box for Show legends and click Plot. Results are shown in Figure 5.28.

14. To build an App for this model, launch COMSOL if not already running. From File menu select New. In the New window select Applications Wizard. In the Select Model for Application window click on Browse; locate and open Example 5.3. mph. The New Form window will appear. In this window, select all entries, except the last four ones, under Parameters, from the Inputs/outputs tab and move them to Selected window. Click on Graphics tab, select and move Temperature (ht) to Selected window. Click on the Buttons tab; select and move Compute Study 1 to Selected window. See Figure 5.29. Click on Done. Form Editor desktop window appears. Save the file as App_Example 5.3.

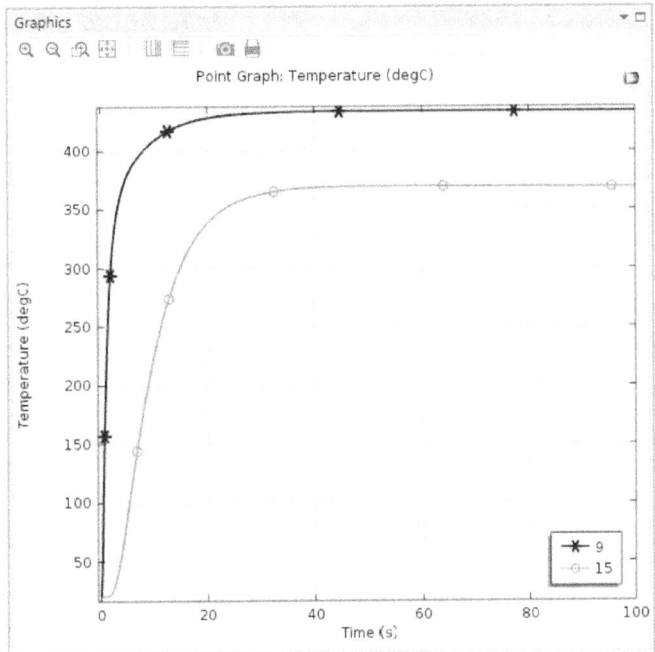

FIGURE 5.28 Temperature variation at the fin base (x) and its tip (o) versus time.

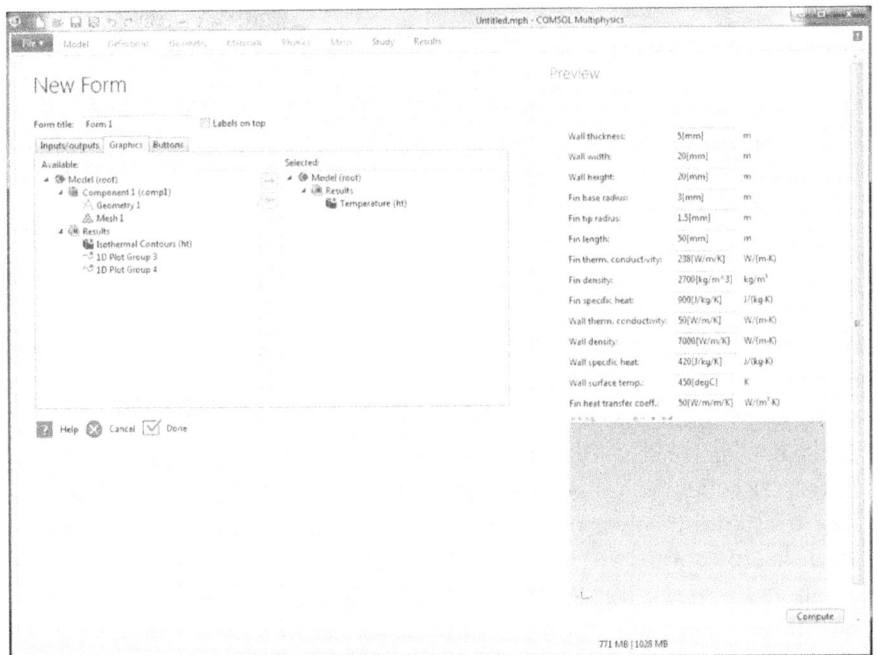

FIGURE 5.29 Application settings for Example 5.3.

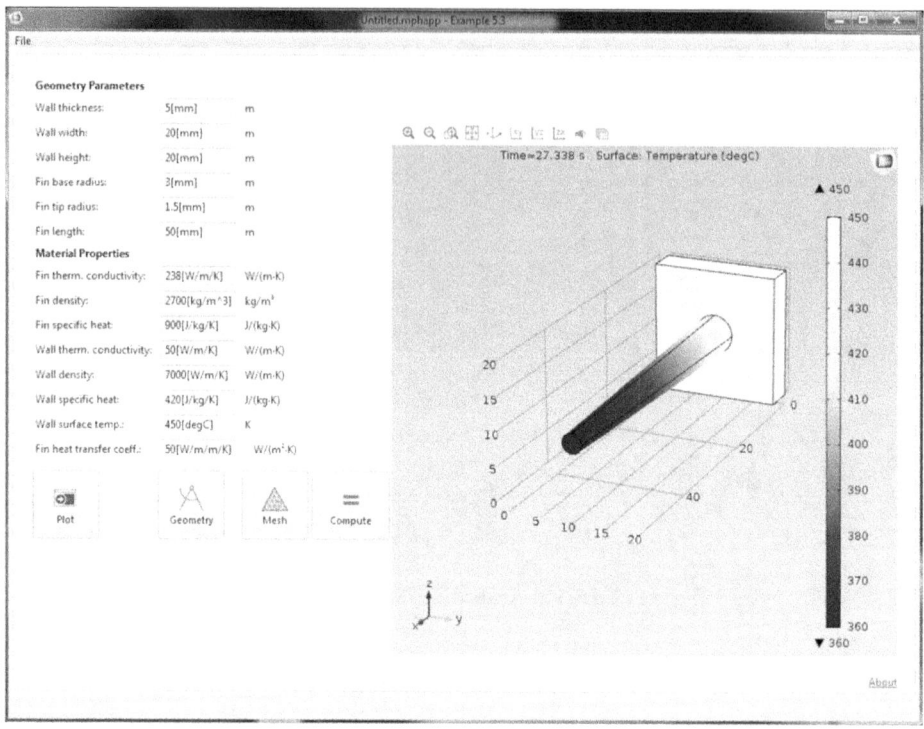

FIGURE 5.30 Application GUI for Example 5.3.

15. Several editing tools are available in the Form Editor window. We use some of these tools in order to lay out the App interface. Click on Grid from the ribbon bar to relocate the objects in the Form 1. Use the Settings for each object to modify the size, affiliate a picture, and specify the function. Refer to detailed instructions given for Example 2.1, for a guide. Final results for the App interface are shown in Figure 5.30.

Example 5.4: Heat conduction through a multilayer wall with contact resistance

In this example, we model heat conduction through a multilayer wall while considering thermal resistance between two of its layers. Contact resistance exists between dissimilar layers at their interface. The physics of contact resistance involves dimensions at micro scale (or even nano scale) level in order to capture the surface roughness of the two adjacent corresponding layers. For thermal engineering calculations, we consider an average (or statistically averaged) thermal resistance to represent the thermal contact resistance effects. However, the question is, "What is the thickness of this equivalent layer?" As mentioned above, this layer is very thin as compared with other actual corresponding layers' thicknesses. Therefore, if we create a very thin layer to represent the contact resistance, it requires very small mesh size to build finite elements inside this layer, and therefore increases the number of elements and also creates unnecessary difficulties for the computations and numerical convergence of a model. One possible solution to this problem is to use conservation of energy (first law of thermodynamics) to conserve the heat flux going through the contact boundary. Therefore, the temperature will behave like a step function or will at least vary with very steep slope across the fictitious layer or the boundary. In COMSOL 5, this tool is available as a type of boundary condition called Thin Layer (see COMSOL 5 manual for details). Thin Layer replaces Thin Thermally Resistive Layer and Highly Conductive Layer features that were available in previous versions. The data required, for a resistive layer, are an estimate of the layer thickness and optional thermal conductivities. If data are available, then users can enter it; otherwise the material properties of the adjacent layers will be used for calculations. Having this tool makes the calculations and inclusion of thermal resistance very simple as well as effective. However, users should note that realistic results require good data or estimates for the contact resistance interface layer and corresponding layers material properties.

In this example, we build a multilayer flat wall that includes a thin layer to represent contact resistance between two adjacent layers and calculate heat conduction and temperature profile through the wall section.

Solution:

1. Launch COMSOL and in New window click on Model Wizard. Save the file as Example 5.4.

2. In the Select Space Dimension window, select 2D.

3. From the Select Physics list, expand Heat Transfer and select Heat Transfer in Solids (ht). Click on Add, and then click on the Study arrow button.

4. From the Select Study list, click on Stationary and then click on Done.

5. In the Geometry window, select mm from the list for Length unit.

6. Draw a rectangle, using the tools under Geometry tab, with dimensions Width 100 and Height 400, and set the Corner at x = y = 0. Rename the rectangle Brick. Similarly, draw four more rectangles with

Widths 70, 30, 100, and 13, attached to each other with equal Heights of 400. Rename these layers Rain Screen, Air, Concrete, and Gypsum, respectively, as shown in Figure 5.31.

7. Add materials to the wall model. Right-click on the Materials node in the Model Builder window, and select Add Material. In the corresponding window, expand Built-In, select Brick, and add it to the model by clicking on Add to Component. Assign Layer 1 to the settings window for Brick. Similarly, for subsequent layers select Acrylic plastic, Air, and Concrete from the list and add them to the model. For the Gypsum layer, click on the Materials tab from the toolbar and select Blank Material. In the corresponding settings window, choose the Gypsum layer from the Graphics window and add it to the Selection list. Enter the following data in the Material Contents section under Value: k = 0.276, rho = 711, Cp = 1017. Rename this node to Gypsum. If needed, material property values can be changed for each layer by entering the desired value under the relevant Value column. Make sure each material is associated with the right corresponding layer.

8. Define boundary conditions. Click on Physics tab from the toolbar and select Boundaries > Temperature from the list. In the Temperature settings window, add the left edge of the wall (boundary 1) to the Selection list. For Temperature, enter −30°C. Similarly, create another Temperature boundary condition and select and add the right edge of the wall (boundary 16) to its corresponding Selection list. For its Temperature, enter 20°C.

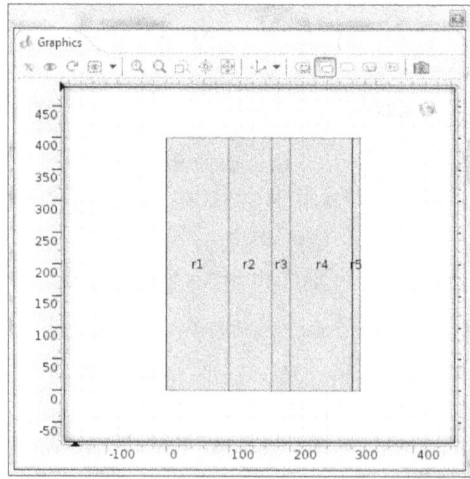

FIGURE 5.31 Window showing multilayer wall geometry.

9. Next we define the interface between the Brick and Rain screen layers as a thin layer with contact resistance. Click on Boundaries button and select Thin Layer. In the corresponding settings window, add boundary 4 to the Selection. Locate the Thin Layer section and enter the data as shown in Figure 5.32.

10. To build a mesh, click on the Mesh tab from the toolbar and select Mapped. From the ribbon list select Distribution. In the Distribution settings window, select Boundary 2 (i.e. thickness of Brick layer) and choose Predefined distribution type and enter 10 for Number of elements. Similarly, create four more Distribution nodes and assign

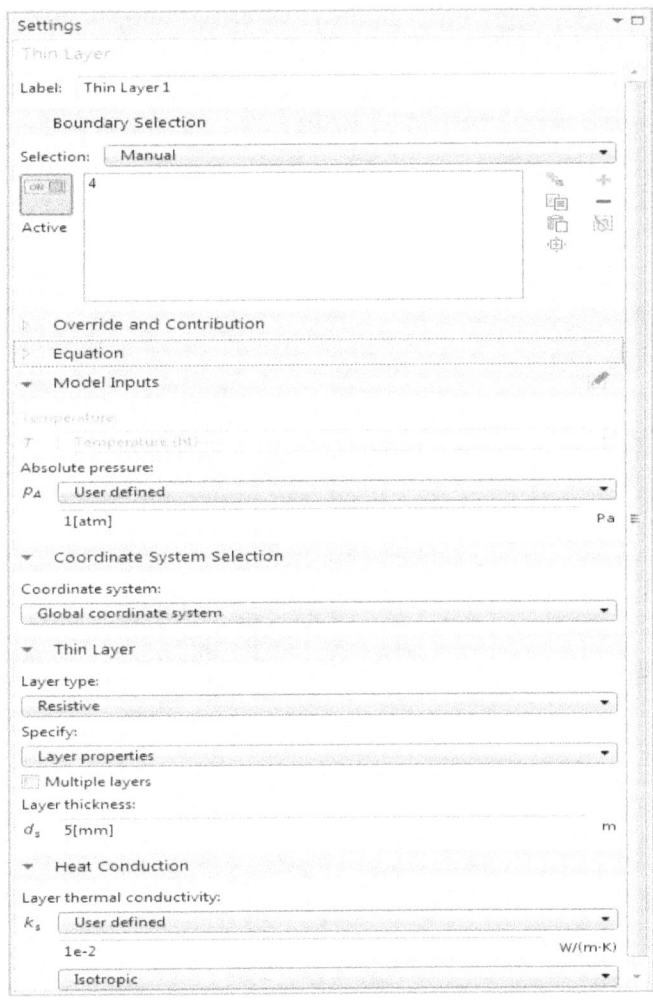

FIGURE 5.32 Settings window for resistive Thin Layer parameters.

10, 5, 15, and 3 for Number of elements. These distribution nodes should have their corresponding layer assigned. Finally, create a Distribution for the Height of the wall and set 20 for its Number of elements. Click Build All. See Figure 5.33.

11. To run the model, right-click on Study 1 node in the Model Builder window and select Compute. Wait for computations to finish.

12. Default results for surface temperature and Isothermal contours appear in the model tree. Expand the Temperature (ht) node and click on Surface 1. In the Surface settings window, change Unit to °C and click Plot. Similarly, expand the Isothermal Contours (ht) node and click on Contour 1. In the Contour settings window, change Unit to °C. We add a plot for showing the heat flux vectors. Right-click on Isothermal Contours (ht) node and select Arrow Surface and click Plot. Results are shown in Figure 5.34.

13. It would be useful to make a graph of temperature variations across the wall. Click on the Results tab from the toolbar and select Cut Line 2D. In the corresponding window, locate Line Data and enter (0, 200) for

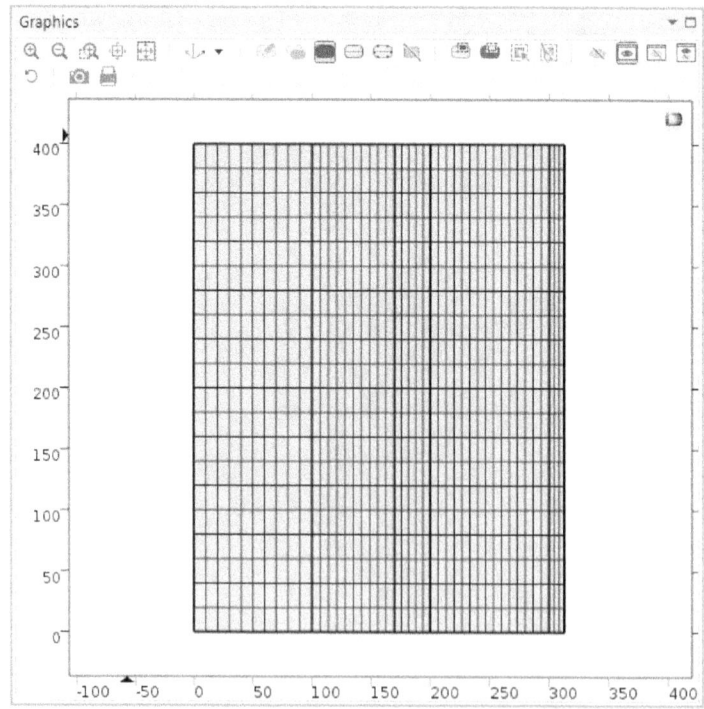

FIGURE 5.33 Finite elements Mapped mesh for multilayer wall.

Point 1 and (313, 200) for Point 2. From the ribbon, click on 1D Plot Group. From the ribbon, again select Line Graph. In the Line Graph settings window, change Unit to °C and click Plot. Result, see Figure 5.35, shows the effect of contact resistance at the interface between Brick and Rain Screen layers, located at 100 mm, which resulted in a sharp change in temperature.

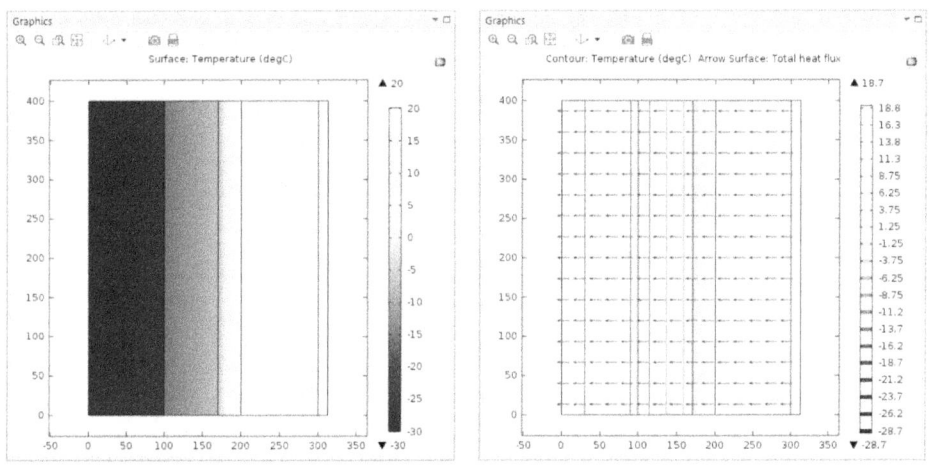

FIGURE 5.34 Windows showing results for temperature and heat flux through multilayer wall.

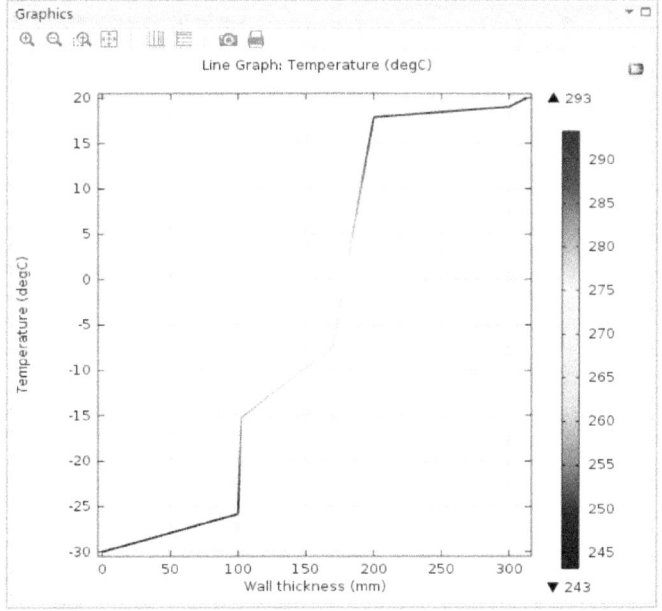

FIGURE 5.35 Results for temperature variations across multilayer wall with contact resistance at 100 mm.

6

MODEL EXAMPLES FOR ELECTRICAL CIRCUITS AND GENERATOR

In this chapter, we use COMSOL modules and 0D features to model some examples of basic electrical circuits. In engineering, we encounter modeling problems that do not have any space or dimension dependency, such as chemical reactions, electric circuits, and Optimization. The 0D feature in COMSOL is used to model such problems.

Models discussed and built in this section include dynamic analysis and parametric study for typical RC and RLC electrical circuits. The main objective is to provide, for users and readers, some solved examples that can be used directly or lead to further solutions of similar or more complex structures using COMSOL. It is assumed that readers are familiar with relevant engineering principles and governing equations [23]. In each example, we provide brief explanations of physics involved along with governing equations and phenomena, as applicable. It is recommended that readers cover Chapters 1 and 2 before attempting examples in this chapter.

Example 6.1: Modeling an RC electrical circuit

In this example, we use a feature that is available in COMSOL (since version 4.3), 0D (zero-dimension). We consider a simple resistor-capacitor electrical network, as shown in Figure 6.1. We would like to calculate the voltage and current as a function of time.

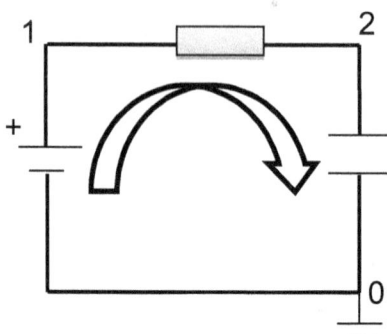

FIGURE 6.1 Sketch for the RC circuit and reference nodes used in the model.

Solution:

1. Launch COMSOL and in New window click on Model Wizard. Save the file as Example 6.1.

2. In the Select Space Dimension window, select 0D.

3. From the Select Physics list, expand AC/DC, select Electrical Circuit (cir), and add it to the list by clicking Add. Selected. Then click on the Study arrow button.

4. From the Select Study list, click on Time Dependent and then click Done. Make sure the file is saved.

For voltage calculation in an electrical network, a reference point is needed (similar to mechanical pressure). This point is referred to as Ground node. We set the Ground as node 0. For the purpose of identifying the components of the network as they connect to one another, we use data provided in Table 6.1, according to the sketch shown in Figure 6.1.

5. Click on the Physics tab from the toolbar and select Voltage Source. In the corresponding settings window, enter data for Voltage and Node Connections as shown in Figure 6.2.

6. Again, select Resistor. In the Resistor settings window, enter Node Connections (1 for p and 2 for n) and 5000[ohm] for Resistance.

7. To define the capacitor, select Capacitor. In the Capacitor settings window, enter Node Connections (2 for p and 0 for n) and 10E-6 for Capacitance. Let Initial capacitor voltage be at the default value of zero (i.e. no initial electric charges exit). The convention for node connections should be consistent and a closed one (i.e. starting node and finishing node is the same physical point in the circuit).

Component	Start node	End node
Battery (voltage source)	0	1
Resistor	1	2
Capacitor	2	0

TABLE 6.1 Data for Building the RC Circuit Nodes

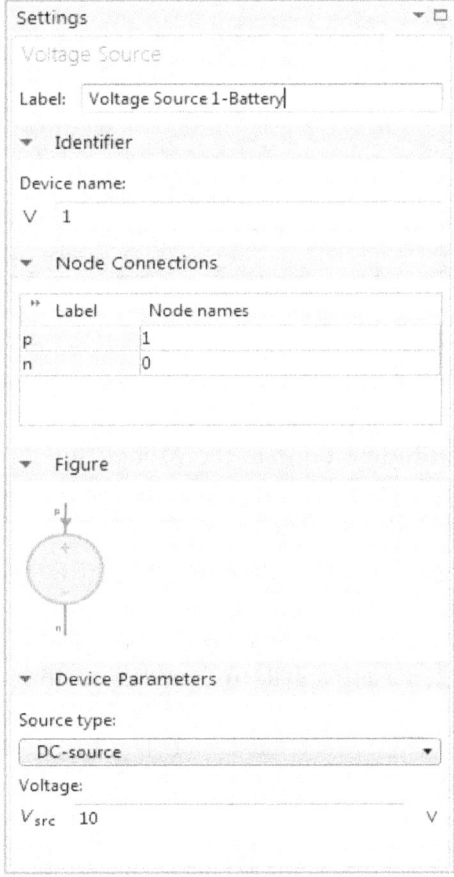

FIGURE 6.2 Voltage source data and nodes.

The time step for this problem needs to be set. In an RC circuit, the time scale is the value of $R \times C$ (resistance times capacitance). Based on data provided, we have 0.05 seconds. Therefore, we run the model with a time step of 0.001 sec. and up to about five times the time scale, or 0.25 sec.

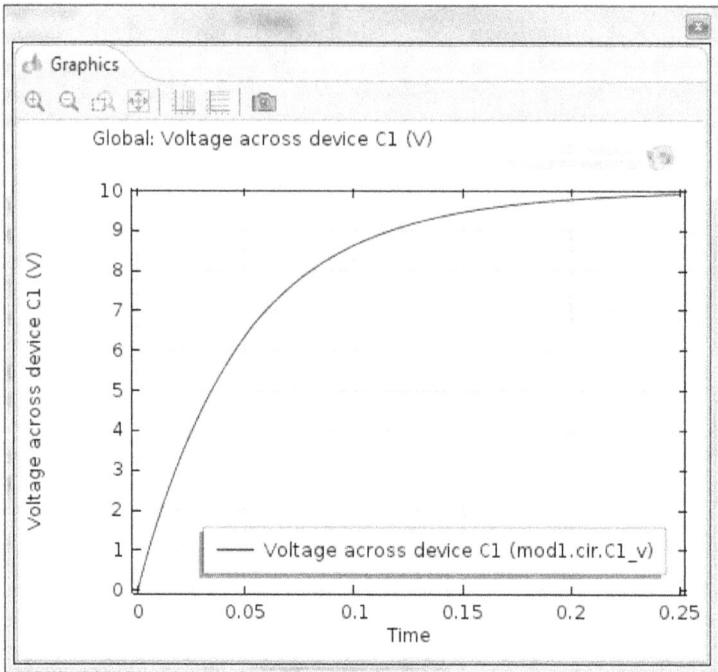

FIGURE 6.3 Results for voltage variation across capacitance device-C1.

8. Expand the Study 1 node in the Model Builder window and click on Step 1: Time Dependent. In the corresponding settings window, enter range(0, 0.001, 0.25), click Compute.

9. Right-click the Results node in the Model Builder and select 1D Plot Group. In the corresponding settings window, expand the Legend section and choose Lower right for Position. From the ribbon, under 1D Plot Group 1, select Global. In the Global settings window under y-Axis Data section, click on (+) and select Electrical Circuit > Two pin devices > Voltage across device C1. Click Plot. The results, as shown in Figure 6.3, show the variation of voltage versus time, which shows that the capacitor has been fully charged. Graphs for voltage and currents for other circuit components can be plotted by choosing the corresponding variables for the list.

Example 6.2: Modeling an RLC electrical circuit

An electrical circuit with resistor, capacitor, and inductor will be modeled in this example, using the 0D feature available in COMSOL. An RLC circuit behaves in oscillation like a mass-spring-damper mechanical system. To be exact, mass m is analogous to inductance L, spring stiffness k to inverse of capacitance C, damping coefficient γ to resistance R when we write the governing differential equation for electric current $i(t)$, which is analogous to mass displacement $x(t)$. A similar analogy can be drawn for voltage $v(t)$, as well. Corresponding ODEs are as follows:

$$m\frac{d^2 x}{dt^2} + \gamma\frac{dx}{dt} + kx = f(t)$$

$$L\frac{d^2 i}{dt^2} + R\frac{di}{dt} + \frac{1}{c}i = g(t)$$

$$C\frac{d^2 v}{dt^2} + \frac{1}{R}\frac{dv}{dt} + \frac{1}{L}v = h(t)$$

The damping term is defined as $\alpha = \dfrac{R}{2L}$ for series and $\alpha = \dfrac{1}{2RC}$ for parallel RCL circuits.

Natural fundamental frequency for an RCL circuit is defined as $\omega_0 = (LC)^{-0.5}$. When $D = \alpha^2 - \omega_0^2$ is positive, the circuit is over damped; when equal to zero, the circuit is critically damped; and when negative, the circuit is under-damped.

For this example, we consider a simple resistor-capacitor-inductor electrical network, as shown in Figure 6.4. We would like to calculate the voltage and current as a function of time.

The circuit sketch for this model is shown in Figure 6.4.

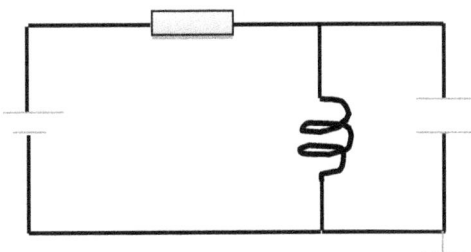

FIGURE 6.4 Sketch for the RCL circuit and reference nodes used in the model.

Solution:

1. Launch COMSOL and in New window click on Model Wizard. Save the file as Example 6.2.

2. In the Select Space Dimension window, select 0D.

3. From the Select Physics list, expand AC/DC, select Electrical Circuit (cir), and add it to the list by clicking Add. Selected. Then click on the Study arrow button.

4. From the Select Study list, click on Time Dependent and then click Done. Make sure the file is saved.

For calculating voltage in an electrical network, a reference point is needed (similar to mechanical pressure). This point is referred to as Ground node. We set the Ground as node 0. For the purpose of identifying the components of the network as they connect to one another, we use data provided in Table 6.2, according to the circuit sketch.

5. Click on Physics tab from the toolbar and select Voltage Source. In the corresponding settings window, enter the data for Voltage and Node names as shown in Figure 6.5.

6. Again, select Resistor. In the Resistor window, enter Node Connections (1 for p and 2 for n) and 3000[ohm] for Resistance.

7. To define the capacitor, select Capacitor. In the Capacitor settings window, enter Node Connections (2 for p and 0 for n) and 1E-6 for Capacitance. Set Initial capacitor voltage at the default value of zero.

8. To define the inductor, select Inductor. In the Inductor settings window, enter Node Connections (2 for p and 0 for n) and 100[mH] for Inductance. Set Initial inductor current at the default value of zero.

Component	Start node	End node
Battery (voltage source)	0	1
Resistor	1	2
Inductor	2	0
Capacitor	2	0

TABLE 6.2 Connection Nodes Used for the RCL Circuit Model

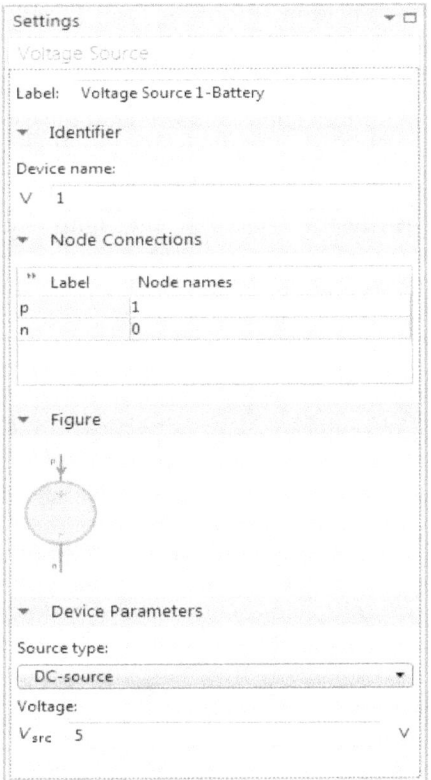

FIGURE 6.5 Voltage source data and nodes.

The time step for this problem needs to be set. In an RLC circuit, there are two parameters that define its time scale. The resonant frequency of the circuit is ω_0; for this circuit the value is about 3163 rad/s. Therefore time scale is about 2E-3 seconds (inverse of $3163/2\pi$). Another time scale is set for the damping coefficient α. For this circuit the value is about 166.7 s^{-1} (=1/2RC, since capacitor and inductor are parallel). Therefore, the time scale is about 6E-3 seconds. We choose the smaller value for time step equal to 0.002 s. Therefore, we run the model with a time step of 0.0001 sec. and up to more than eight times the larger time scale or 0.05 sec.

9. Expand the Study 1 node in the Model Builder window and click on Step 1: Time Dependent. In the corresponding window, enter range (0, 1e-4, 0.05) and click Compute.

10. Click Results tab from the toolbar and select 1D Plot Group. In the corresponding settings window, expand the Legend section and choose

Upper right for Position. Right-click on 1D Plot Group 1 and select Global. In the Global settings window under y- Axis Data section, click on (+) and select Electrical Circuit > Two pin devices > Voltage across device R1. Click Plot. The results show the variation of voltage versus time for the resistor. Graphs for voltage and currents across other circuit components can be plotted by choosing the corresponding variables for the list. Results are shown in Figure 6.6.

We perform a parametric analysis for studying the effect of capacitance on the voltage output of the circuit.

11. Click on the Model tab from the toolbar and select Parameters. In the Parameters settings window, enter C under Name and 10[nF] under Expression.

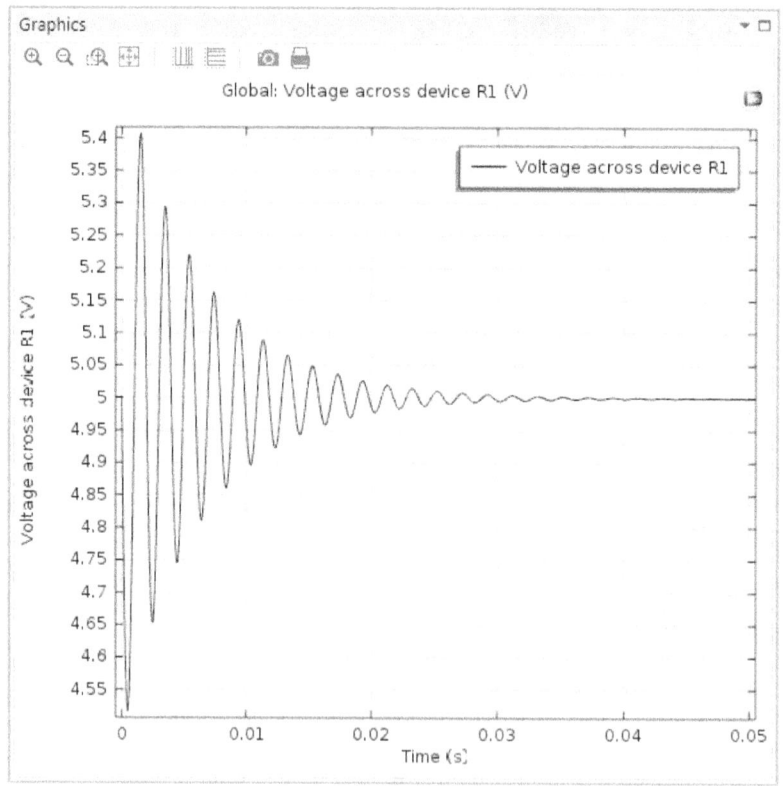

FIGURE 6.6 Voltage variation versus time across resistor device-R1.

12. Click on the Study tab from the toolbar and select Parametric Sweep. In the corresponding settings window, click on the plus icon and select C from the list under Parameter names. Enter 10[nF], 100[nF], 1000[nF] for Parameter value list. Click on Compute. Wait for computations to finish.

13. To show the results, click on the Results tab from the toolbar and select 1D Plot Group. In the corresponding settings window, expand the Legend section and choose Lower right for Position. From the ribbon tools under 1D Plot Group 2 select Global. In the Global window under Data, select Study1/Parametric Solution 1 from the list for Data set. Under y- Axis Data section, click on (+) and select Electrical Circuit > Two pin devices > Voltage across device L1. Click Plot. Results are shown in Figure 6.7. These results show the variation of voltage versus time. Graphs for voltage and currents across other circuit components can be plotted by choosing the corresponding variables for the list.

14. To build an app for this model, launch COMSOL if not already running. From File menu select New. In the New window select Applications Wizard. In the Select Model for Application window click on Browse; locate and open Example 6.2.mph. The New Form window

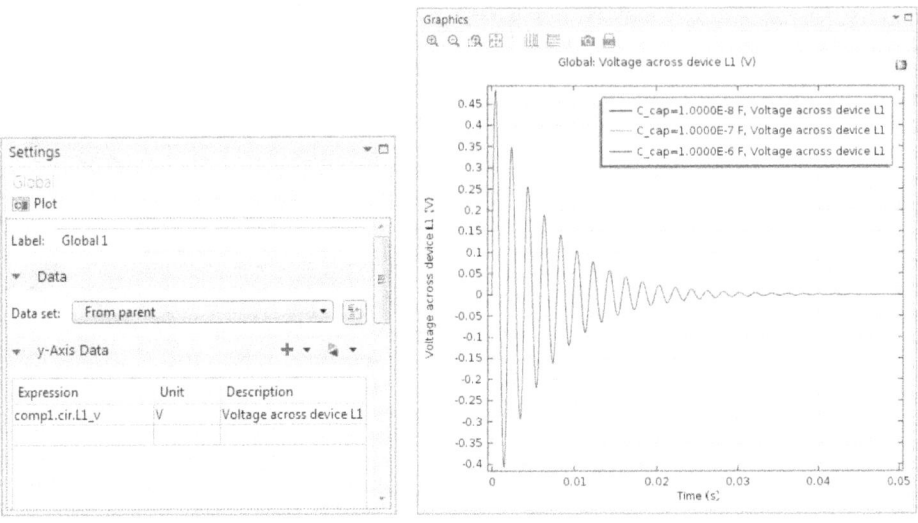

FIGURE 6.7 Windows showing voltage variation versus time across resistor device-L1 for several values of capacitances.

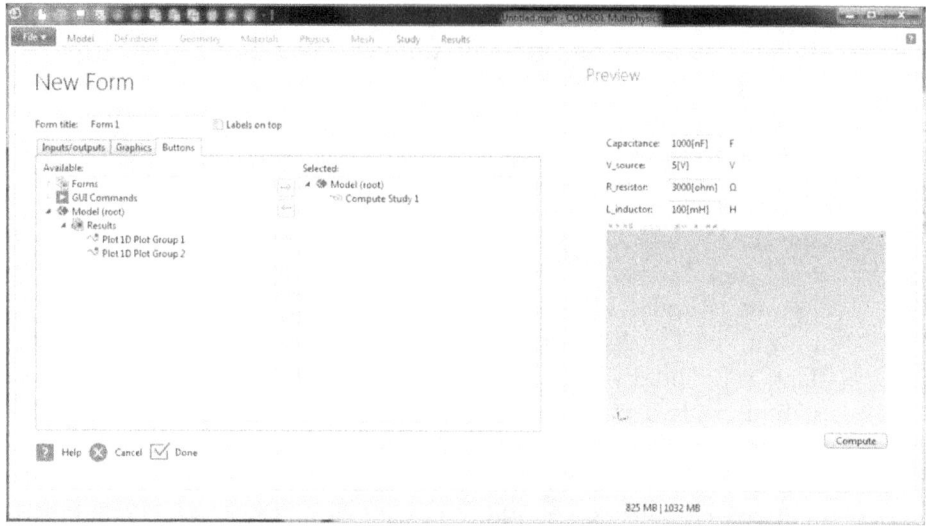

FIGURE 6.8 Application settings for Example 6.2.

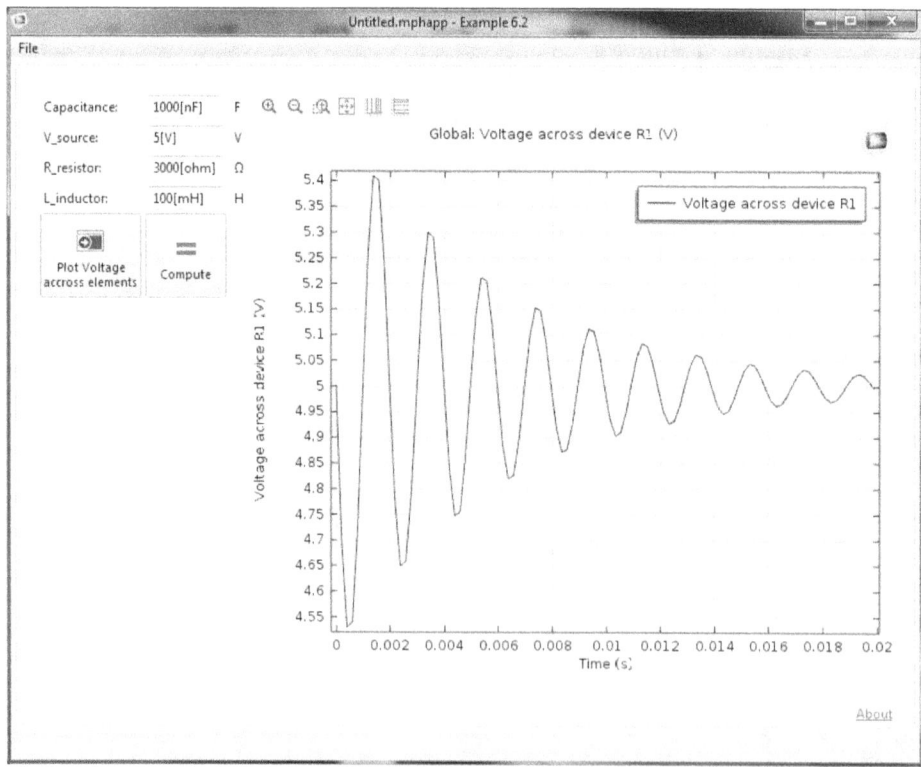

FIGURE 6.9 Application GUI for Example 6.2.

will appear. In this window, select entries Capacitance, V_source, R_ resistor, and L_inductor under Parameters, from the Inputs/outputs tab, and move them to Selected window. Click on the Graphics tab; select and move 1D Plot Group 1 to Selected window. Click on the Buttons tab; select and move Compute Study 1 to Selected window. See Figure 6.8. Click on Done. Form Editor desktop window appears. Save the file as App_Example 6.2.

15. Several editing tools are available in the Form Editor window. We use some of these tools in order to lay out the App interface. Click on Grid, from the ribbon bar, to relocate the objects in the Form 1. Use the Settings for each object to modify the size, affiliate a picture, and specify the function. Refer to detailed instructions given for Example 2.1, for a guide. Final results for the App interface are shown in Figure 6.9.

Example 6.3: Modeling a permanent magnet generator

For this example we build a 2D model of an alternator or synchronous revolving-field generator. The generator rotor is connected to a shaft and has permanent magnets attached to it, hence revolving magnetic field. The stator (or armature) has wire windings and is attached to external network or load. The rotation of the rotor creates the variable magnetic field, which induces voltage/current at the wires of the stator, according to Faraday's law of induction. For this type of generator the mechanical shaft rotation speed, N (in rpm) is directly related to the frequency of the voltage generated, f (in Hz) through the following equation:

$$f = \frac{P}{120} N$$

where P is the number of magnets or poles (including north and south poles) attached to the rotor. The induced voltage is directly proportional to the length of wire wounds in the stator and velocity of the rotor. A schematic of the generator is shown in Figure 6.10. This example is also available from COMSOL 5 Model Library (ACDC_Module/Motors_and_Actuators/ generator_2d).[5]

---

[5]Model made using COMSOL Multiphysics® and is provided courtesy of COMSOL. COMSOL materials are provided "as is" without any representations or warranties of any kind including, but not limited to, any implied warranties of merchantability, fitness for a particular purpose, or noninfringement.

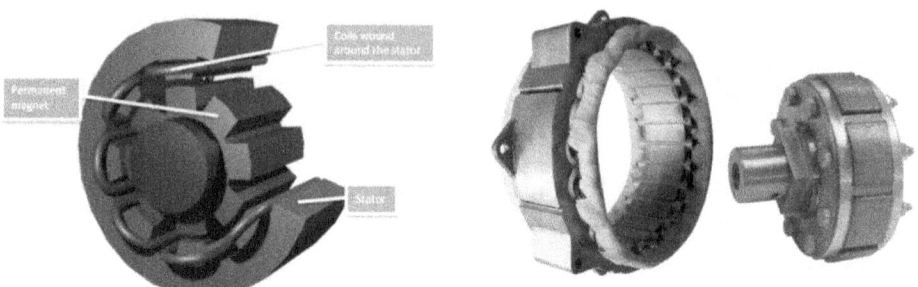

FIGURE 6.10 3D schematics and actual picture of a generator.

Solution:

1. Launch COMSOL and in New window click on Model Wizard. Save the file as Example 6.3.

2. In the Select Space Dimension window, select 2D.

3. From the Select Physics list, expand AC/DC, select Rotating Machinery, Magnetic (rmm), and add it to the list by clicking on AddSelected. Then click on the Study arrow button.

4. From the Select Study list, click on Stationary and then click Done. Make sure the file is saved.

We should point out that the stationary solution selection refers to the steady state of the generator, after initial transient mode is passed on.

5. Define the parameters by clicking on the Parameters button, in the ribbon toolbar. In the corresponding settings window enter the data as shown in Figure 6.11.

6. The geometry of the generator either can be built in COMSOL or imported. We import the geometry from the library. Click on the Geometry tab from the toolbar and select Import. In the corresponding settings window, locate Import section and click on Browse. Find the file generator_2d.mphbin. This file should be available in the COMSOL folder. Click on the Import button. The geometry of the generator appears in the Graphics window, as shown in Figure 6.12. Note that the generator geometry consists of two parts, the rotor and the stator (including the air gap).

7. We make an assembly by uniting the two parts. Click on Form Union (fun) node in the Model Builder window. In the corresponding settings

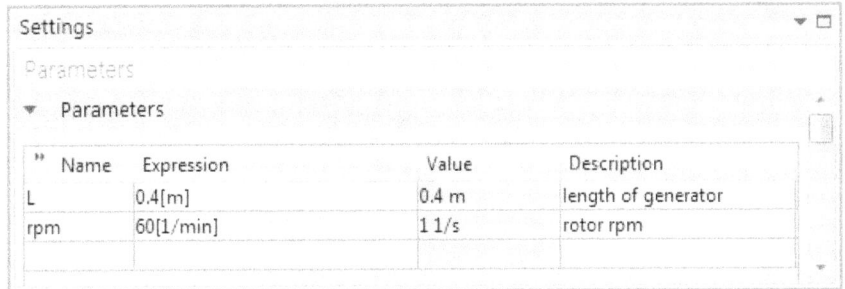

FIGURE 6.11 Parameters data for the model.

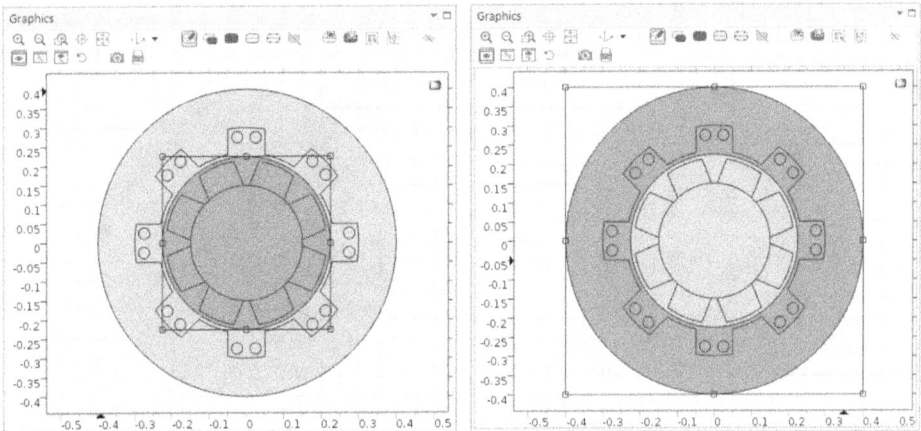

FIGURE 6.12 Two-part generator geometry; rotor (left) and stator (right).

window select Form an assembly, from the list under Action. Click on Build All. Now, if you click on the geometry in the Graphics window, all the geometry will be highlighted, as an assembly. Another feature, selected by default, is the Identity pair, which makes it possible to connect the physics fields across the two parts of the assembly (i.e. rotor and stator boundaries). This feature is crucial for this model for satisfying the continuity of magnetic field.

8. To facilitate the selection of the geometrical entities with their labels, expand Definitions node, under Component 1 (comp1) node, in the Model Builder window. Click on View 1. In the corresponding settings window locate View section and check the Show geometry labels box. The label number for surfaces of the generator should appear in the Graphics window, as shown in Figure 6.13.

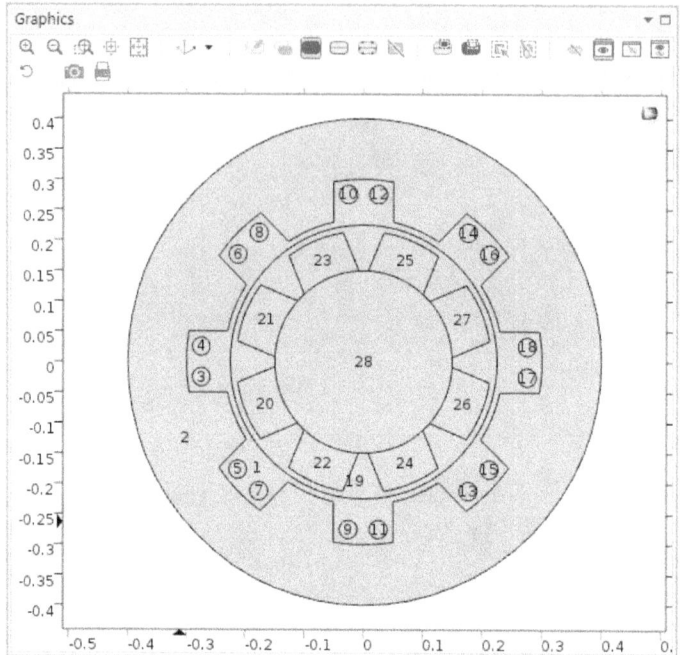

FIGURE 6.13 Geometry of the generator showing each entity label.

9. Now we define the materials for different parts of the generator: air for the gap, soft medium carbon (with relatively high permeability) for the rotor core and stator ring. Click on Materials node from the toolbar and select Add Material. In the Add Material window, type in Air in the search space. Find Air in the search results, click on it, and then click on Add to Component. Similarly, search for Soft Iron (without losses) and add it to the model component. In the corresponding settings window, locate Geometric Entity Selection section and assign domains 2 and 28 to the Selection list. Locate Materials Contents and enter 1 for Relative Permeability.

10. Now we define the rotor rotational velocity. Click on the Physics tab from the toolbar and select Domains > Prescribed Rotational Velocity. In the corresponding settings window add domains 19–28 to the Selection list. In the same window, locate Prescribed Rotational Velocity section and enter rpm for rps. This defines the rotor angular speed, as defined in the Parameters.

11. To define the radial magnetic field of the permanent magnets, we use a local cylindrical system of coordinates. Click on the Definitions tab from the toolbar and select Coordinate Systems > Cylindrical

Systems. Vector potential fields (electric and magnetic) are defined using Ampere's Law tool in COMSOL. Click on the Physics tab from the toolbar and select Domains > Ampere's Law. In the corresponding settings window add domains 20, 23, 24, and 27 to the Selection list. Locate Coordinates system section and select Cylindrical System 2 (sys2) from the list. To define the constitutive relations, to relate magnetic flux density **B** to the magnetic field **H**, locate Magnetic Field section and select Remanent flux density from the list. For r-component of $\mathbf{B_r}$, enter 0.84[T]. See Figure 6.14. The domains selected are the

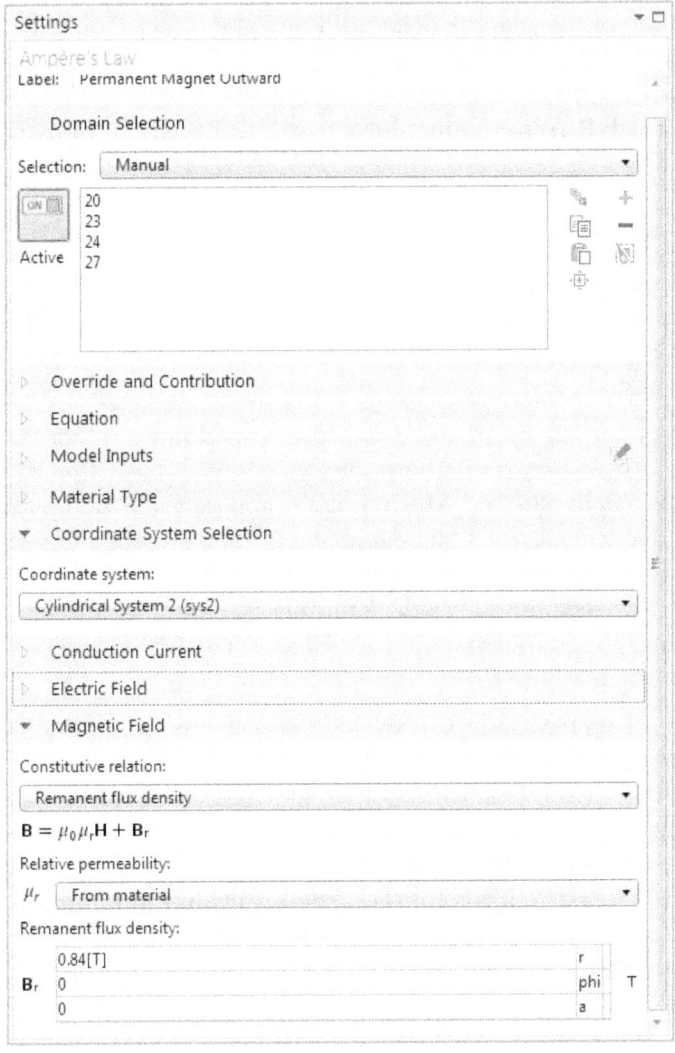

FIGURE 6.14 Magnetic flux and field settings for permanent magnets.

outward magnets; we rename the Ampere's Law node to Permanent Magnet Outward.

12. Similarly, assign the other four magnets (i.e. domains 21, 22, 25, and 26) as inward types. To do this right-click on Permanent Magnet Outward node and select Duplicate. Rename the new node created to Permanent Magnet Inward. In the corresponding settings window select domains 21, 22, 25, and 26 and add them to the selection list. For r-component of $\mathbf{B}_r$, enter −0.84[T].

13. Now we define the magnetic field for the iron core of the rotor and casing of the stator. Click on Physics and select Domains > Ampere's Law. In the Corresponding settings window select domains 2 and 28 and add them to the Selection. Locate Magnetic Field section and select **HB** from the list, for Constitutive relation. Rename this node to Iron.

14. To satisfy the continuity of the magnetic field between rotor and stator, through air gap, we use the Continuity tool. Click on Pairs from the ribbon toolbar and select Continuity. In the corresponding settings window locate Pair Selection and select Identity Pair 1 (ap1). This Pair was created when we build an assembly of the geometry parts imported.

15. To create the mesh, click on Mesh 1 node in the Model Builder window. In the corresponding settings window select Finer for Element size. Right-click on Mesh 1 node and select Edit Physics-Induced Sequence. Click on Size node and in the corresponding settings window locate Element Size Parameters and enter 2 for Resolution of narrow regions. This will determine the number of elements in the narrow region of the air gap. Click Build All. The resulted mesh is shown in Figure 6.15.

16. To set up the study, we would like to see the solution plotted during the computation. We then need to have the default results. Click on the Study tab from the toolbar and select Get Initial Value. Expand Data Sets and click on Study 1/Solution 1 node. In the corresponding settings window locate Solution section and select Spatial (x, y, z) for the Frame. This will show the results in the reference frames for rotor and stator, during the solution. From the tools under Magnetic Flux Density (rmm), select Contour. In the corresponding settings window locate expression and type Az for the Expression. This is

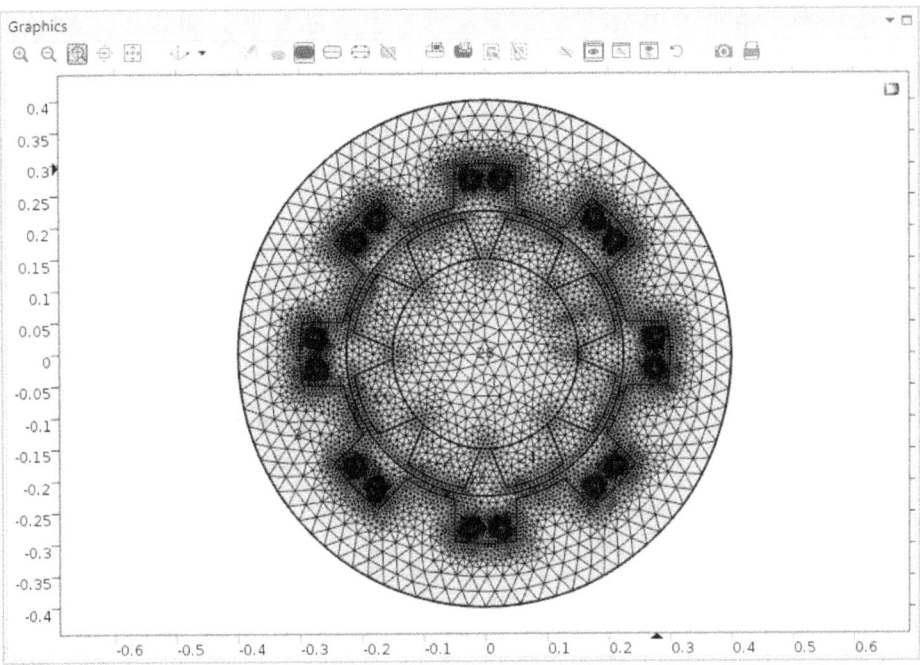

FIGURE 6.15 Custom-made mesh for the generator, including air gap.

the z-component of the magnetic vector potential. In the Coloring and Style section select Uniform for Coloring, Black for Color, and uncheck the box for Color legend.

17. We now set the transient solution. Click on the Study tab from the toolbar and select Study Steps > Time Dependent. In the corresponding settings window locate Study Settings and enter range(0, 0.01, 0.25), for Times. This is the minimum run time required, since the rotor speed (RPM) is 1 and we have 4 magnets (or 8 poles) hence, 1/4 = 0.25. More run time than 0.25 s will basically create the periodic solution for the voltage. Users may try to increase the total time for the solution and experiment this. To help the solution convergence, click on the Study tab from the toolbar and select Show Default Solver. In the Model Builder window, under Study 1, expand Solver Configurations > Solution 1 > Time Dependent Solver 1 and click on Fully Coupled 1 node. In the corresponding settings window expand Method and Termination section. Select On every iteration, for the Jacobian update and enter 1e-3, for Tolerance factor. Expand Results While Solving

section and check the box for Plot. To watch the solution plots during computation more clearly, Float the Graphics window and use full-screen mode. Run the model by clicking on Compute. Results at times 0.2 s and 0.25 s are shown in Figure 6.16.

Now we calculate the voltage induced in the windings of the stator. The voltage can be calculated by taking the line integral of the electric field over the length of the wires in the windings. For this model, a 2D geometry, we approximate the voltage by calculating the average electric field $\bar{E}$ over a wire cross-sectional area using the z-component of the electric field E_Z, as $\bar{E} = \dfrac{\int E_Z dA}{A}$. Now multiplying the average voltage by the length of the generator L and sum it up over all wires cross-sectional areas in a winding, we get the voltage induced in a winding as $\sum_{all\,x\text{-}sections} L\,\bar{E}$, which gives the total induced voltage as $V_i = N_w \sum_{all\,x\text{-}sections} L\,\bar{E}$. Where Nw is number of turns is the winding.

18. To set up these calculations in this model, we create several integral definitions. Click on the Definitions tab from the toolbar and select Component Coupling > Integration. In the corresponding settings window select domains 5–8 and 13–16 and add them to the selection. These represent half of the wires cross sections in the stator. Duplicate Integration 1 by right-clicking on it and selecting Duplicate. In the corresponding settings window, clear all selections and add domains 3, 4, 9–12, and 17–18. These represent the other half of the wires

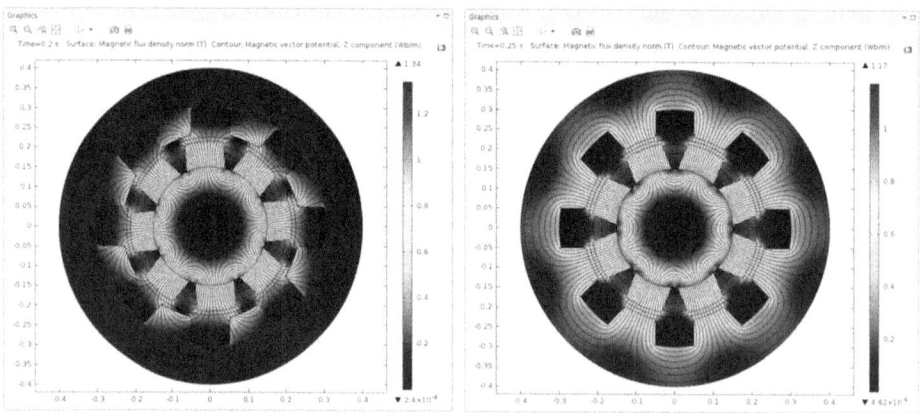

FIGURE 6.16 Magnetic flux density and z-component vector potential contours at 0.2 s (left) and 0.25 s (right) for the generator.

cross sections in the stator. Create another integration node and assign domain 8 to it. The last one is just used for calculation the cross-sectional area of the winding.

19. To set up the global variables for calculating the induced voltage, click on the Model tab from the toolbar and select Variables > Global Variables. In the corresponding settings window enter the variables as shown in Figure 6.17.

20. Click on the Study tab from the toolbar and select Update Solution. Now we want to plot the induced voltage variation versus time. Click on Results tab and select 1D Plot Group. Then select Global from the ribbon. In the Global settings window locate y-axis data and change the type in Vi under Expression. This should be exactly as defined in the Variables, or alternatively could be found by clicking on Replace Expression and search for Vi. Click Plot. Result is shown in Figure 6.18. As mentioned previously and shown here, the period of 0.25 s was sufficient to get a full cycle of the induced voltage.

21. To build an app for this model, launch COMSOL if not already running. From the File menu select New. In the New window select Applications Wizard. In the Select Model for Application window click on Browse; locate and open Example 6.3. mph. The New Form window will appear. In this window, select entries length of generator and rotor rpm under Parameters, from the Inputs/outputs tab, and move them to Selected window. Click on Graphics tab; select and move Magnetic Flux Density (rmm) to Selected window. Click on the Buttons tab; select and move Compute Study 1 to Selected window. See Figure 6.19. Click on Done. Form Editor desktop window appears. Save the file as App_Example 6.3.

22. Several editing tools are available in the Form Editor window. We use some of these tools in order to lay out the App interface. Click on Grid,

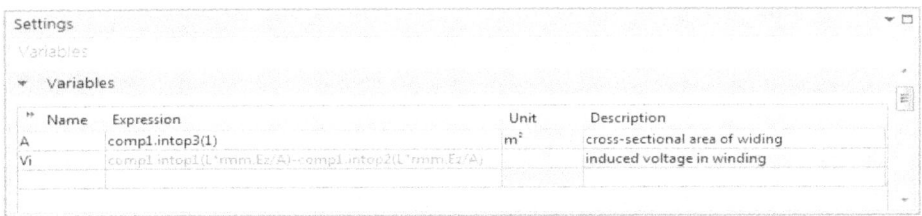

FIGURE 6.17 Global variables settings for calculating induced voltage in the windings.

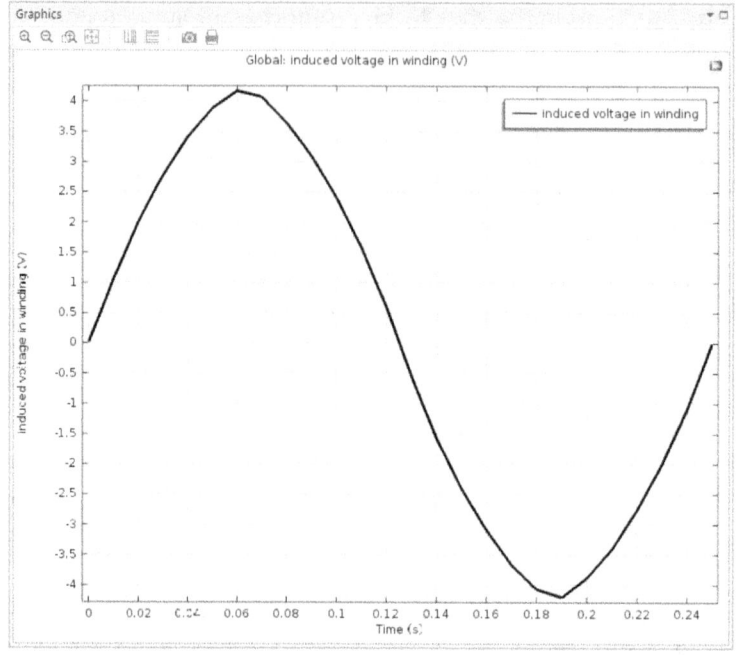

FIGURE 6.18 Voltage variations versus time of the generator.

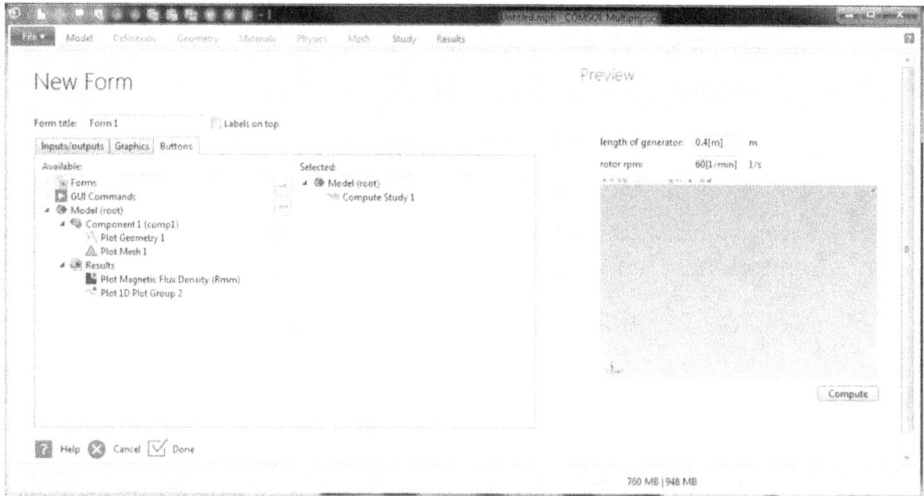

FIGURE 6.19 Application settings for Example 6.3.

from the ribbon bar, to relocate the objects in the Form 1. Use the Settings for each object to modify the size, affiliate a picture, and specify the function. Refer to detailed instructions given for Example 2.1, for a guide. Final results for the App interface are shown in Figure 6.20.

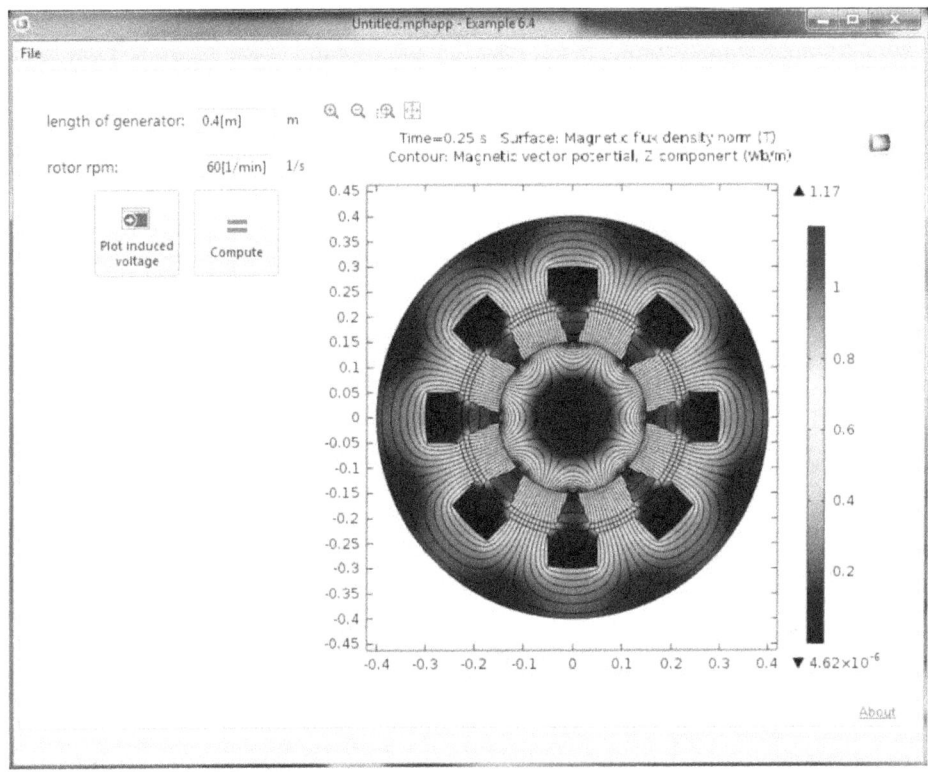

FIGURE 6.20 Application GUI for Example 6.3.

MODEL EXAMPLES FOR COMPLEX-MULTIPHYSICS SYSTEMS

In previous chapters, we used COMSOL features to model problem examples that involved several different physical phenomena but mainly with one phenomenon dominating. Modern engineering involves problems that require solutions of multiphysics phenomena [25]. A multiphysics problem is defined as one that has two or more competing physics happening simultaneously. For example, when we have a fluid flowing over a relatively hot plate, heat is transferred to it and the temperature variations in the plate may change its thermal conductivity as well as that of the fluid. This interactive effect should be accounted for in stress analysis of the plate and flow velocity and pressure calculations. Another example may be a cantilever beam in a fluid stream. The deflection of the beam, as a result of its stiffness, will affect the fluid's velocity and pressure and vice versa. Analyzing and modeling multiphysics problems can be done using COMSOL, which has many features and modules enabling users to "simply" add physics to a problem at hand.

In this chapter, we also solve complex problems for which users may have the relevant mathematical models (like ODEs and PDEs) developed and would like to have the solutions using FEM with features available in COMSOL, such as a multibody dynamics problem or multiphysics problem for which COMSOL modules may not yet be available. We demonstrate the applications of FEM to solve such problems through some examples. For examples given in this chapter, previous knowledge (or review) of differential equations (both ordinary and partial) and Lagrangian mechanics [26], would help users to benefit fully from their obtained solutions.

Example 7.1: Stress analysis for an orthotropic thin plate

This example extends the model built in Example 3.1. We consider an orthotropic material (i.e. material properties different along x and y axes). The plate carries stationary mechanical loads. It is recommended that readers review or build the model described in Example 3.1 before studying this example.

The plate geometry (12 in. by 8 in.) is given as shown in the Figure 7.1. All other dimensions are as those given in Example 3.1. We would like to calculate the displacement and stress field under the applied loads. We consider a plane stress case, which means that the stress component along the z-axis (normal to the plane of the plate) is zero, but not necessarily the strains. Plate thickness is 0.4 in. and properties of the material (for example, Kevlar-epoxy) are given in Table 7.1. The boundaries on the left side of the plate are constrained while the other plate sides are let free.

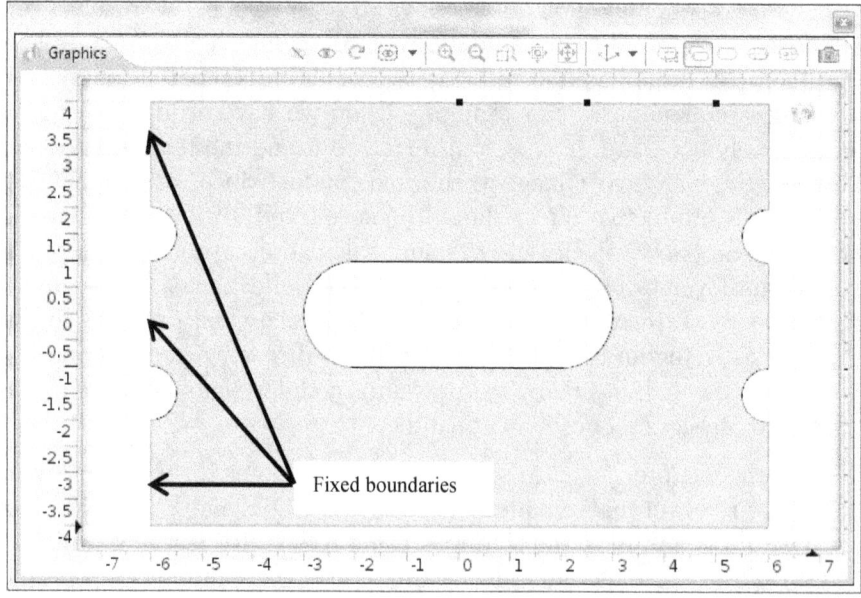

FIGURE 7.1 Geometry, dimensions, and boundary conditions.

Modulus of elasticity	Poisson's ratio	Shear modulus	Specific gravity
$E_x = 80$ GPa	$v_{xy} = 0.31$	$G_{xy} = 2.1$ GPa	1.44
$E_y = 5.5$ GPa	$v_{yx} =$(calculated)		

TABLE 7.1 Material Properties for Orthotropic Kevlar-Epoxy.

Before starting the model construction, we explain the material properties, since we are using an orthotropic material. From Table 7.1 we see that we have two moduli of elasticity. Obviously, this material is relatively stiffer in x-direction than in y-direction. We also have two Poisson's ratios. If we recall from isotropic materials, we usually have one modulus of elasticity and one Poisson's ratio. In order to explain the significance of orthotropic material properties, we use the stress-strain relationship resulting from general theory of elasticity. For our 2D plane stress problem at hand, the strains (ε's and γ) are related to stresses (σ's and τ) as:

$$\begin{Bmatrix} \varepsilon_x \\ \varepsilon_y \\ \gamma_{xy} \end{Bmatrix} = \begin{bmatrix} 1/E_x & -v_{yx}/E_y & 0 \\ -v_{xy}/E_x & 1/E_y & 0 \\ 0 & 0 & 1/G_{xy} \end{bmatrix} \begin{Bmatrix} \sigma_x \\ \sigma_y \\ \tau_{xy} \end{Bmatrix}$$

Because of symmetry we should have $-v_{yx}/E_y = -v_{xy}/E_x$, which is used in COMSOL for calculation of v_{yx} ($= 0.31 \times 5.5/80$) in this example. For further readings, see [26] and [16].

Solution:

1. Open COMSOL and open the file Example 3.1.mph. If you have not built the model for Example 3.1, you can obtain it from the attached CD. Save the new file as Example 7.1 by clicking on File > Save As in the toolbar.

   In the Model Builder window, open the tree under Component 1 (*comp 1*). Under Plate (*plate*), click on the Linear Elastic Material 1 node. In the Linear Elastic Material settings window, locate the Linear Elastic Material section and select Orthotropic from the list under Solid model. Under Young's modulus section, select User defined and enter the data for *E* as given in Table 7.1. Similarly, enter data for Poisson's ratio, Shear modulus, and density (1.44E3).

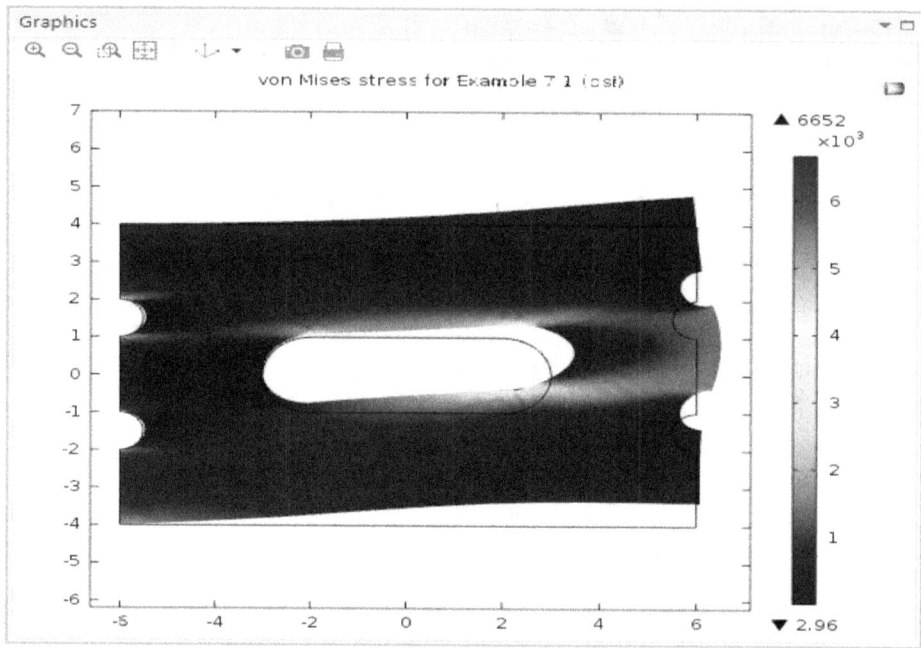

FIGURE 7.2 Results for von Mises stress and deformed plate.

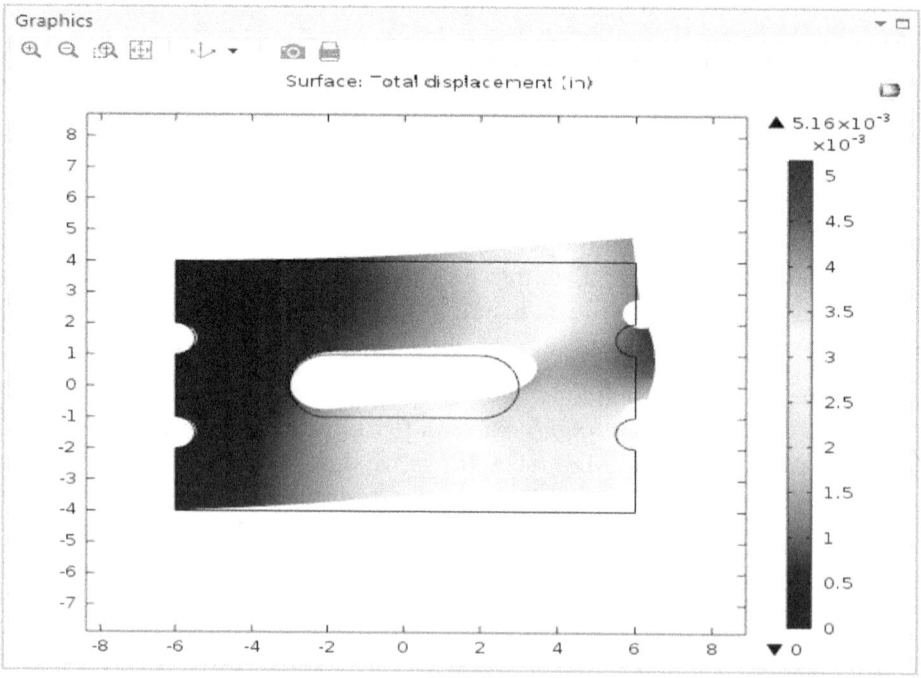

FIGURE 7.3 Results for total displacement for the orthotropic plate.

2. To run the model, Click on the Study tab from the toolbar and select Compute. The default result for von Mises stresses will appear in the Graphics window.

3. To manipulate the results, click on the Surface 1 node under the Stress Top (plate) node in the Model Builder window. This will open the Surface settings window. In this window, open the list under Unit in the Expression section and choose psi, then expand the Title section and choose Manual from the Title type and enter von Mises stress for Example 7.1 (psi). Click on the Plot icon. Periodically, you may need to zoom in/zoom out to fit the results in the Graphics window. To show the displacement results, click on Surface 2 node and click Plot. These results are shown in Figure 7.2 and Figure 7.3.

Example 7.2: Thermal stress analysis and transient response of a bracket

In this example, we extend the model built in Example 3.3, the bracket. The first feature we consider is extending the model to calculate thermal stresses due to temperature variations at the boundaries. The second feature is modeling transient response of the bracket structure due to a transient load. Users can also add these features directly to the file from Example 3.3.

Solution:

1. Launch COMSOL and click on Model Wizard. Save the file as Example 7.2.

2. In Select Space Dimension window select 3D. From Select Physics list select Structural Mechanics > Thermal Stress and click on Add. Click on the Study arrow button.

3. In Select Study window select Stationary and click on Done.

4. Now we import the geometry of the model from Example 3.3. Click on the Geometry tab from the toolbar and select Insert Sequence. Locate Example 3.3 and open it (users can find the Example 3.3 file in the accompanied CD). In the corresponding settings window click on Build All Objects. The bracket geometry should appear in the Graphics window. Make sure the file is saved.

5. **Thermal stresses:** Click on the Material tab from the toolbar and select Add Material. In the Add Material window locate Built-In > Structural Steel and click on Add to Component. Close the Add Material window.

6. To group the boundaries, click on the Geometry tab and select Selections >Explicit Selection from the ribbon bar. In the corresponding settings window select Boundaries for Geometric entity level. Select boundaries associated with the bolt holes (18–21 and 31–34). Rename the Explicit Selection 1 (sel1) node to Bolt holes. Similarly create another selection for the shaft holes and assign boundaries (4, 5, 42, 43) to it. Rename it to Shaft holes. Results are shown in Figure 7.4.

7. To define the boundary conditions for the solid mechanics, click on the Physics tab and select Boundaries > Fixed Constraint. In the corresponding settings window locate Boundary Selection section select Bolt holes from the list. This will constrain the bolts connectors to the bracket structure.

8. To define the boundary conditions for the heat transfer, click on the Physics tab and change the physics to Heat Transfer in Solids. Select

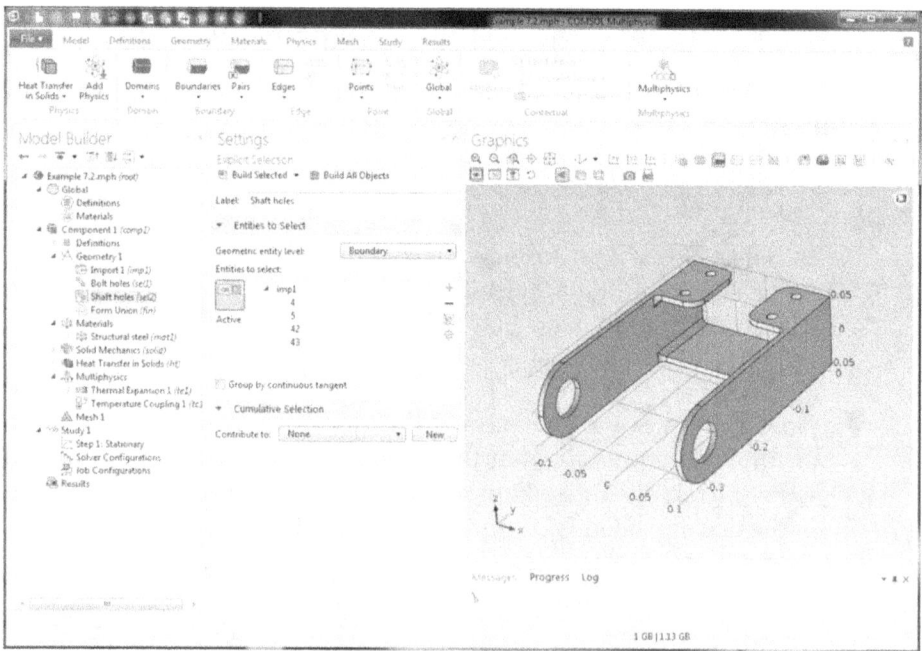

FIGURE 7.4 Geometry and boundary selections for the bracket.

Boundaries > Heat Flux. In the corresponding settings window locate Boundary Selection section; select All Boundaries from the list. Locate the Heat Flux section and check the Convective heat flux box; enter 10 for h and 15 °C for T_{ext}. This will assign convective cooling conditions for the bracket. To set the temperature boundaries, select Boundaries > Temperature. In the corresponding window locate Boundary Selection section and select Bolt holes from the list. Enter 20 °C for the Temperature T_0. Similarly create another temperature boundary condition for the Shaft hole with a temperature of 100 °C. See Figure 7.5.

9. Click on the Mesh node in the Model Builder. In the Mesh window, select Normal for the Element size and click Build All.

10. Run the model by right-clicking on the Study 1 node and select Compute. After the computation is finished, 3D plots for stress and temperature are available. Click on the Stress (solid) node under Results in the Model Builder window. The Graphics window shows the results for von Mises thermal stresses. Click on the Temperature (ht) node to view the results for temperature in the Graphics window. For locating the Legend and changing the Units, click on the

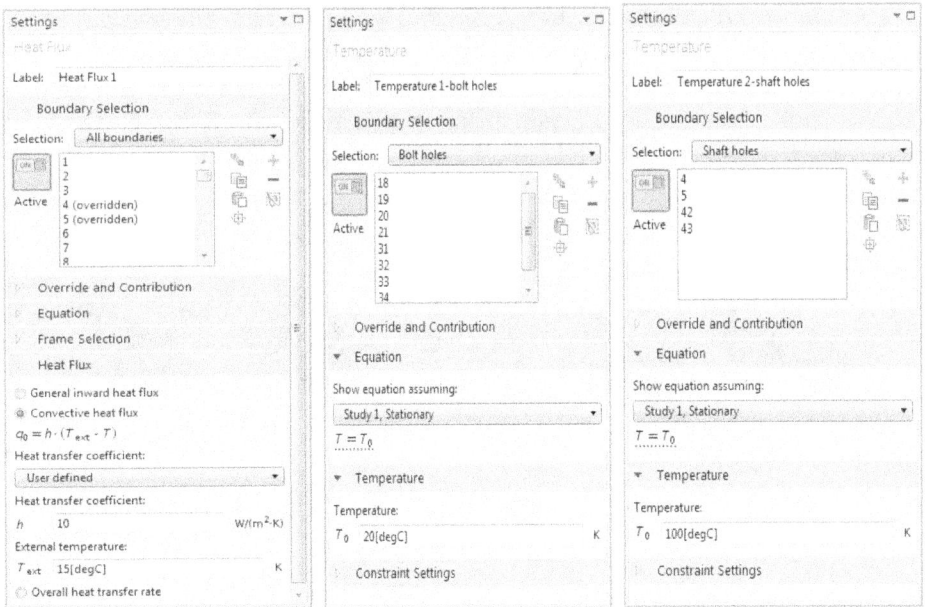

FIGURE 7.5 Windows showing boundary conditions setting for thermal stress calculations.

corresponding 3D Plot Group (or Surface) node to select the desired options. See Figure 7.6.

Transient response: Now we would like to model the bracket response to a set of given dynamic time-varying loads. Dynamic modeling of a structure requires definition for damping (the way a structure dissipates absorbed energy). In COMSOL, damping can be added to the model using Rayleigh damping or Loss Factor damping for isotropic and anisotropic, including orthotropic, materials (see COMSOL manual). We use the Rayleigh damping method in this model. The Rayleigh method relates the damping coefficient c to the mass m (in general, mass matrix) and stiffness k (in general, stiffness matrix) using two parameters α_{dM} and β_{dK} as:

$$c = \alpha_{dM}m + \beta_{dK}k$$

11. Expand the Solid Mechanics (*solid*) node, right-click on the Linear Elastic Material node, and select Damping. In the Damping settings window, locate the Damping Settings section and choose Rayleigh damping from the list under Damping type. Enter 100 for α_{dM} and 3E-4 for β_{dK}, as shown in Figure 7.7.

12. We also change the bolt connections boundary condition from fixed to the type that can include the elasticity of the bolts material. Right-click on Fixed Constraints 1 node, in the Model builder window, and select Disable. Click on Physics tab from the toolbar and select Boundaries > Spring Foundation. In the corresponding settings window select Bolt holes from the list. Locate Spring section and choose Diagonal from the

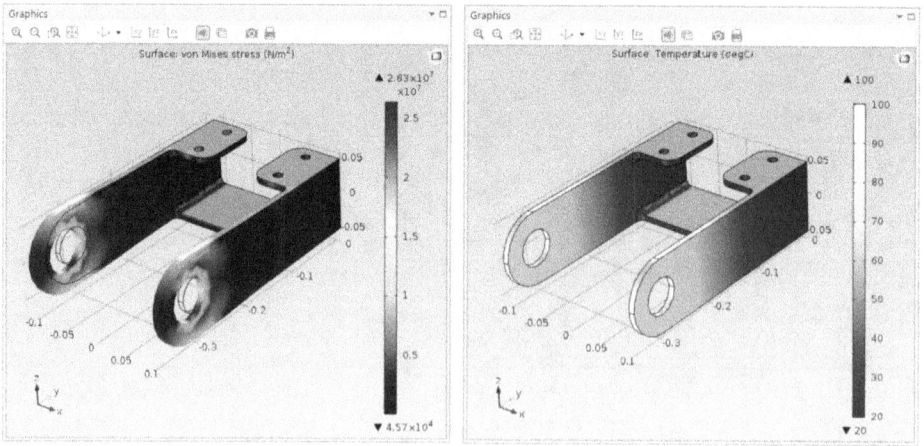

FIGURE 7.6 Results for thermal von Mises stress and temperature variations.

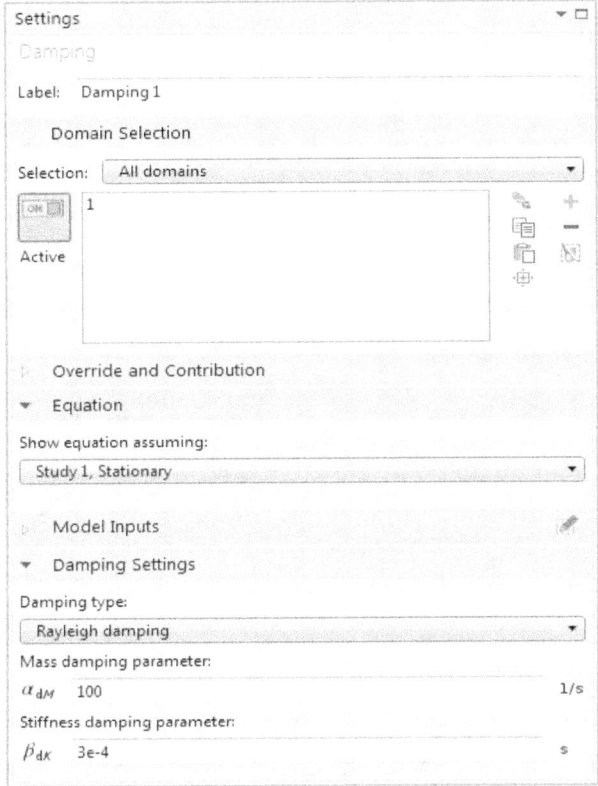

FIGURE 7.7 Rayleigh damping parameters setting.

list. Enter 1E7 for diagonal values in the matrix, as shown in Figure 7.8. This will assign spring type boundary conditions for displacements in x and y directions (not z-directions).

13. The load is through a shaft to the bracket. Therefore, the shaft holes should be connected such that their boundary conditions are compatible with each other. We assume a relatively rigid virtual shaft connecting the two holes. Click on the Physics tab and select Boundaries > Rigid Connector. In the corresponding settings window locate Boundary Selection section and select Shaft holes from the list. This will represent the shaft that connects the two arms of the structure, as shown in Figure 7.9.

14. The loads are applied on the shaft connectors. To enter these forces, click on Physics tab and select Attributes > Applied Force. In the

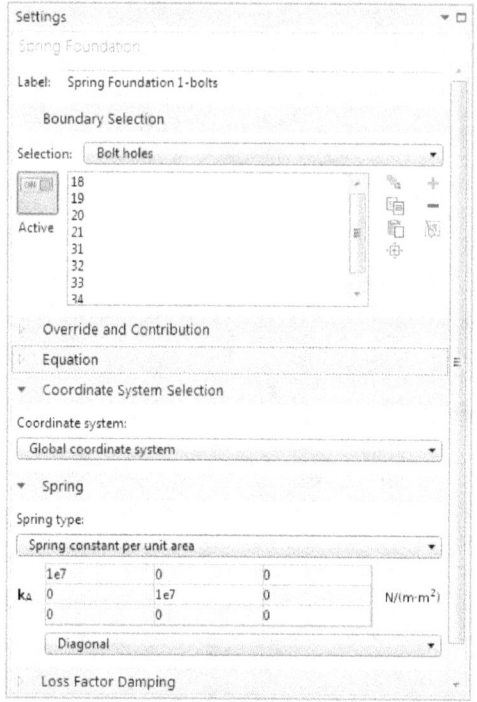

FIGURE 7.8 Spring Foundation boundary conditions settings for bolt holes.

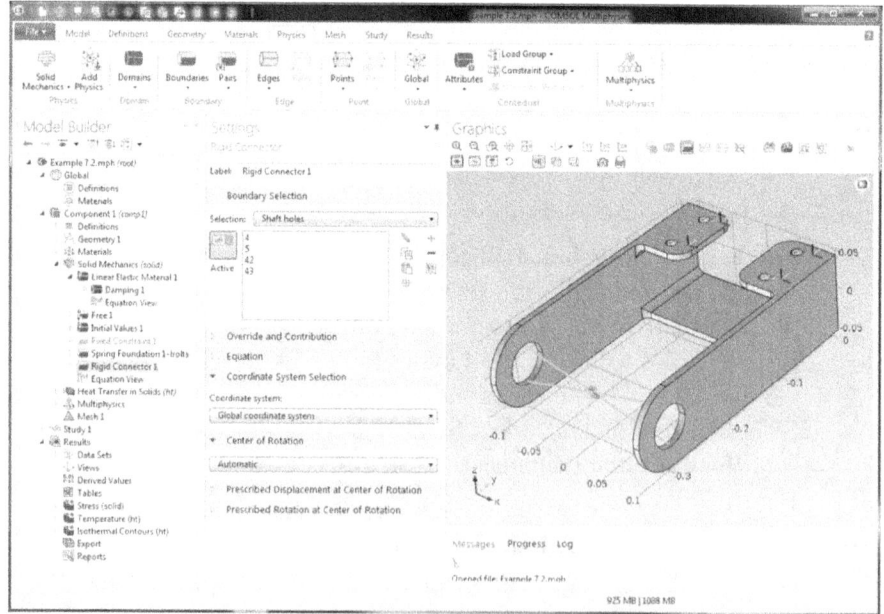

FIGURE 7.9 Rigid Connector boundary conditions setting for the bracket.

corresponding settings window locate Applied Force section and enter $1e2*\sin(2*pi*t*200\,[1/s])$ for x, -11000 for y, and $700*\sin(2*pi*t*100\,[1/s])$ for z.

15. To disengage the thermal stress analysis, right-click on the Heat transfer in Solids (*ht*) node and select Disable. Nevertheless, the previous solution for thermal stress is still available after running the transient model.

16. Click on the Study tab from the toolbar and select Add Study. From the list select Time Dependent and click on Add Study, in the Add Study window. Close Add Study window.

17. We set up a range of values for time variable *t*, for saving the results. Click on Study 2 >Step 1: Time Dependent. In the Time Dependent settings window, enter range (0, 0.25 [ms], 25 [ms]) in the space provided for Times under Study Settings. This will tell the solver to store the results for 25 ms at 0.25 ms intervals. Click the box in front of Relative tolerance and enter 0.001 in the space provided. This will ensure that the solver takes small time intervals during the transient solution. Also expand Results While Solving and check the box for Plot. See Figure 7.10.

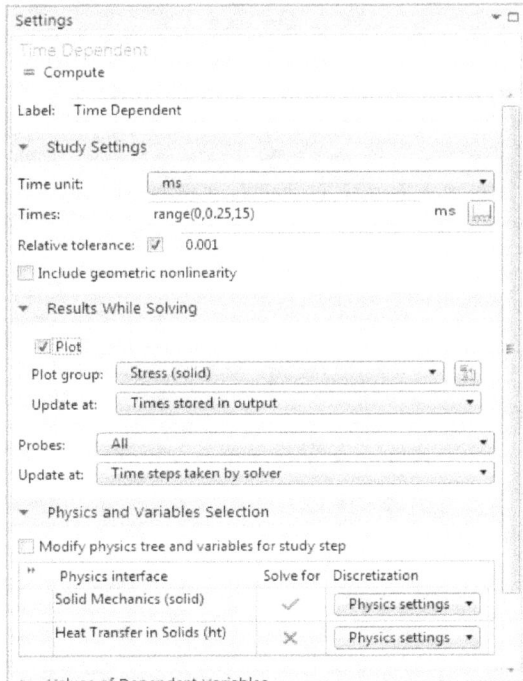

FIGURE 7.10 Time steps settings for transient solver.

18. Click on the Show icon in the Model Builder toolbar and select Advanced Study Options. In the Model Builder window, right-click Study 2 and choose Show Default Solver. Expand Study 2 > Solver Configurations > Solution 2 > Dependent Variables 1. Click on Displacement field (Material) (comp 1.u) and in the Settings window locate the Scaling section and enter 1e-4 for Scale, as shown in Figure 7.11.

19. Click on the Study tab and select Compute. Wait until computations are done.

20. Click on Stress (solid) 1 under Results node in the Model Builder window and rename it to Stress (solid) 1-transient. In the corresponding settings window, locate Data section and select Study 2/Solution 2 from the Data set: list and 0.015 (in second) from the Time list (notice the time unit selection). Click Plot. The results show von Mises stress due to applied loads at time equal to 0.015 seconds, as shown in Figure 7.12. It would be useful to animate the displacements. Click on the Results tab from the toolbar and select Player in the ribbon toolbar. In the Settings window locate Scene and select Stress (solid) 1-transient.

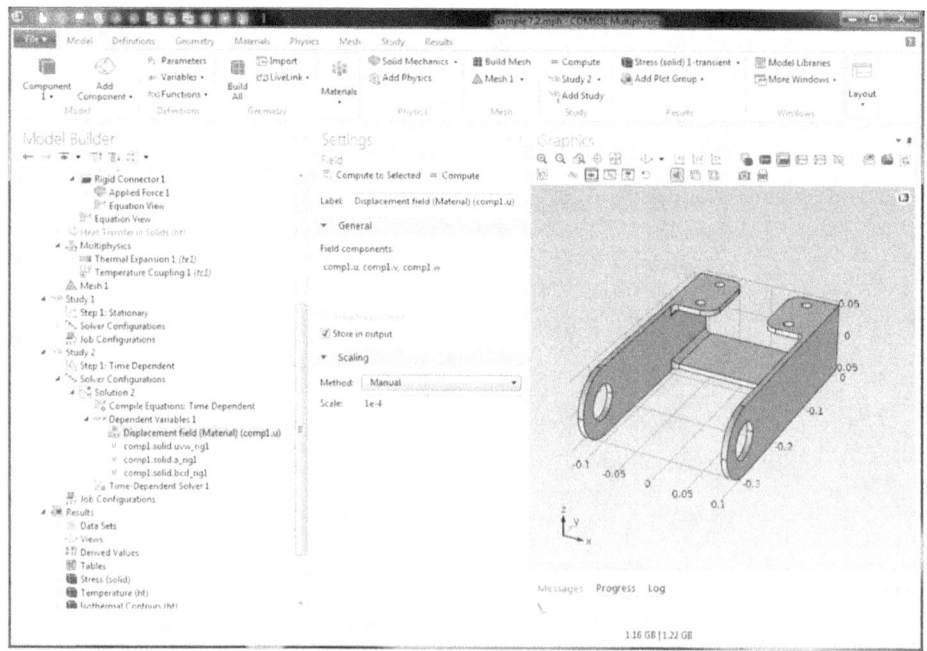

FIGURE 7.11 Transient Solver parameters setting.

Locate the Playing section and enter 0.1 for Display each frame for and check the box for Repeat. Click on the Play icon, from the toolbar in the Graphics window. A movie of the von Mises mapped on displacements will be displayed in the Graphics window. Click on the Player 1 node in the Model tree to change the parameters of the animation.

21. To graph the displacements, right-click on Results and select 1D Plot Group. In the corresponding window locate Data section and select Study 2/Solution 2 from the Data set: list. In the same window, expand Legend and select Lower left from the Position list. From the ribbon toolbar, under 1D Plot Group, select Global. In the corresponding settings window enter solid.rig1.u, solid.rig1.v, solid.rig1.w in the table under y-Axis Data section. Click Plot. For comparison the effects of bolt holes boundary conditions on the displacements are shown in Figure 7.13.

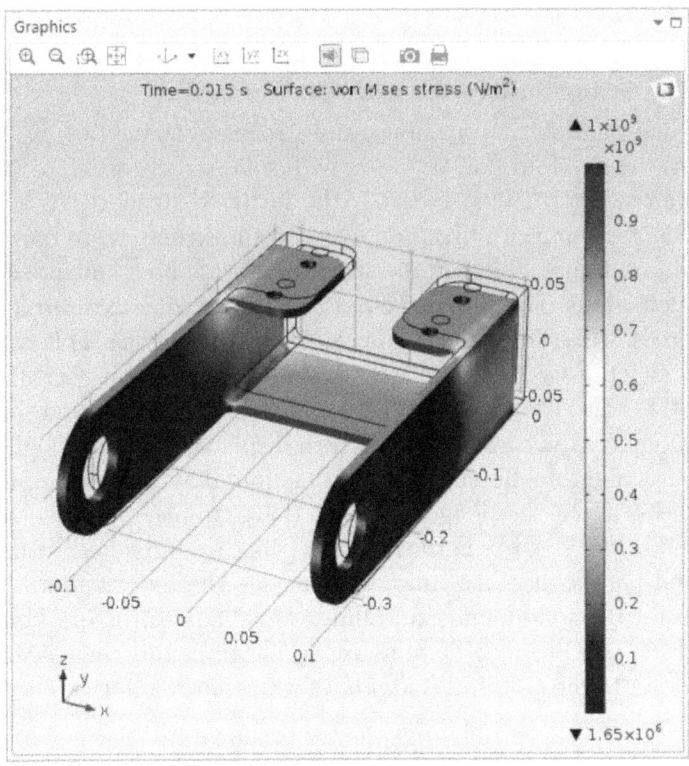

FIGURE 7.12 von Mises stress due to applied loads at time = 0.015 seconds.

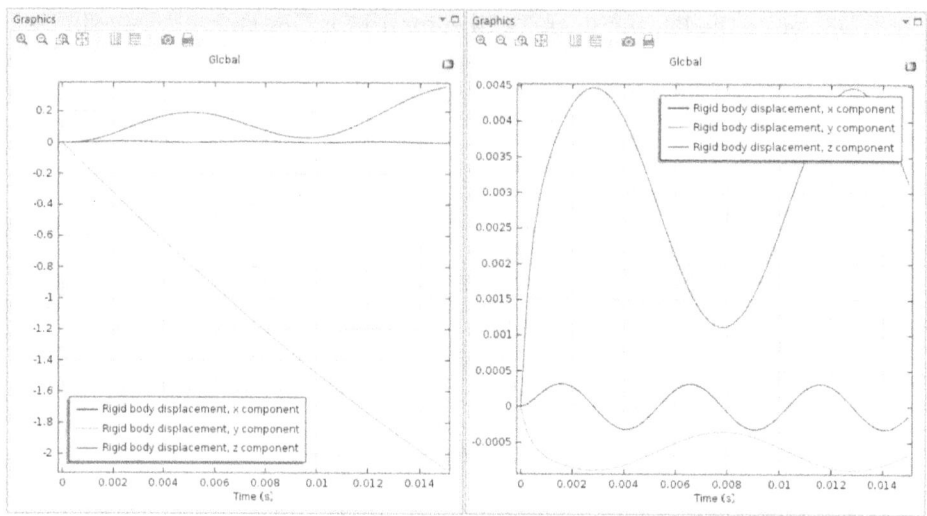

FIGURE 7.13 Displacement components due to applied loads versus time (s), for flexible (left) and (rigid) bolts boundary conditions.

Example 7.3: Static fluid mixer with flexible baffles

In many industrial machines and processing facilities, we may have cases in which a fluid flow interacts with solid structures or parts. We usually, as an approximation, consider the involved structure as a rigid body and calculate quantities like drag force on the structure. In many practical applications, the flexibility of the structure should be taken into account for such calculations. Therefore, the deformed structure, as a result of exerted fluid forces, will change the fluid velocity and vice versa. This type of modeling is called *fluid-structure interaction*, abbreviated in COMSOL as *fsi*. COMSOL 5 also has a feature that includes fluid-structure interaction with fixed geometry for cases that structure displacements are negligible. The calculation of interaction of a solid with a fluid flow is, in principle, an iterative process and a very lengthy one. In this example, we model a 2D static mixer with two flexible baffles (beams) that are attached to the walls of the mixer and behave like cantilever beams. The deformations, von Mises stresses for the solid beams, and fluid velocity and pressure fields are calculated. The governing equations are equilibrium and Navier-Stokes equations. Refer to the *COMSOL Manual* for further readings.

A sketch of the problem geometry is shown in Figure 7.14. We build this geometry using CAD tools available in COMSOL.

FIGURE 7.14 Sketch of the geometry for fluid mixed with baffles.

Solution:

1. Launch COMSOL and in New window select Model Wizard. Save the file as Example 7.3.

2. Select 2D from the Select Space Dimension window. In the Select Physics window, expand the list for Fluid Flow and select Fluid-Structure Interaction (*fsi*). Click on Add and then click on the Study arrow button. In Select Study window select Stationary and click Done.

3. To build the geometry of the mixer, use the following steps:

   3.1. Click on the Geometry 1 node in the Model Builder window (if not already highlighted). In the Geometry window under Units, change the Length unit to cm from the list (open the list to see the selection options). Draw a box in the Graphics window by clicking on the Geometry tab from the toolbar and selecting Rectangle. The Rectangle settings window will open, or click on Geometry 1 >Rectangle (r1) node in the Model Builder window. In this window, enter 50 for Width and 15 for Height in the Size section, and 0 for x: and 0 for y: under Position. Make sure that Corner is selected for Base. Click on the Build Selected icon. The mixer box geometry will appear in the Graphics window.

   3.2. Similarly, draw two more boxes for the beams/baffles. Right-click on the Rectangle 1 node and select Duplicate. In the Rectangle settings window, enter 1 for Width and 10 for Height in the Size section, and 18 for x: and 0 for y: under Position. Make sure that Corner is selected for Base. Click on the Build Selected icon. Again, right-click on the Rectangle 1 node in the Model Builder window and select Duplicate. In the Rectangle settings window, enter 1 for Width and 10 for Height in the Size section, and 32 for x: and 5 for y: under Position. Make sure that Corner is selected for Base. Click on the Build Selected icon. Finally, to smooth the tips of the beams just created, click on the Geometry tab from the toolbar and select Fillet. In the Fillet settings window, select the points on the tips of the beams/baffles and add them to the Selection window. Enter 0.2 for Radius, and then click on the Build Selected icon.

3.3. Rename the nodes under the Geometry 1 node as mixture, baffle 1, and baffle 2, as shown in Figure 7.15.

4. To assign materials properties, we start with the baffles/beams. To define the baffles as solid media beams, expand the Fluid-Structure Interaction (*fsi*) node in the Model Builder and click on the Linear Elastic Material 1 node. From the Graphics window, select and add the domains (2, 3) for the baffles to the Selection list in settings window.

To assign the baffles material properties, locate Linear Elastic Material section and enter the following data for the beams properties: Young's modulus of elasticity (5.6e6), Poisson's ratio (0.4), and density (1e3). See Figure 7.16.

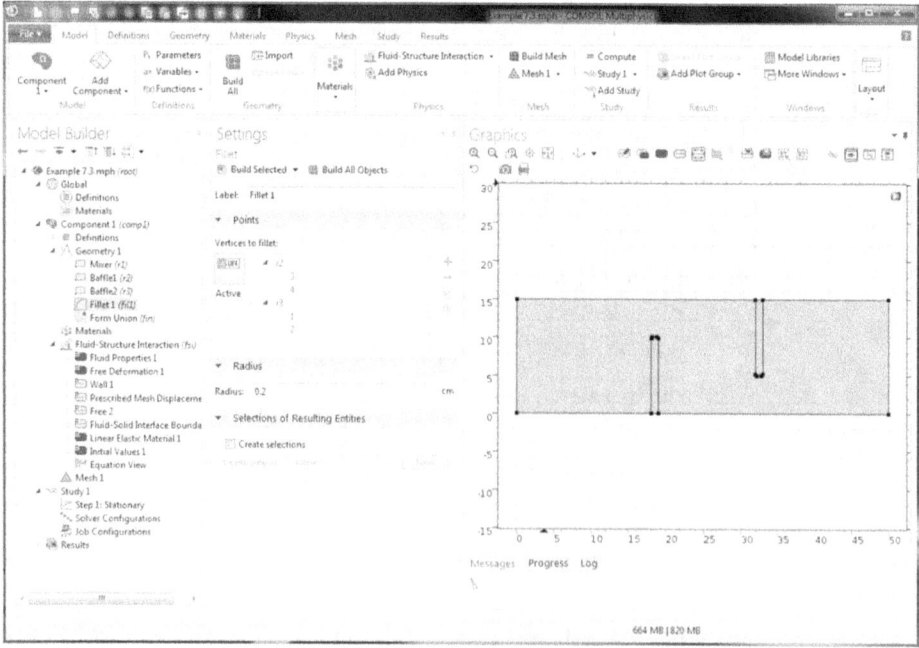

FIGURE 7.15 Building the mixer geometry and baffles with fillets.

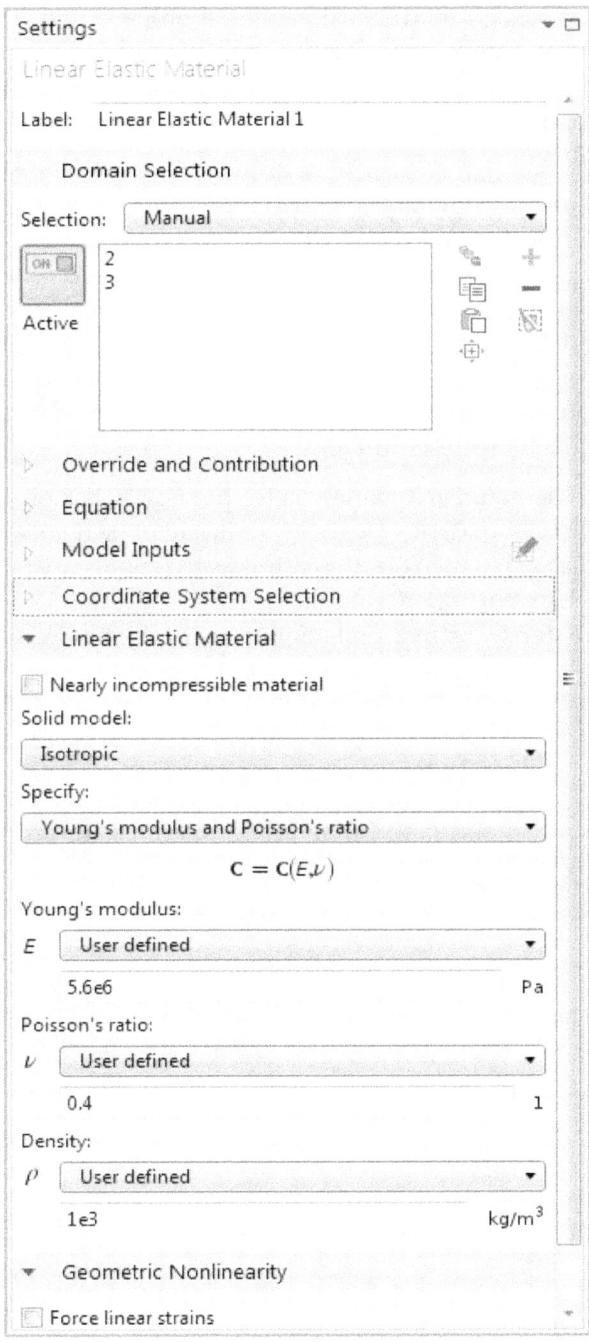

FIGURE 7.16 Material properties setting for elastic baffles.

Similarly, assign fluid domain material properties. Click on the Fluid Properties 1 node in the Model Builder window; make sure that only domain 1 is listed, as it represents the fluid flow domain. Locate Fluid Properties section and enter 1E3 for fluid density and 1 for dynamic viscosity.

5. To assign the boundary conditions, we first assign those related to the beams/baffles. Click on the Physics tab from the toolbar and select Boundaries > Fixed Constraint. From the Graphics window, select the boundaries (5, 10) at the base of the beams where they are attached to the mixer wall, and add them to the Selection list in the Fixed Constraint settings window. This step is very crucial in finding a converged solution for this model, and in general fluid-structure interaction models. See Figure 7.17.

6. Assign the boundary conditions for the flow. Click on the Physics tab and select Boundaries > > Inlet. In the Inlet settings window, add the boundary (1) to the Selection list. Locate Boundary Condition section, select Laminar inflow, and enter Vin for U_{av} and 10 for L_{entr}. Check the box for Constrain endpoints to zero. Similarly for create outlet

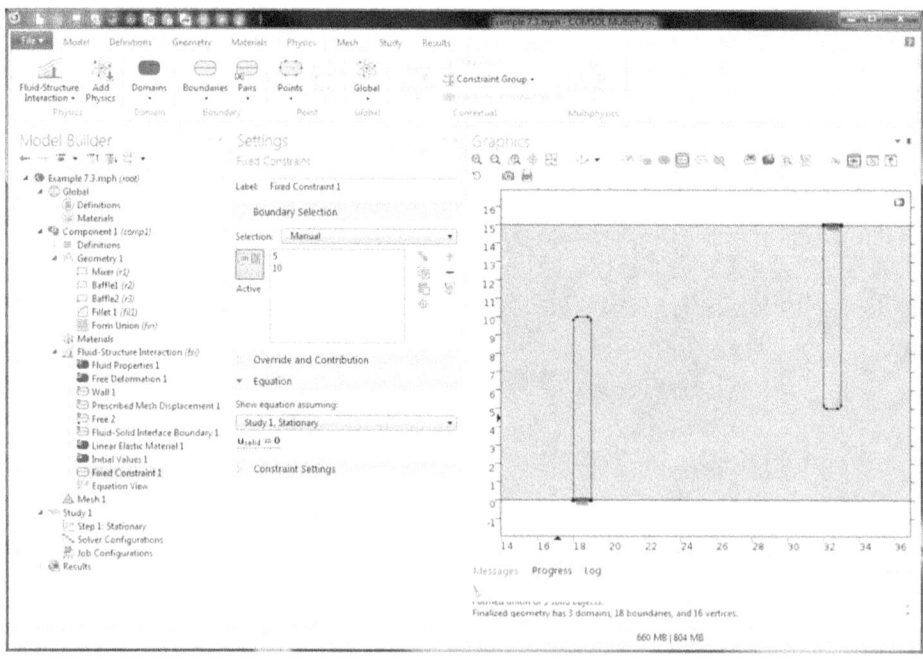

FIGURE 7.17 Assigning fixed-type boundary conditions for the beams/baffles.

boundary condition, select Boundaries > Outlet. In the Outlet window, add the boundary (14) to the Selection list. We leave the pressure as zero, for reference.

7. To define the parameter Vin, the fluid velocity at the inlet, click on Model tab from the toolbar and select Parameters. In the Parameters settings window, enter Vin for Name and 0.2 [m/s] for Expressions. You may want to add Fluid Velocity under Description (optional).

At this stage, we have defined all the physics and boundary conditions for this model. Note that COMSOL will assign a moving mesh that moves according to the fluid flow and beams displacements. This feature is very useful for calculating the interactions and is done automatically. It is recommended that at this point users click on each of the nodes listed under the Fluid-Structure Interaction node in the Model Builder window and study their definitions.

8. We change the default discretization or the order of function/polynomials for finite elements for fluid flow. We want to use second-order elements for velocity and first-order elements for pressure. This is defined as P2 + P1. To do this, click on the Show button, located in the Model Builder window toolbar, and select Discretization. Now click on the Fluid-Structure Interaction node and in the corresponding settings window locate the Discretization section and expand it. From the list under Discretization of fluids, select P2 + P1.

9. To build the mesh, click on the Mesh 1 node in the Model Builder window. In the Mesh settings window select Coarse for Element size and click Build All. Zoom on the mesh to see the hybrid mesh, which consist of boundary-layer quadrilateral and main domain triangular elements. Also note that the beam/baffle solid domain is meshed, as shown in Figure 7.18.

10. The model is now ready to run. Click on the Study tab and select Compute. After the calculations are complete, the default results appear in the Graphics window, showing the fluid velocity and the von Mises stress in the beams/baffles for the assigned value of the inlet velocity 0.2 m/s.

11. It is useful to do a parametric analysis for this model based on the inlet velocity. Click on the Study tab from the toolbar and select Parametric Sweep. In the corresponding settings window, locate the Study Settings section and click on the plus icon (+); the Vin (Inlet fluid

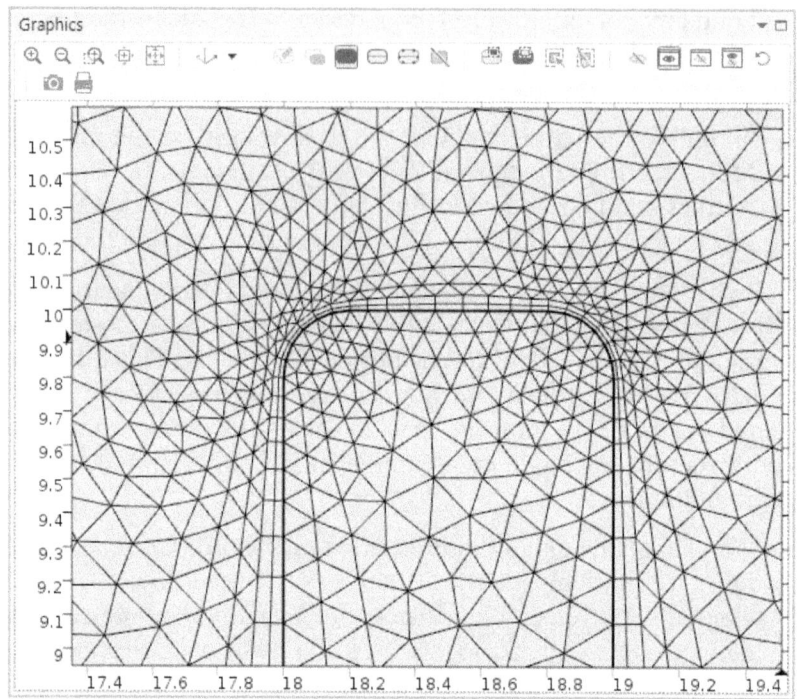

FIGURE 7.18 Built hybrid mesh, zoomed around the first baffle tip.

velocity) will appear in the table. This is the same as it was defined in the Global definitions > Parameters (see Step 7). In the Table under the Parameter value list, enter 0.2, 0.4, 0.6, 0.8. The model will assign these values to the inlet velocity, in turn, and calculate the corresponding results. To see the results while being solved, check the Plot box under Output While Solving. This will slow down the solution process because it requires more computer resources.

12. Again, click on the Study tab and select Compute. The results will appear in the Graphics window. To see each set, click on the Flow and Stress node in the Model Builder window and in the corresponding settings window select the desired value for Parameter value (Vin), under the Data section.

13. The results for Vin = 0.2, 0.4, 0.6, and 0.8 m/s are shown in Figure 7.19. Note the progressive deflections of the beams/baffles as the fluid flow moves faster.

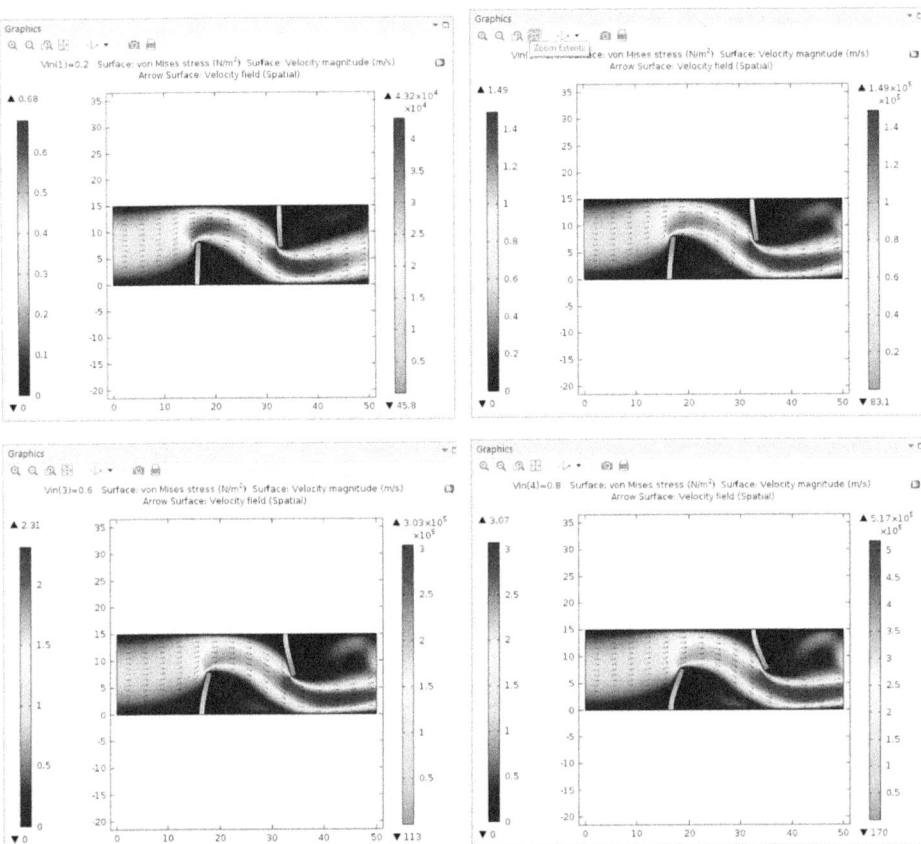

FIGURE 7.19 Windows showing results for fluid-beam interactions for different values of fluid inlet velocity.

14. It would be useful to draw the displacements of the tips of the baffles/beams versus the inlet velocity of the fluid. Right-click on the Results node and select 1D Plot Group. A new node, 1D Plot Group 3, will be created. Right-click on it and select Point Graph. In the corresponding settings window, select the points (4, 13) representing the tip points of the baffles and add them to the Selection list. In the y-axis Data section, enter fsi.disp. This variable is the total displacements as a result of beam deflection. Change the unit to mm, by selecting it from the Unit list. To draw the results, click on Plot. See Figure 7.20.

15. To build an app for this model, launch COMSOL if not already running. From File menu select New. In the New window select

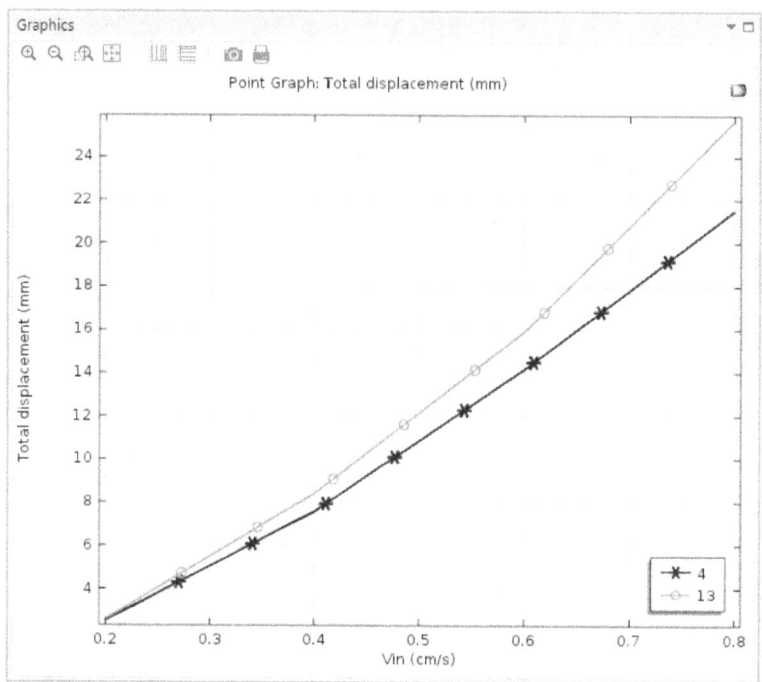

FIGURE 7.20 Results for tip displacements of the beam/baffles versus inlet velocity.

Applications Wizard. In the Select Model for Application window click on Browse; locate and open Example 7.3.mph. The New Form window will appear. In this window, select entries Inlet fluid velocity, Mixer width, and Mixer height under Parameters from the Inputs/outputs tab and move them to Selected window. Click on the Graphics tab; select and move Flow and Stress (*fsi*) to Selected window. Click on the Buttons tab; select and move Compute Study 1 to Selected window. See Figure 7.21. Click on Done. Form Editor desktop window appears. Save the file as App_Example 7.3.

16. Several editing tools are available in the Form Editor window. We use some of these tools in order to lay out the App interface. Click on Grid, from the ribbon bar, to relocate the objects in the Form 1. Use the Settings for each object to modify the size, affiliate a picture, and specify the function. Refer to detailed instructions given for Example 2.1, for a guide. Final results for the App interface are shown in Figure 7.22.

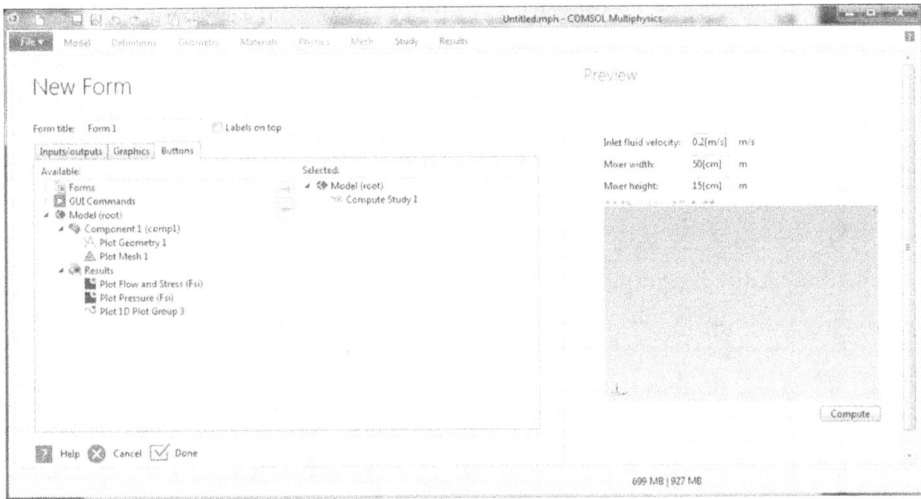

FIGURE 7.21 Application settings for Example 7.3.

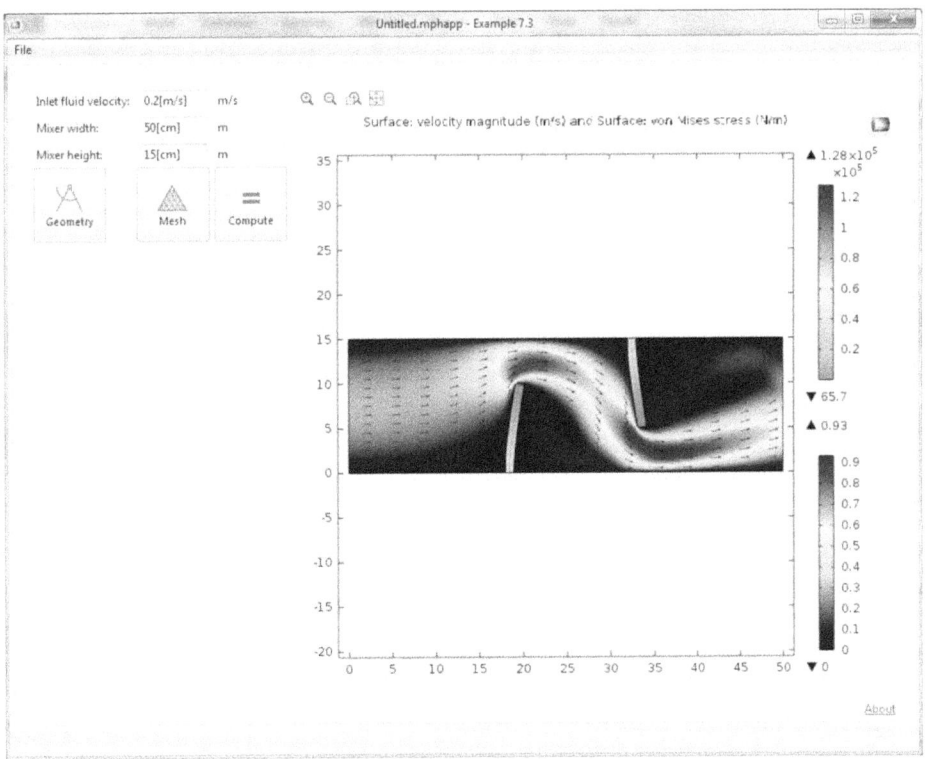

FIGURE 7.22 Application GUI for Example 7.3.

Example 7.4: Double pendulum motion: Multibody dynamics

In this example, we consider motion of a double pendulum. This is a classic problem in applied mechanics under multibody dynamics topic. As shown in Figure 7.23, the double pendulum system consists of two masses (m_1 and m_2) hanging by two strings with lengths L_1 and L_2. We assume that strings are massless, relatively. Vibration is in the x-y plane. We are mainly interested in finding the variations of angles θ_1 and θ_2, as functions of time.

The governing equations for motions of masses m_1 and m_2 are derived using Lagrangian mechanics [25]. The Lagrangian function is defined as $L = K - V$, where K is the kinetic energy and V potential energy of the system. For the double pendulum, we have

$$K = \frac{1}{2}m_1(\dot{x}_1^2 + \dot{y}_1^2) + \frac{1}{2}m_2(\dot{x}_2^2 + \dot{y}_2^2)$$

$$V = -m_1 g L_1 \cos(\theta_1) - m_2 g(L_1 \cos\theta_1 + L_2 \cos\theta_2)$$

where $x_1 = L_1 \sin\theta_1$, $x_2 = x_1 + L_2 \sin\theta_2$, $y_1 = -L_1 \cos\theta_1$, $y_2 = y_1 - L_2 \cos\theta_2$ with reference to the coordinates system shown in Figure 7.23. Dot-symbols are used for time derivatives of variables. By substituting K and V into L, after simplification we have:

$$L = \frac{1}{2}(m_1 + m_2)L_1^2\dot{\theta}_1^2 + \frac{1}{2}m_2 L_2^2\dot{\theta}_2^2 + m_2 L_1 L_2 \dot{\theta}_1 \dot{\theta}_2 \cos(\theta_1 - \theta_2)$$
$$+ (m_1 + m_2)gL_1 \cos\theta_1 + m_2 g L_2 \cos\theta_2$$

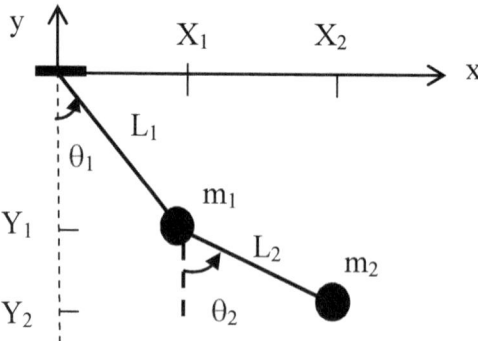

FIGURE 7.23 Geometry sketch and coordinates for a double pendulum.

The Euler-Lagrange equations are:

$$\frac{\partial L}{\partial \theta_1} - \frac{d}{dt}\left(\frac{\partial L}{\partial \dot\theta_1}\right) = 0$$

$$\frac{\partial L}{\partial \theta_2} - \frac{d}{dt}\left(\frac{\partial L}{\partial \dot\theta_2}\right) = 0$$

We now apply Euler-Lagrange equations using the resulting equation for L. The final result is a system of nonlinear ODEs, as given below:

$$(m_1 + m_2)L_1^2\ddot\theta_1 + m_2 L_1 L_2 \ddot\theta_2 \cos(\theta_1 - \theta_2) + m_2 L_1 L_2 \dot\theta_2^2 \sin(\theta_1 - \theta_2)$$
$$+ (m_1 + m_2)L_1 g \sin\theta_1 = 0$$

$$m_2 L_2^2 \ddot\theta_2 + m_2 L_1 L_2 \ddot\theta_1 \cos(\theta_1 - \theta_2) - m_2 L_1 L_2 \dot\theta_1^2 \sin(\theta_1 - \theta_2) + m_2 L_2 g \sin\theta_2 = 0$$

After simplification, we get;

$$(m_1 + m_2)L_1\ddot\theta_1 + m_2 L_2 \ddot\theta_2 \cos(\theta_1 - \theta_2) + m_2 L_2 \dot\theta_2^2 \sin(\theta_1 - \theta_2) + (m_1 + m_2)g \sin\theta_1 = 0$$

$$L_2 \ddot\theta_2 + L_1 \ddot\theta_1 \cos(\theta_1 - \theta_2) - L_1 \dot\theta_1^2 \sin(\theta_1 - \theta_2) + g \sin\theta_2 = 0$$

Solution:

To solve the nonlinear system of ODEs, we let m_1 = 50 g, m_2 = 100 g, L_1 = 20 cm, L_2 = 10 cm. The initial conditions are $\theta_1 = \dot\theta_1 = \dot\theta_2 = 0$ and $\theta_1 = \pi/6$, $\theta_2 = \pi/4$.

1. Launch COMSOL and select Model Wizard. Save the file as Example 7.4.

2. In Select Space Dimension window select 0D. In the Select Physics window expand the Mathematics node and select ODE and DAE Interfaces > Global ODEs and DAEs (ge). Click Add and then click on the Study arrow button.

3. From the Select Study list, select Time Dependent and click Done.

4. For this example, we don't have geometry to build (hence the 0D option selected). Expand the Component 1 (*comp 1*) node in the Model Builder window and click on the Global Equations 1 node. In the corresponding settings window, enter the governing equations derived for θ_1, θ_2, and their initial conditions. See Figure 7.24. Note

that the variable names should be entered under column Name, and
they should be consistent with the variables used in the equations. We
used theta 1 for θ_1 and theta 2 for θ_2. Note that in COMSOL derivatives are available. For example, the time derivative of theta 1 is theta
1t, and so on.

5. To enter the data, we enter them as parameters into the model. Click
   on the Model tab in the toolbar and select Parameters. In the corresponding window, enter the data as shown in Figure 7.25.

6. Expand the Study 1 node in the Model Builder window and click on
   Step 1: Time Dependent. In the corresponding window, enter range
   (0, 0.05, 10).

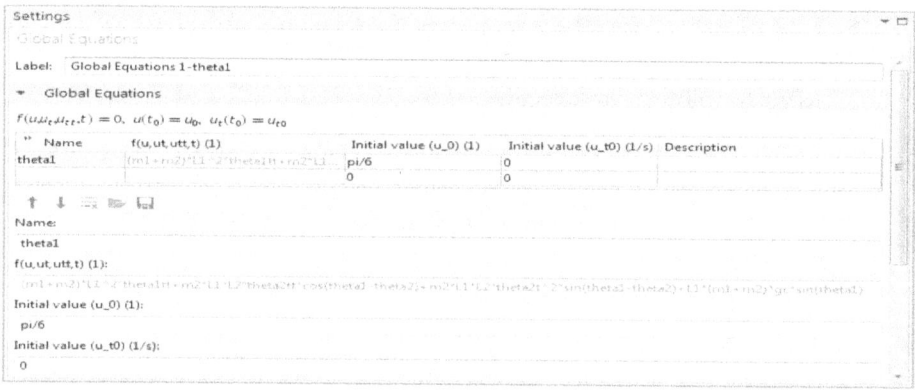

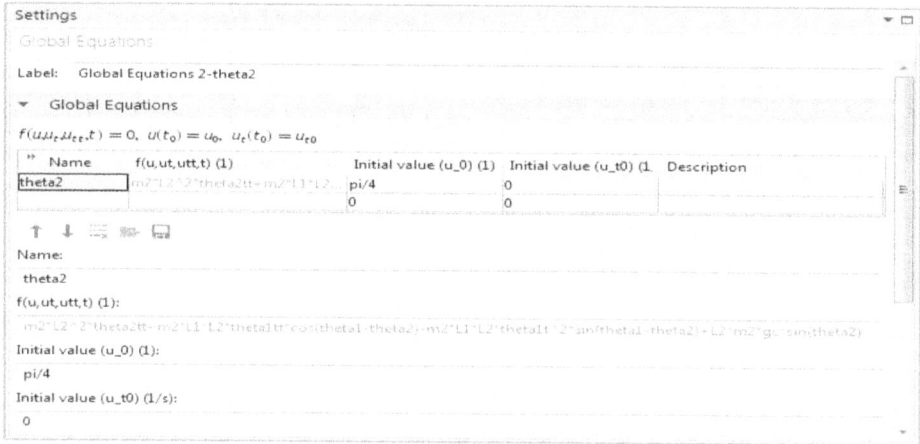

FIGURE 7.24 Entering system of ODEs, as Global Equations.

7. Click on the Study tab from the toolbar and select Compute. Default results show line graphs for variations of theta 1 and theta 2 with respect to time, as shown in Figure 7.26.

8. It is useful to calculate the trajectories of mass 1 $(x_1(t), y_1(t))$ and mass 2 $(x_2(t), y_2(t))$, as well. Create another 1D Plot Group. Select Global and in the corresponding settings window find the Data section. In the table,

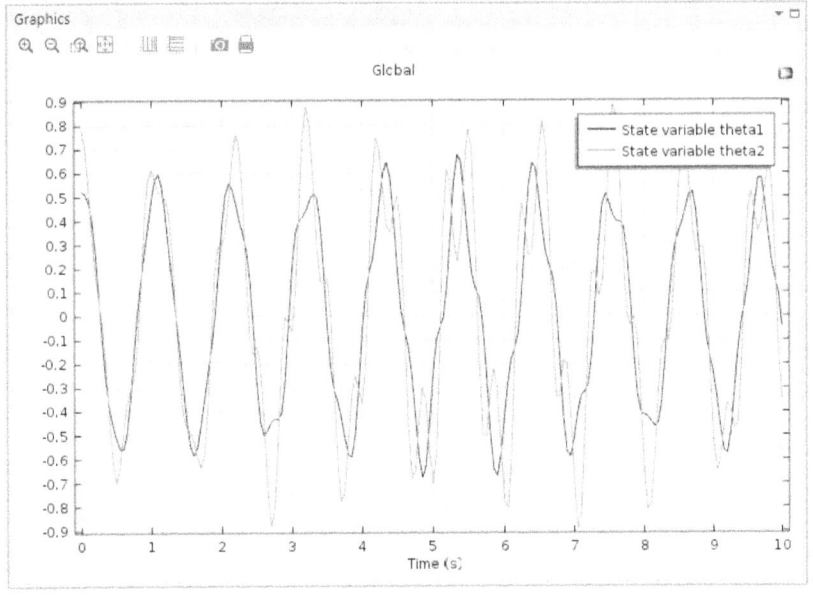

Settings

Parameters

▼ Parameters

Name	Expression	Value	Description
m1	50[g]	0.05 kg	mass 1
m2	100[g]	0.1 kg	mass 2
L1	20[cm]	0.2 m	lenght of string 1
L2	10[cm]	0.1 m	length of string 2
gc	9.81[m/s^2]	9.81 m/s²	grav. accelelarion

FIGURE 7.25 Data for the double pendulum, as Parameters.

FIGURE 7.26 Variations of θ_1 and θ_2 w.r.t. time for the double pendulum.

under y-axis Data, type-L1*cos (comp 1.theta 1). Locate x-Data section select Expression for Parameters and type L1*sin (comp 1.theta 1). These are expressions for Y_1 and X_1, respectively. Click on the Plot icon. Results are shown in Figure 7.27.

Similarly, trajectory of mass 2 can be obtained.

It well known that double pendulums may exhibit periodic, quasi-periodic, or chaotic behavior depending on the amount of initial energy input to the system. Double pendulum is a complex system. In order to analyze its dynamics, several graphs usually are needed (e.g. velocity versus position, trajectories, angular velocities, angles). COMSOL allows users to make graphs of these quantities.

9. Create another 1D Plot Group. Select Global and in the corresponding settings window find the Data section. Expand y-Axis Data and enter comp 1.theta 1 in the first row under Expression. Expand x-Axis Data and from the Parameters list choose Expression and enter comp 1.theta 2. Expand the Coloring and Style section and select Point for Marker and In data Points for Positioning. Enter the axes and graph title, if desired, and click Plot. Results are shown in Figure 7.28. As shown in

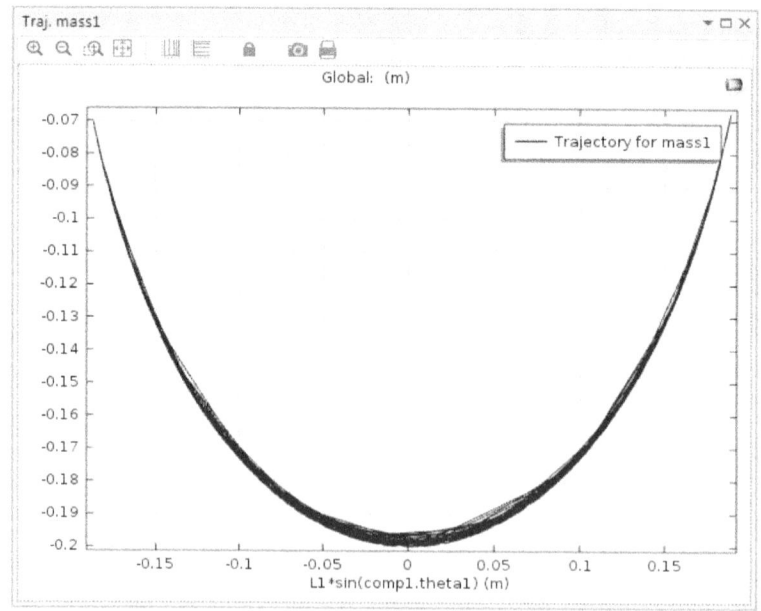

FIGURE 7.27 Trajectory of the double pendulum mass1 w.r.t. time.

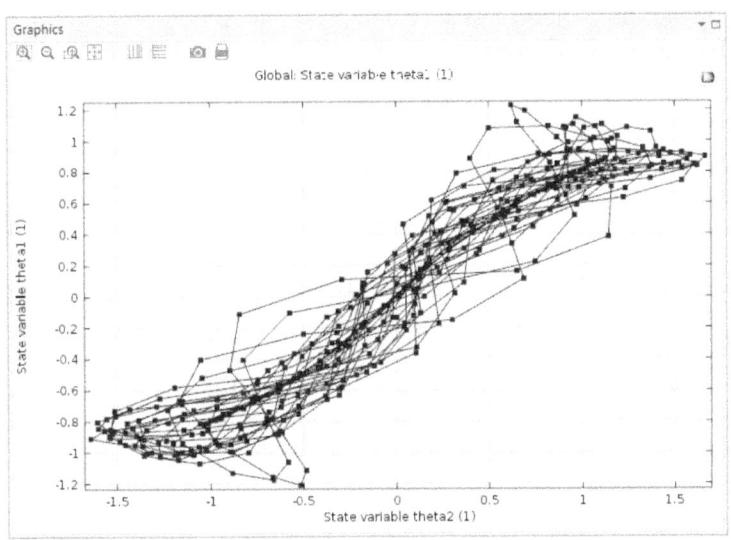

FIGURE 7.28 Angle θ_1 vs. θ_2 for double pendulum mass 1 and mass 2.

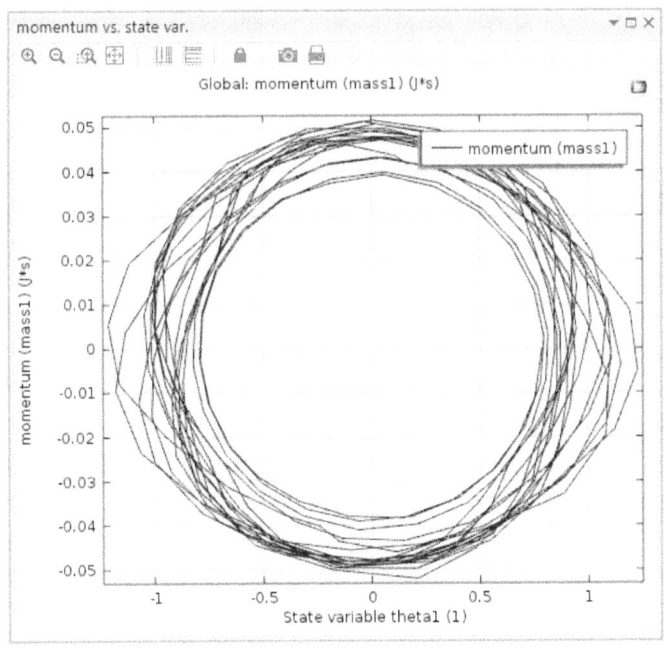

FIGURE 7.29 Momentum for mass1 vs. state variable θ_1.

this graph, when mass 2 is along the y-axis, hence theta 2 = 0, mass 1 is at a different known location on the x-y plane. From this graph users may extract the locations of mass 1 and mass 2 for different values of theta 1 and theta 2.

10. Another useful result is the momenta versus state variables θ_1 or θ_2. The momenta can be calculated using $p_{\theta_i} = \dfrac{\partial L}{\partial \dot{\theta}_i}$. Therefore, for example, $p_{\theta_1} = (m_1 + m_2)L_1^2\dot{\theta}_1 + m_2 L_1 L_2 \dot{\theta}_2 \cos(\theta_1 - \theta_2)$. Create another 1D Plot Group. Select Global and in the corresponding settings window find the Data section. For y-Axis Data, under Expression, enter (m1 + m2)*L1^2*comp 1.theta 1t + m2 * L1 * L2 * comp 1.theta 2t * cos (comp 1.theta 1-comp 1.theta 2). Locate x-Axis Data section and select Expression from the Parameters list and enter comp 1.theta 1. Click Plot. Result is shown in Figure 7.29.

Example 7.5: Multiphysics model for thermoelectric modules

Thermoelectric (TE) materials can be used for power generation and refrigeration, as shown in Figure 7.30. In this example, we model a TE module system. The TE modules in a system are connected in series and in parallel to produce useful power and required electrical voltage and current.

When two dissimilar materials are connected together (electrically in series and thermally in parallel) and a differential temperature is applied at their two ends, an electric voltage is generated. This is known as the Seebeck effect, which states that generated voltage is proportional to the temperature difference. When an electrical load is connected to the module, then electric current running through generates electric power. There are also other effects, such as Peltier and Thomson effects, associated with current going through thermoelectric module [27].

The Peltier effect appears when an external electric power is applied instead of the electric load and as a result a temperature difference occurs, as shown in Figure 7.30. TE refrigeration uses the Peltier effect.

The efficiency of a TE module (e.g. related to power generation mode) is referred to as figure-of-merit. The figure-of-merit is given as $Z = \dfrac{S^2\sigma}{k}$, where S is the Seebeck coefficient (V/K), σ electrical conductivity ($\Omega^{-1}\mathrm{m}^{-1}$), and k thermal conductivity (W/m.K). The product $S^2 \sigma$ is referred to as the electrical power factor. The figure-of-merit for a TE device is sometimes given in its dimensionless form, as ZT, where T is absolute temperature.

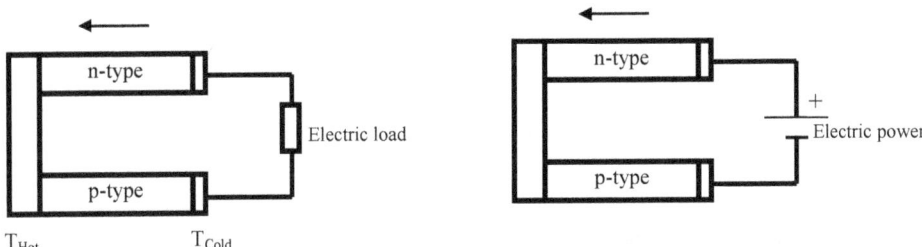

FIGURE 7.30 Seebeck effect: electric power generated by providing $\Delta T = T_{Hot} - T_{Cold}$ (left diagram) and Peltier effect: refrigeration by absorbing heat on the T Cold side providing electric power input (right diagram).

The main phenomena associated with a TE are heat flux and electric current density. Heat flux is proportional to temperature gradient (Fourier's law) and generates electric current (Seebeck effect). Electric current by itself heats up the materials (Joule heating effect). The electric current absorbs heat reversibly (Peltier effect). A combination of these phenomena constitutes the governing equation for heat transfer in a TE (in vector/tensor format):

$$q = \Pi \cdot J - k \cdot \nabla T$$
$$\nabla \cdot q = \dot{Q}$$

where q is heat flux (W/m²), Π Peltier coefficient matrix (V), J current density (A/m²), and T temperature (K). $\dot{Q} = J \cdot E$ is the heat source due to the Joule heating effect and $E = -\nabla V$ is the electric field due to electric voltage potential V.

The electric current density is due to Seebeck voltage and should satisfy Ohm's law. Hence we have the governing equation for current density as:

$$J = \sigma \cdot (E - S \cdot \nabla T)$$
$$\nabla \cdot J = 0$$

In the last equation ($\nabla \cdot J = 0$), which governs convergence of current density, we neglect electric displacement (or time variation of electric flux density). This assumption is valid for TE.

By combining and manipulating the above equations (see [28], [29]), we have:

$$\begin{cases} \nabla \cdot [-(S^2 \sigma T + k) \nabla T] + \nabla \cdot (-S \sigma T \nabla V) = \sigma[(\nabla V)^2 + S \nabla T \cdot \nabla V)] \\ \nabla \cdot (S \sigma \nabla T) + \nabla \cdot (\sigma \nabla V) = 0 \end{cases}$$

These equations are the mathematical model for TE. Note that these are coupled nonlinear partial differential equations in vector notation format.

We can use the Mathematics module in COMSOL for modeling, specifically the Coefficient Form for solving the above equations. However, COMSOL 5 (since version 4.4) has a thermoelectric module (i.e. Thermoelectric Effect) included in the Heat Transfer module. In COMSOL definition, the Coefficient Form of a PDE reads:

$$e_a \frac{\partial^2 u}{\partial t^2} + d_a \frac{\partial u}{\partial t} + \nabla \cdot (-c\nabla u - \alpha u + \gamma) + \beta \cdot \nabla u + au = f$$

where u is the dependent variable vector and coefficients are defined independently. Therefore, after comparing our mathematical model with the Coefficient Form, we have:

$$u = \begin{pmatrix} T \\ V \end{pmatrix}, \quad c = \begin{pmatrix} k + S^2\sigma T & S\sigma T \\ S\sigma & \sigma \end{pmatrix}, \quad f = \begin{pmatrix} \sigma[(\nabla V)^2 + S\nabla T \cdot \nabla V)] \\ 0 \end{pmatrix}$$

All other coefficients (e_a, d_a, α, β, γ, a) are zero. We will discuss the boundary conditions during model construction.

Solution:

1. Launch COMSOL and in New window select Model Wizard. Save the file as Example 7.5.

2. Select 3D from the Select Space Dimension window.

3. From the Select Physics list, expand the Δu Mathematics node and select Δu PDE Interfaces > Δu Coefficient Form PDE (c). Click Add and locate the Dependent Variables section (on the right side of the same window) and enter 2 in the space for Number of dependent variables and T and V for Dependent variables. Leave Field name as default. Click on Study arrow button.

4. From the Select Study list, select Stationary and click Done.

5. To build the TE modules geometry, click on the Geometry tab from the toolbar and select Block. In the corresponding settings window, enter 1.24 [cm] for Width and 1 [cm] for both Depth and Height. For Position enter 0.1 [cm] for z. Click the Build Selected icon on the toolbar of the Block settings window. Similarly, build more blocks with the dimensions provided in Table 7.2.

6. Click on the Geometry tab and select Transforms > Copy. In the corresponding settings window, add Blocks 1, 3, and 4 to the selection list. In the Displacement section, enter 3.24 [cm] for x: and click on Build Selected. The final result for the geometry is shown in Figure 7.31.

	Width	Depth	Height	Position (x, y, z)
Block 2	1.24 [cm]	1 [cm]	0.1 [cm]	$(0, 0, 0)$
Block 3	2.74 [cm]	1 [cm]	0.1 [cm]	$(0, 0, 1.1 \,[cm])$
Block 4	1 [cm]	1 [cm]	1 [cm]	$(1.74 \,[cm], 0, 0.1 \,[cm])$
Block 5	2.74 [cm]	1 [cm]	0.1 [cm]	$(1.74 \,[cm], 0, 0)$
Block 6	1 [cm]	1 [cm]	0.1 [cm]	$(4.98 \,[cm], 0, 0)$

TABLE 7.2 Dimensions for Building Geometry of TE Modules.

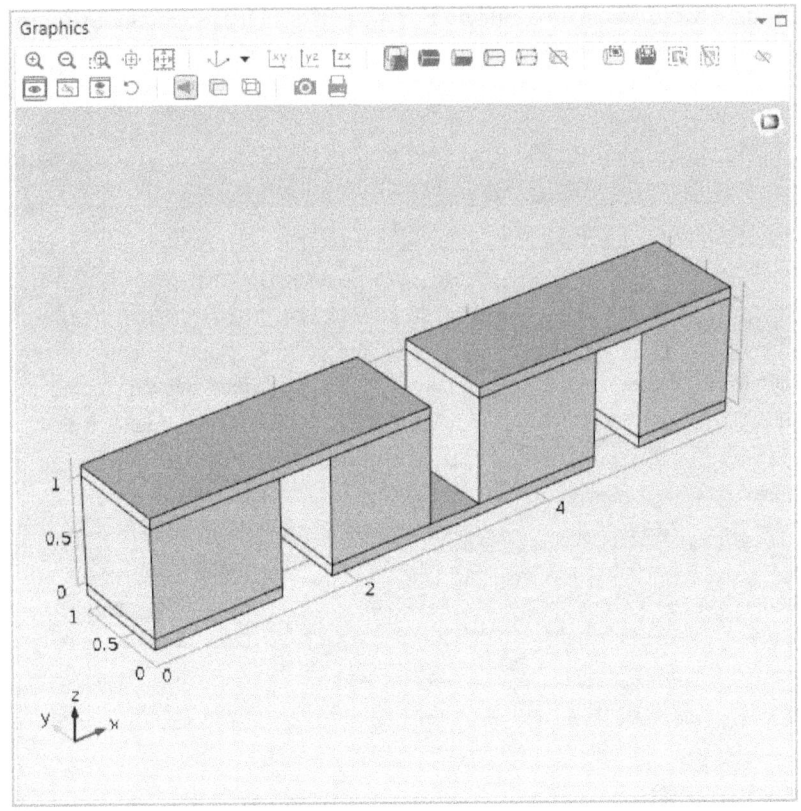

FIGURE 7.31 Built geometry for two TE modules connected in series.

7. We use the Parameters feature for entering material properties. As mentioned, p-type (Block 1) and n-type (Block 4) silicon-based materials are common in industry for TE modules. In addition, electrodes (Blocks 2 and 3) are usually made out of copper. Click on Model tab from the toolbar and select Parameters. In the corresponding settings window, enter the data as shown in Figure 7.32. These are typical material properties for Bismuth-Telluride and copper.

For this example, we consider constant material properties. However, the model is capable of using materials with varying properties, such as functions of temperature.

8. To enter the model equations, click on the Coefficient Form PDE 1 node in the Model Builder window. Rename this node to Coefficient Form for TEp. In the corresponding settings window, we will enter expressions for Coefficients c and f only and make the rest as zeros, as discussed previously. Expand the Diffusion Coefficient and Source Term sections and enter the expressions for c and f matrices for the p-type TE (Domains 2 and 6), as shown in Figure 7.33. The

Pi Parameters

▼ Parameters

Name	Expression	Value	Description
alphaTEp	230e-6[V/K]	2.3000E-4 V/K	Seebeck coeff-p type
kTEp	1.2[W/m/K]	1.2000 W/(m·K)	therm. cond.-p type
sigmaTEp	(1/(1.75e-5))[S/m]	57143 S/m	elec. cond.-p type
alphaTEn	-195e-6[V/K]	-1.9500E-4 V/K	Seebeck coeff-n type
kTEn	1.4[W/m/K]	1.4000 W/(m·K)	therm. cond.-n type
sigmaTEn	(1/(1.35e-5))[S/m]	74074 S/m	elec. cond.-n type
alphac	6.5e-6[V/K]	6.5000E-6 V/K	Seebeck coeff-copper
sigmac	5.9e8[S/m]	5.9000E8 S/m	elec. cond. -copper
kc	350[W/m/K]	350.00 W/(m·K)	therm. cond. - copper

FIGURE 7.32 Parameters for material properties for TE modules, p-type, n-type, and copper electrodes.

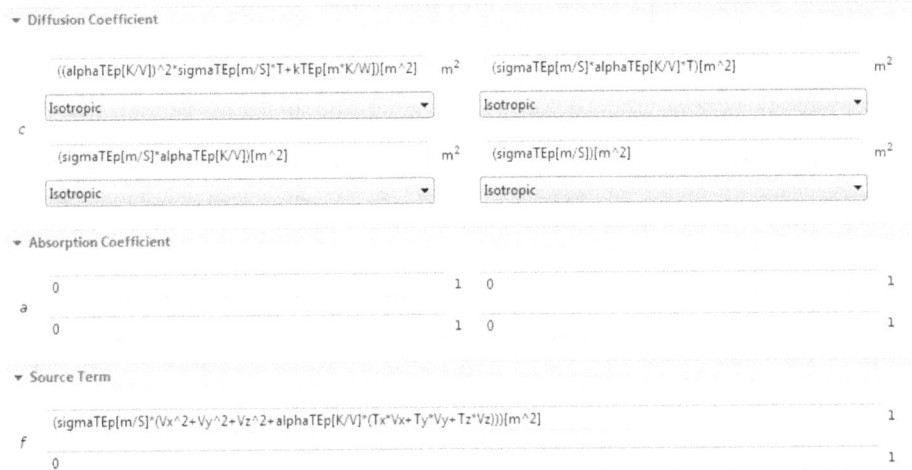

FIGURE 7.33 Coefficient expressions for p-type TE blocks.

variables should be exactly the same as those defined as Parameters (see Figure 7.32) and they are case sensitive. Users should note that variables are made dimensionless by applying the inverse of their corresponding units. The color of expressions entered in the space holders should turn black when units are consistent. The dimensions of the elements of the coefficient c entered are m^2 in order to have the entire equation dimensionless after gradient operators are applied twice, i.e. the term $\nabla.(-c\nabla\mathbf{u})$.

9. Similarly, we define coefficients for n-type TE modules. Right-click on the Coefficient Form PDE (c) node in the Model Builder window, and select Coefficient Form PDE from the list. Coefficient Form PDE 2 node will appear; rename it to Coefficient Form for TEn. In the corresponding settings window, select and add domains 5 and 9 to the list under Domain Selection. Expand the Diffusion Coefficient section and enter the expressions for c and f matrices for the n-type TE (Domains 5 and 9), as shown in Figure 7.34. The variables should be exactly the same as those defined as Parameters (see Figure 7.32) and they are case sensitive. Nondimensionalization of the expressions for coefficient c is applied similar to Step 9 above.

10. Similarly, we define coefficients for copper electrodes. Click on the Physics tab from the toolbar and select Domains > Coefficient Form PDE from the list. Coefficient Form PDE 3 node will appear; rename

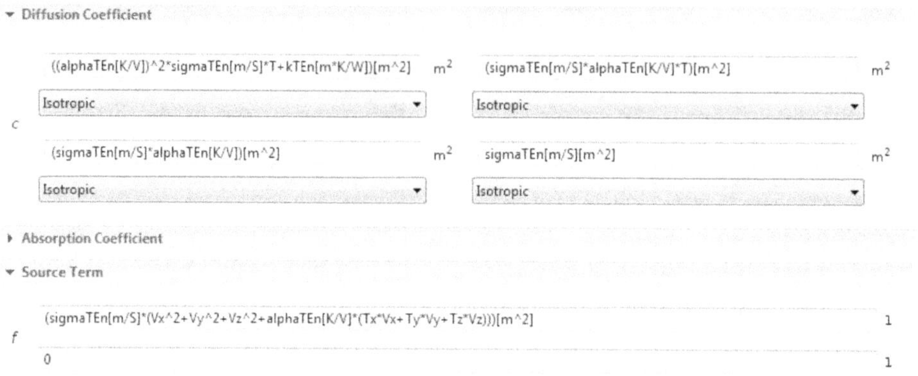

▼ Diffusion Coefficient

$((alphaTEn[K/V])^2*sigmaTEn[m/S]*T+kTEn[m*K/W])[m^2]$	m^2	$(sigmaTEn[m/S]*alphaTEn[K/V]*T)[m^2]$	m^2
Isotropic ▼		Isotropic ▼	

c

$(sigmaTEn[m/S]*alphaTEn[K/V])[m^2]$	m^2	$sigmaTEn[m/S][m^2]$	m^2
Isotropic ▼		Isotropic ▼	

▶ Absorption Coefficient

▼ Source Term

$(sigmaTEn[m/S]*(Vx^2+Vy^2+Vz^2+alphaTEn[K/V]*(Tx*Vx+Ty*Vy+Tz*Vz)))[m^2]$	1
f	
0	1

FIGURE 7.34 Coefficient expressions for n-type TE blocks.

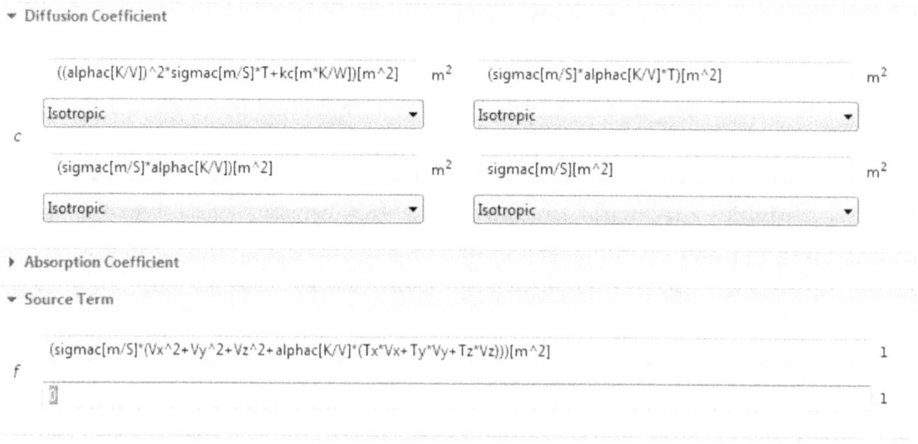

▼ Diffusion Coefficient

$((alphac[K/V])^2*sigmac[m/S]*T+kc[m*K/W])[m^2]$	m^2	$(sigmac[m/S]*alphac[K/V]*T)[m^2]$	m^2
Isotropic ▼		Isotropic ▼	

c

$(sigmac[m/S]*alphac[K/V])[m^2]$	m^2	$sigmac[m/S][m^2]$	m^2
Isotropic ▼		Isotropic ▼	

▶ Absorption Coefficient

▼ Source Term

$(sigmac[m/S]*(Vx^2+Vy^2+Vz^2+alphac[K/V]*(Tx*Vx+Ty*Vy+Tz*Vz)))[m^2]$	1
f	
0	1

FIGURE 7.35 Coefficient expressions for copper electrode blocks.

it to Coefficient Form for Copper Electrodes. In the corresponding settings window, select and add domains 1, 3, 4, 7, and 8 to the list under Domain Selection. Expand the Diffusion Coefficient section and enter the expressions for c and f matrices for the electrodes, as shown in Figure 7.35. The variables should be exactly the same as those defined as Parameters (see Figure 7.32) and they are case sensitive.

11. This model can be used for modeling Seebeck-type thermoelectric power generation or Peltier-type thermoelectric refrigeration. For this example we apply a set of boundary conditions for power generation. Therefore, we should apply temperatures at the electrodes

to calculate the resulting electrical voltage. From the ribbon, under the Physics tab, select Boundaries > Constraint. Constraint 1 node will appear in the model tree list. Select boundaries corresponding to electrodes on the top of the modules (i.e. 10, 35) from the geometry and add them to the Selection list in the corresponding settings window. Expand the Equation section. From the equation we can see that we should define values for R, corresponding to the $0 = R$ equation. In the Constraint section, enter 180-T and 0 into values for R. This will set the temperature at the upper electrodes at 180° C. Similarly, create another constraint boundary condition for the remaining electrode surfaces (i.e. boundaries 3, 19, 43) and enter 10–T and –V into the values for R. This will set the temperature at the lower electrodes at 10° C and ground the voltage for reference at zero.

12. Now the model is ready for meshing. Click on Mesh 1 and, in the corresponding window, accept default inputs and click on Build All.

13. To run the model, click on the Study tab from the toolbar, and then click on Compute.

14. The results appear in the Graphics window. The default is the Slice 1. Click on Slice 1, under 3D Plot Group 1, and in the corresponding window under the Expression, type T and check the results. Then type V and check the results. Number of slices and their orientations can be modified by changing the parameters in the Plane Data section. Other variables, such as electric field, could be visualized by typing $V_x + V_y + V_z$ for the Expression. To add another graph, click on the Results tab and select 3D Plot Group and select Surface from the ribbon. The 3D Plot Group 2 node will appear in the model tree as well as Surface 1 node. In the corresponding settings window enter the desired variable in the Expression space, such as V for visualization; click Plot. See Figure 7.36.

The numerical values for voltage, after validation, could be used for generated power. Validation for this model could be performed when experimental results are available. We leave this task open for those readers who may have access to the experimental setup. It can be shown that changing the temperature at the upper electrode surfaces will change the resulting voltage generated. This will suggest that we could use temperature-dependent material properties to get more accurate results. Readers are referred to [30] for the data and further reading.

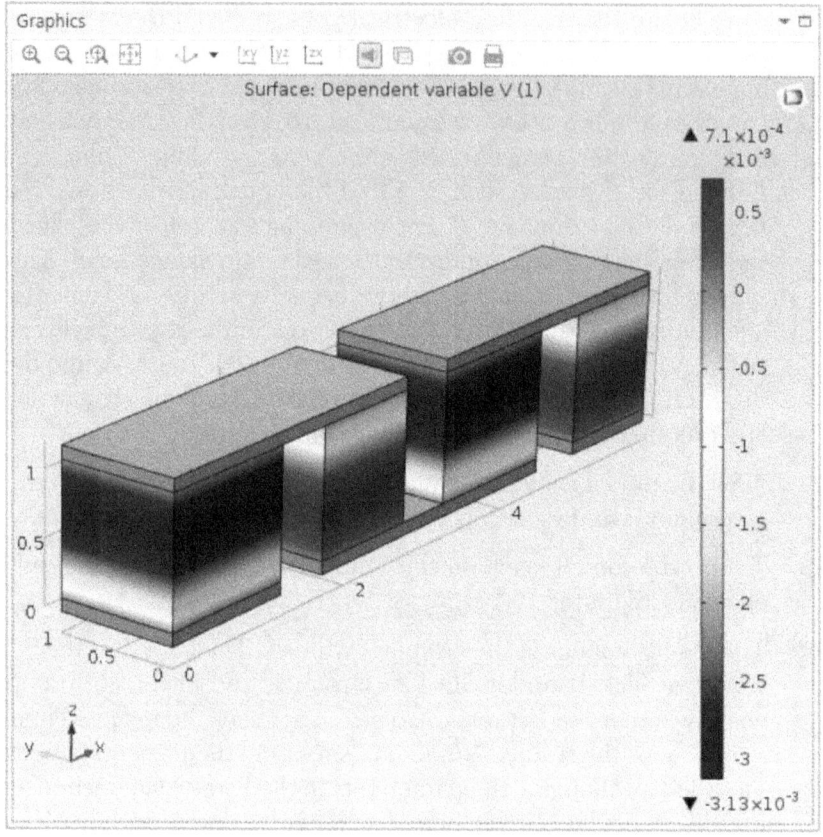

FIGURE 7.36 Results for voltage variations for TE modules.

Example 7.6: Acoustic pressure wave propagation in an automotive muffler

In this example, we use the COMSOL Multiphysics module to model and analyze pressure wave propagation and attenuation in a muffler.

Sound travels through air as a longitudinal wave; that is, the material (for example, air molecules) displacement and the sound wave propagation are in the same direction. When a sound wave is created by a source, such as a speaker or human vocal cords, the extra pressure due to compression of air molecules propagates as a wave. The extra pressure or acoustic pressure is in addition to the background ambient pressure or atmospheric pressure in the air or the medium involved. It is important to understand that the velocity of the air molecules is different from the velocity of the sound wave generated. Speed of sound wave V_s in a medium is, in general, proportional to the square root of medium stiffness over its density. For air as an ideal gas, $V_s \cong 20\sqrt{T}$ where T is the absolute temperature of air. Typical sound wave velocity at room temperature is 343 m/s depending on the source; sound wave length λ and frequency f are defined and their product should be equal to the generated sound wave velocity $V_s = \lambda f$. Acoustic pressure, or the generated sound wave amplitude p_s depends on the energy released from the sound source. The flux of this energy (i.e. energy per unit area per unit time) is called sound Intensity I, which is also equal to acoustic power per unit area and is proportional to p_s^2. If r is the distance from the sound sources, then sound intensity variation is proportional to $1/r^2$ and p_s to $1/r$ (see [31]). The human ear, as a mechanical sound receiver device, is sensitive to sound frequency (20Hz to 20kHz) and can detect sound intensities as low as $I_0 = 10^{-12}$ W/m^2, which corresponds to a sound pressure equal to about 2×10^{-5} Pa. I_0 is used as a scale to define sound level or "loudness" equal to $10 \log_{10}(I/I_0)$ as decibels and acoustic sound pressure level (SPL). Humans' hearing comfort sound level is about 50 dB.

To model the sound pressure-wave propagation as a result of combustion in a car engine, through a muffler, we use the COMSOL Model Library and relevant data, after some modifications. For more details and mathematical models, refer to Acoustics_Module/Industrial_Models/absorptive_muffler.[6] The main objective of a muffler is to attenuate the sound level to a bearable level for human hearing. In this model, the muffler is considered as a resonator with empty space.

Solution:

1. Launch COMSOL and in New window select Model Wizard. Save the file as Example 7.6.

2. In the Select Space Dimension window select 3D.

3. From the Select Physics list, expand Acoustics and select Pressure Acoustics > Pressure Acoustics, Frequency Domain (acpr), and then click Add. Locate Dependent variables and notice that Pressure p is defined as the sound pressure. Click on the Study arrow button.

4. From the Select Study list, select Frequency Domain and click Done.

5. To define the model data as parameters, click on the Model tab from the toolbar and select Parameters. In the corresponding settings window,

---

[6]Model made using COMSOL Multiphysics® and is provided courtesy of COMSOL. COMSOL materials are provided "as is" without any representations or warranties of any kind including, but not limited to, any implied warranties of merchantability, fitness for a particular purpose, or noninfringement

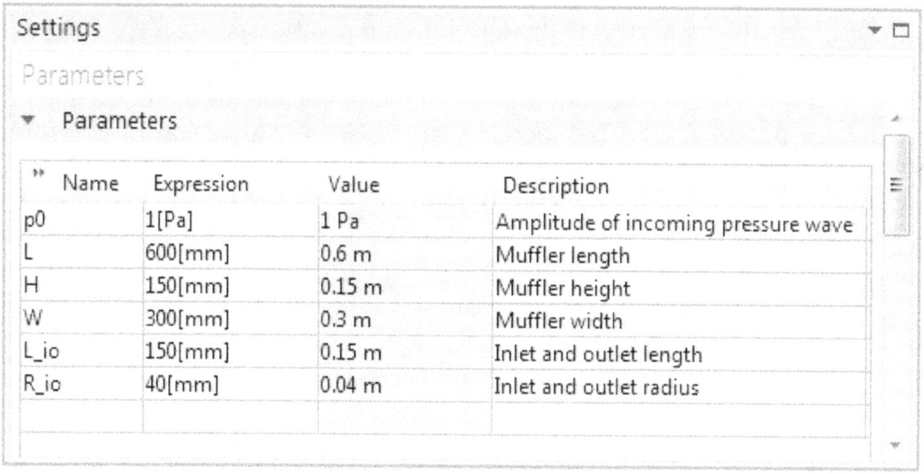

Name	Expression	Value	Description
p0	1[Pa]	1 Pa	Amplitude of incoming pressure wave
L	600[mm]	0.6 m	Muffler length
H	150[mm]	0.15 m	Muffler height
W	300[mm]	0.3 m	Muffler width
L_io	150[mm]	0.15 m	Inlet and outlet length
R_io	40[mm]	0.04 m	Inlet and outlet radius

FIGURE 7.37 Model data as parameters.

enter the data shown in Figure 7.37. Alternatively, users can use Load from File tool and import the file absorptive_muffler_parameters.txt, usually located in the Models folder where COMSOL is installed.

To build the geometry, we use the CAD tools available in COMSOL. Users may want to build it using their desired CAD software.

6. Click on the Geometry 1 node and in the corresponding settings window change Length to mm. Click on the Geometry tab from the toolbar and select Cylinder from the ribbon. In the corresponding settings window, locate the Size and Shape section and enter R_io for Radius and L_io for Height. For Position, enter –L_io for x. In the Axis section, select Cartesian from the list for Axis type. Enter 1 for x and 0 for both y and z. Then click Build Selected.
7. Similarly, build another cylinder. All data are the same except the Position x value should be L.

These two cylinders are the inlet and outlet pipes to the muffler resonate chamber. We now build this chamber.

8. From the ribbon under Geometry select Work Plane. In the corresponding settings window, select zy-plane from the Plane Definition section. Click Build Selected. Right-click on the Plane Geometry node (created under Work Plane 1) and select Ellipse. In the corresponding settings window, enter H/2 for a-semiaxis, W/2 for b-semiaxis, and

360 for Sector angle. Click Build Selected. Right-click on Work Plane 1 and select Extrude. In the corresponding settings window, enter −L for Distances, under Distances from Plane section. Click Build All Objects and then click on the Form Union (*fin*) node, in the Model Builder window. Final geometry is shown in Figure 7.38.

9. We now add material properties of the air inside the muffler. Click on the Materials tab from the toolbar and select Add Material. In the Add Material window, select Liquid and Gases > Gases > Air and add it to the model by clicking on Add to Component. By default, air is added to all three domains (1, 2, 3) for the muffler geometry. Close Add Material window.

10. Click on the Pressure Acoustic Model 1 node. In the corresponding settings window, extend Equation and study the equation. This

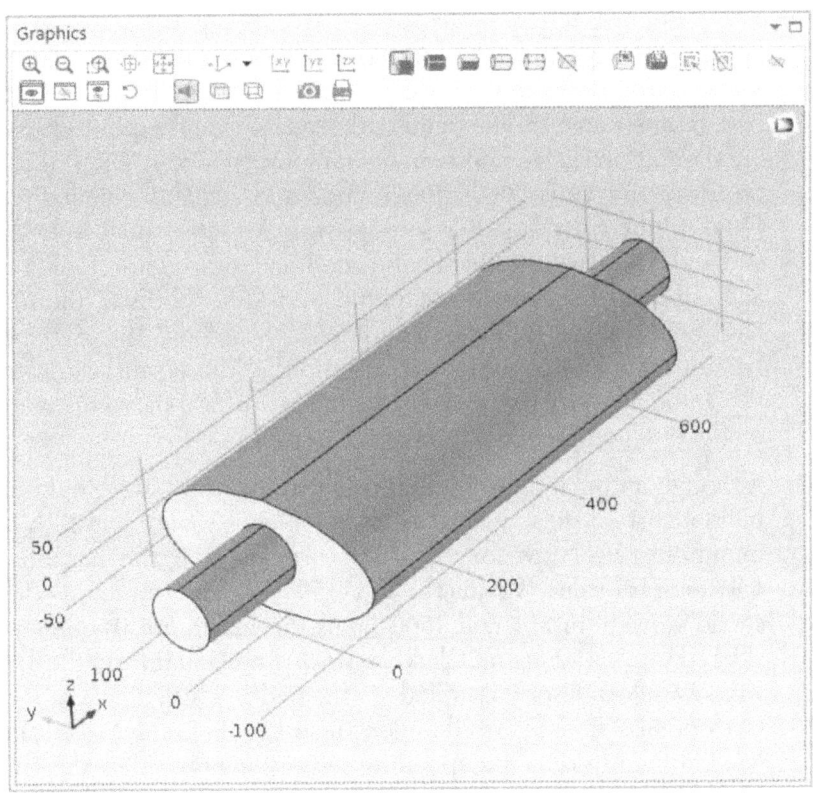

FIGURE 7.38 Graphics window showing muffler built geometry.

equation can be derived from a general wave equation using a transform, like Fourier's, to change the time domain to frequency domain. The result is a "Helmholtz"-type equation that governs the acoustic pressure-wave propagation for different frequencies as shown in the Equation section. Variable p is a function of space and angular velocity ω (which is itself a function of frequency). We accept all default values given under the Model Inputs and Pressure Acoustics Model sections.

11. For the boundary conditions at the inlet and outlet, we have incoming and outgoing plane sound waves. To assign boundary conditions, we first assign explicit names for them. Click the Definitions tab from the toolbar and select Explicit. In the corresponding settings window, select Boundary from the list for Geometric entity level and select and add boundary 1. Right-click on the Explicit 1 node and rename it Inlet. Similarly, create Explicit 2 and rename it to Outlet and assign boundary 18 to it.

12. Click on the Physics tab from the toolbar and select Boundaries > Plane Wave Radiation. In the corresponding settings window, add boundaries 1 and 18 (inlet and outlet surfaces of the pipes) by selecting them from the geometry in the Graphics window. Then right-click on the Plane Wave Radiation 1 node and select Incident Pressure Field. In the corresponding settings window, select Inlet from the list and add boundary 1 to the list. Locate Incident Pressure Field section and enter p0 for Pressure amplitude. All other walls of the muffler have Sound Hard Boundary conditions, by default. By assigning this type of boundary condition we assume that the normal derivative of the sound pressure is zero at the walls. Figure 7.39 shows the result for model tree nodes under Component 1 (*comp 1*), so far.

13. Although automatic tetrahedral mesh would work, we would like to build a custom mesh with maximum element size of one-fifth of the minimum sound wave length, which corresponds to the maximum frequency. This value is equal to 343/(1500*5). The number 343 is the sound velocity in *m/s* and 1500 Hz is the maximum frequency considered for this problem. Click on Mesh 1 node and from the settings window select User-controlled mesh, under Mesh Settings section. Click on Size node, located under Mesh 1 node in the Model Builder window, and in the corresponding settings window click the Custom button. Enter 343 [m/s]/1500 [Hz]/5 for the Maximum element size. Click on the Mesh tab from the toolbar and select Boundary > Free

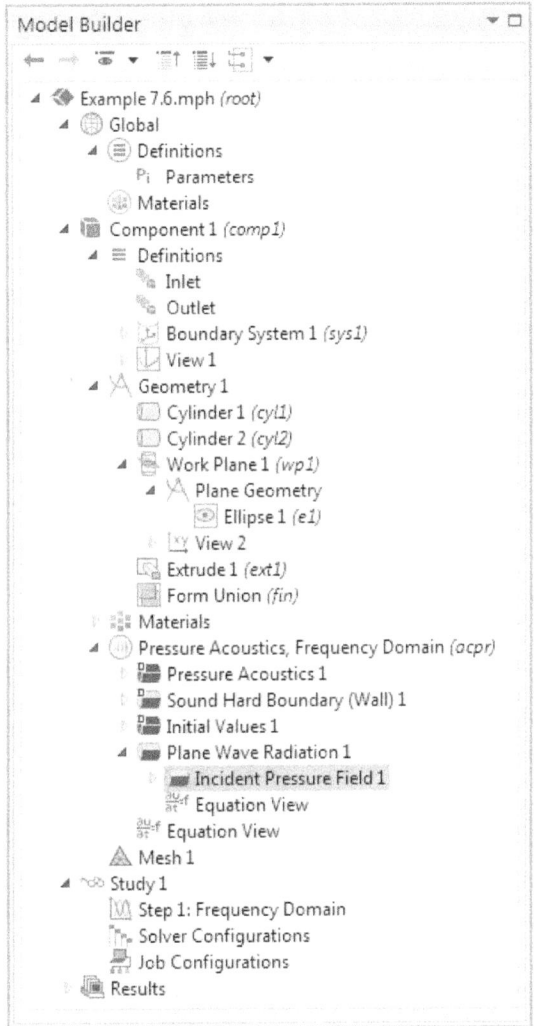

FIGURE 7.39 Model tree for model physics and boundaries.

Triangular. In the corresponding settings window, select boundaries 6 and 9 and add them to the list. These are boundaries at the face of the muffler resonator at the end of the entrance/inlet pipe. Click Build Selected. You may want to click on the Transparency node in the Graphics window toolbar to clearly see these surfaces i.e. boundaries 6 and 9. Also delete any other meshing tools that may have been created, by default.

14. We build the volume elements using the Swept tool. This method is suitable for building a mesh for wave propagation. Click on the Mesh tab and select Swept. In the corresponding settings window, select Domain for Geometric entity level and add Domain 1 (the inlet pipe) to the list and click Build Selected. Similarly, create Swept 2. In the corresponding settings window, select and add Domain 2 to the Selection list. In the Source Faces section, select and add boundaries 6 and 9. In the Destination Faces section, select and add boundaries 12 and 13 to the list. Click Build Selected. Finally, create Swept 3. In the corresponding settings window, click Build All. Figure 7.40 shows the final mesh.

15. To set the range for frequency, click on Step 1: Frequency Domain node under Study 1 in the Model Builder window. In the corresponding settings window, enter range (50, 25, 1500) in the Frequencies space.

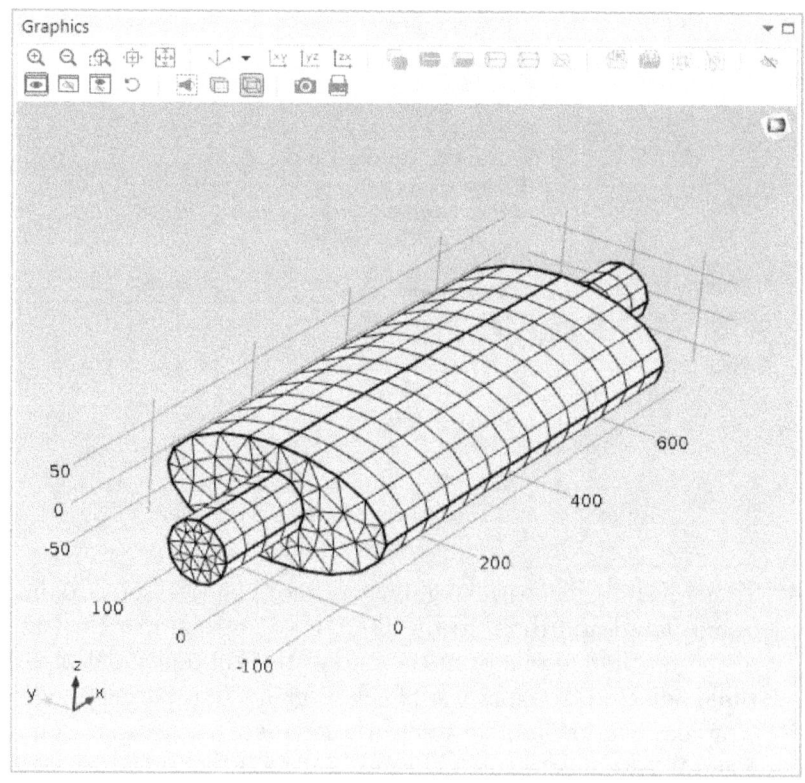

FIGURE 7.40 Custom-built mesh for the muffler geometry.

Expand Results While Solving and check the box for Plot. Click on the Study tab and select Compute. Wait until calculations are finished.

16. The default results for acoustic pressure appear in the Graphics window. Expand the Acoustic Pressure (acpr) node and click on Surface 1. In the corresponding settings window, locate the Expression section. Change the variable to acpr.absp by clicking on Replace Expression and select Pressure Acoustics, Frequency Domain > Pressure and sound pressure level > acpr.absp-Absolute pressure. Click Plot. The results, which are absolute (norm) values of the sound pressure on the surface of the muffler, appear in the Graphics window, as shown in Figure 7.41. Graphs for other frequencies could be obtained by clicking on the Acoustic Pressure (acpr) node and choosing the desired value from the Parameter value (freq (Hz)).

17. Click on the Sound Pressure Level (acpr) node. The graph appears in the Graphics window. The sound pressure is the sound pressure

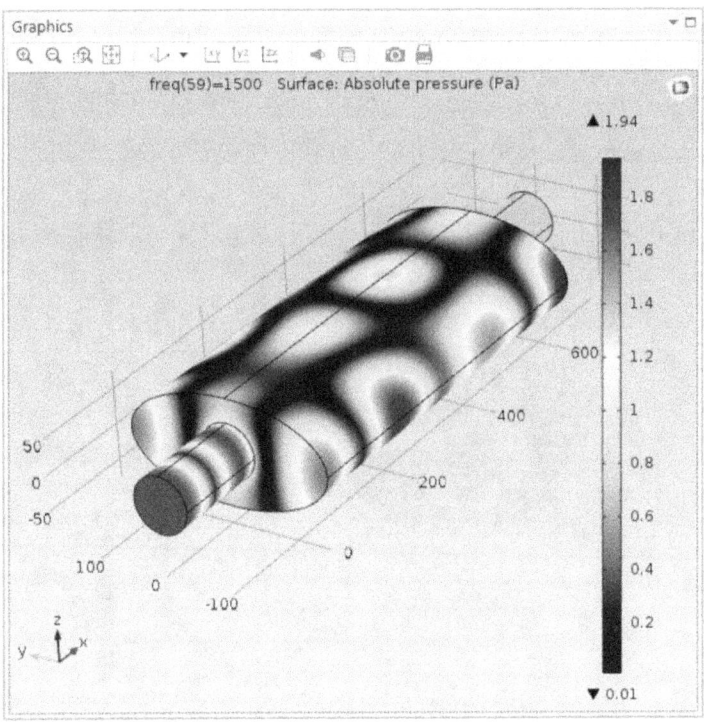

FIGURE 7.41 Total acoustic pressure on the surface of the muffler for frequency of 1500 Hz.

associated with acoustic pressure with reference to the 0 dB, associated with the lowest human hearing sensitivity or 10^{-12} W/m², which is set as 20 μPa in COMSOL. The result is shown in Figure 7.42.

18. Another useful graph is the isosurfaces for acoustic pressure, or locations for constant pressure in the muffler. Click on the Acoustic Pressure, Isosurfaces (acpr). We keep the default Expression acpr.p_t, which is the total acoustic pressure (real part). Click on Plot. Results are shown in Figure 7.43.

19. Attenuation of acoustic power (or transmission loss) is the main function of the muffler. Sound intensity or power per unit area is the quantity of interest given by $p^2/2\rho V_s$. To integrate this quantity over the surface area of the inlet and outlet pipes, we create two integral operations and define the intensity as a variable. Click on the Definitions tab from the toolbar and select Local Variables. In the corresponding settings window, under Variables section enter W_in for Name and int op1(p0^2/(2*acpr.rho*acpr.c)) for Expression. Similarly, in the second row, enter W_out for Name and int op2 (abs (p)^2/(2*acpr.rho*acpr.c)) for Expression. Then create the integral operations (*int op1* and *int op2*)

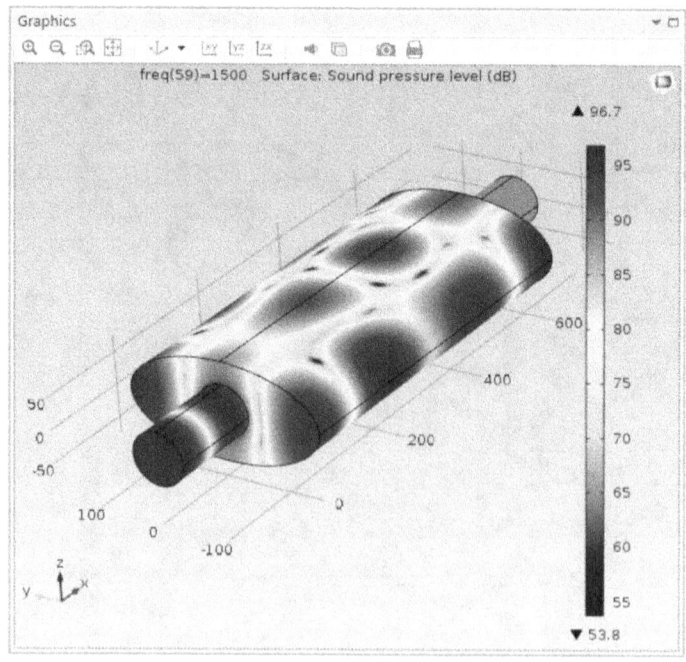

FIGURE 7.42 Sound pressure level (referenced to 0 dB) on the surface of the muffler for frequency of 1500 Hz.

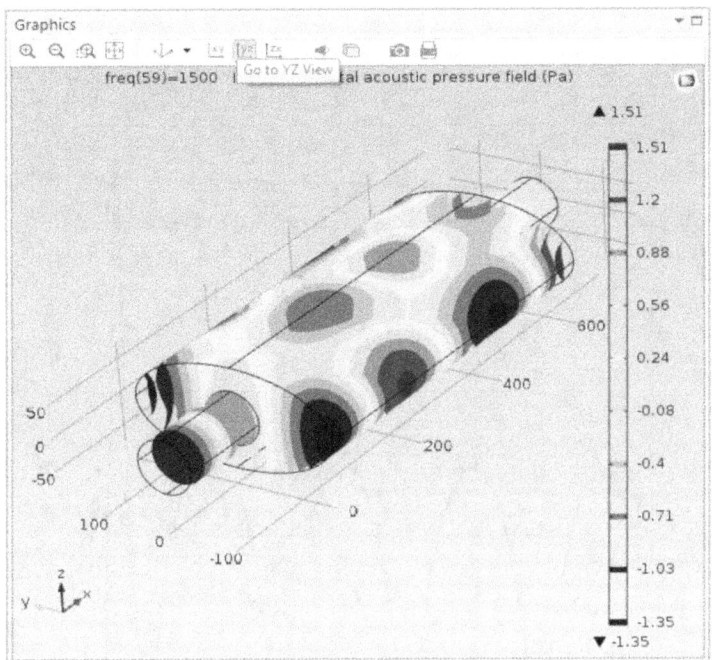

FIGURE 7.43 Total acoustic pressure isosurfaces on the surface of the muffler for frequency of 1500 Hz.

by clicking on the Definitions tab in the toolbar and select Component Couplings > Integration. In the corresponding settings window, select and add boundary 1 (if needed, change the Geometric entity level to Boundary). Similarly, create Integration 2 (*int op2*) and assign boundary 18 to it. Click on Study and select Compute to update all solutions.

20. Now we are ready to extract the integrals of the acoustic power densities from the solution results. Click on the Results tab from the toolbar and select 1D Plot Group. 1D Plot Group 4 appears in the model tree. Select Global. In the corresponding settings window, locate the y_Axis data section and enter 10*log 10 (W_in/W_out). This expression is standard sound level in dB. It shows the ratio (logarithmic) of the incoming sound power over the outgoing one. Results are shown in Figure 7.44. Open Coloring and Style to change the graphing options for the line graph (2 for Width, Point for Marker, and In data points for Positioning).

The model is complete. Further studies or graphs may be built. In practice, a muffler resonate chamber is composed of two shells, instead of one, for damping the acoustic energy. Users may want to modify this model

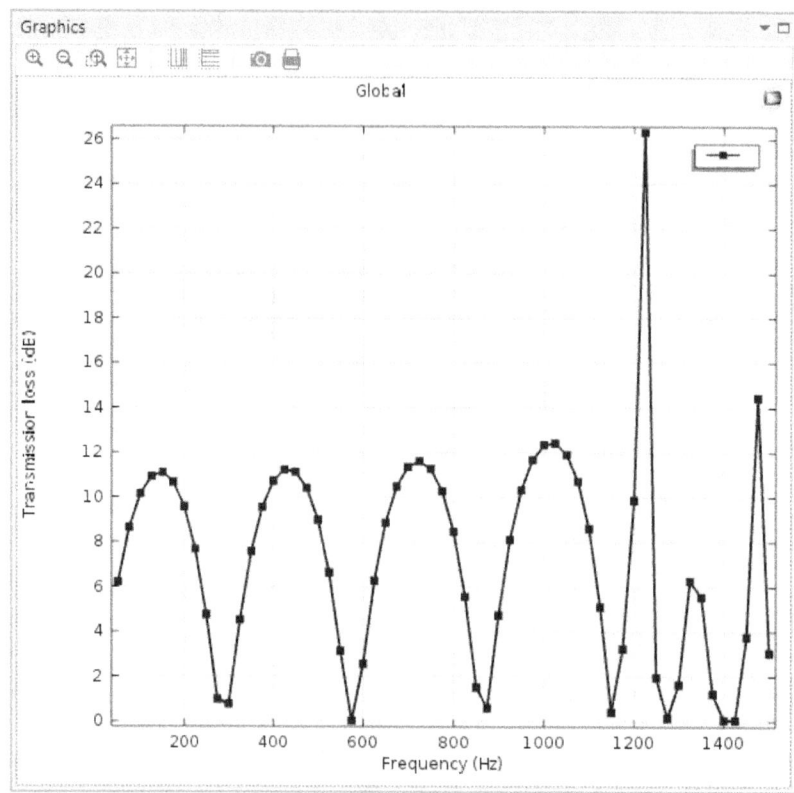

FIGURE 7.44 Power attenuation or transmission loss of the muffler versus frequency.

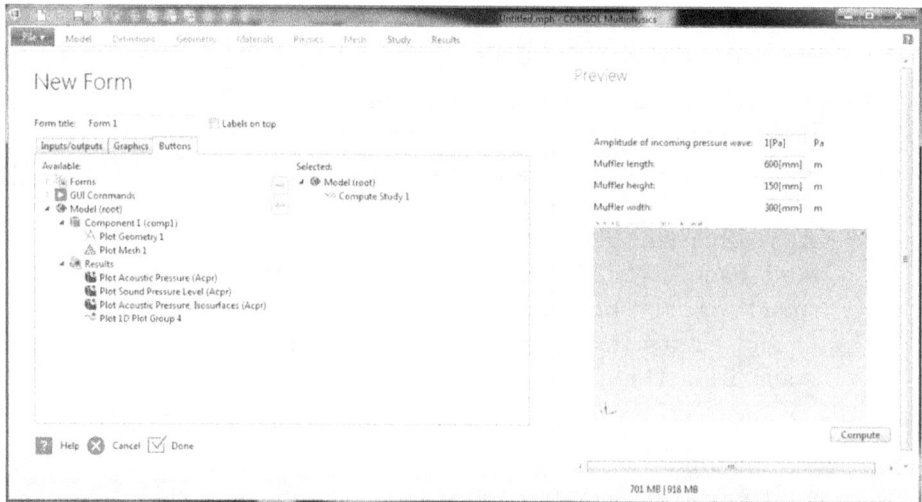

FIGURE 7.45 Application settings for Example 7.6.

by adding another chamber inside the existing one and study its effects as compared with results obtained in this example.

21. To build an app for this model, launch COMSOL if not already running. From File menu select New. In the New window select Applications Wizard. In the Select Model for Application window click on Browse; locate and open Example 7.6.mph. The New Form window will appear. In this window, select entries Amplitude of incoming pressure wave, Muffler length, Muffler height, and Muffler width under Parameters, from the Inputs/outputs tab and move them to Selected window. Click on Graphics tab; select and move Flow and Stress (*fsi*) to Selected window. Click on the Buttons tab; select and move Compute Study 1 to Selected window. See Figure 7.45. Click on Done. Form Editor desktop window appears. Save the file as App_Example 7.6.

22. Several editing tools are available in the Form Editor window. We use some of these tools in order to lay out the App interface. Click on Grid, from the ribbon bar, to relocate the objects in the Form 1. Use the Settings for each object to modify the size, affiliate a picture, and specify the function. Refer to detailed instructions given for Example 2.1, for a guide. Final results for the App interface are shown in Figure 7.46.

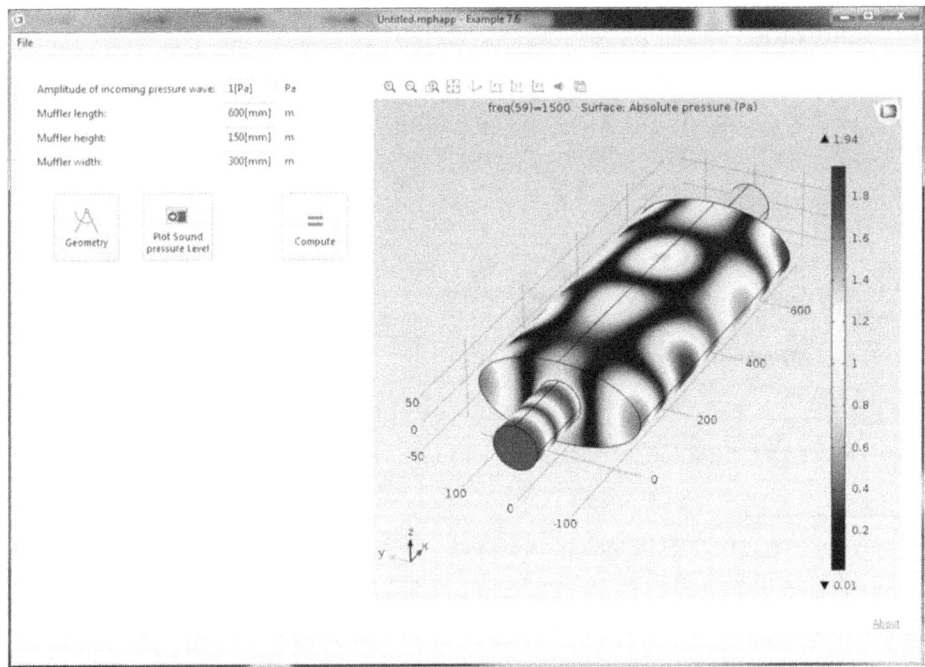

FIGURE 7.46 Application GUI for Example 7.6.

Example 7.7: Viscous fluid damper with conjugate heat transfer

For this example we model a viscous fluid damper. Dampers are used in machines and structures for absorbing energy due to exerted vibrations or extreme loads. Their sizes cover a wide range from small-size dampers to big ones, usually used for larges structures. Figure 7.47 shows a schematic and an actual damper.

The energy exerted on the damper is dissipated to heat through viscous shear force created inside a very narrow gap/orifice between the damper cylinder/chamber and its moving piston head. For modeling we solve unsteady Navier-Stokes and heat transfer equations for the fluid flow inside the chamber. Also we consider heat transfer through the material of the chamber and piston head and moving fluid; hence we consider the multiphysics phenomenon of conjugate heat transfer. For this model we use the Moving Mesh tool available in COMSOL, for modeling the dynamics of the moving boundaries as a result of piston head vibration. We will explain the detail of boundary conditions and material properties used, along with modeling tools during the solution steps. We also make use of axisymmetric geometry of the damper, for modeling. A similar model is available in COMSOL Model Libraries (CFD_Module/Non-sothermal_Flow/fluid_damper).[7] For this example, we modify some of the solution steps given in the Model Library.

Solution:

1. Launch COMSOL and in New window select Model Wizard. Save the file as Example 7.7.

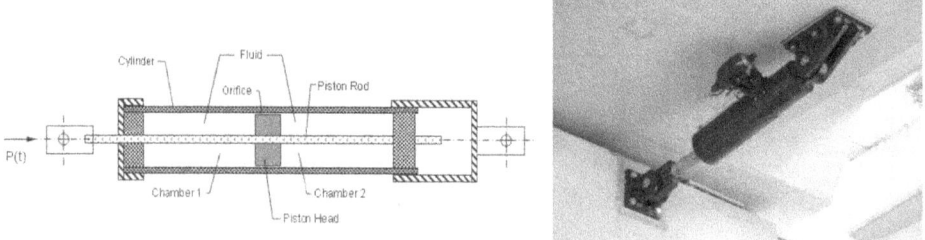

FIGURE 7.47 A schematic of fluid damper (left) and an actual one in operation (right), [32].

---

[7]Model made using COMSOL Multiphysics® and is provided courtesy of COMSOL. COMSOL materials are provided "as is" without any representations or warranties of any kind including, but not limited to, any implied warranties of merchantability, fitness for a particular purpose, or noninfringement.

2. In the Select Space Dimension window select 2D Axisymmetric.

3. From the Select Physics list, expand Heat Transfer and select Conjugate Heat Transfer > Laminar Flow, and then click Add. Also expand Mathematics and select Deformed Mesh > Moving Mesh (ale), and then click Add. Click on the Study arrow button.

4. From the Select Study list, select Time Dependent and click on Done icon.

To define the model data as parameters, click on the Model tab from the toolbar and select Parameters. In the corresponding settings window, enter the data shown in Figure 7.48. Alternatively, users may upload the file Example 7.7-parameters.txt, from the accompanied disc.

To build the geometry, we use a COMSOL tool for importing it from the Model Library. Alternatively, users may want to build the geometry using the CAD tools available in COMSOL, by using the instructions given in the corresponding model document available from the Model Library.

5. Click on the Geometry tab from the toolbar and select Insert Sequence. Locate the file Example 7.7.mph, available from the accompanied disc and upload it. Notice the Rectangle 1–5 nodes created in the Model Builder window, under Geometry node. Click on Build All

Name	Expression	Value	Description
Dr	2.83e-2[m]	0.0283 m	Rod diameter
Dp	8.37e-2[m]	0.0837 m	Piston diameter
Lp	0.0254[m]	0.0254 m	Piston half-length
Dd	0.1128[m]	0.1128 m	Damper outer diameter
Hw	1.37e-2[m]	0.0137 m	Damper wall thickness
Ld	U0+Lp	0.1778 m	Damper seal position
U0	0.1524[m]	0.1524 m	Damper chamber height
p0	1[atm]	1.0133E5 Pa	Reference pressure
T0	300[K]	300 K	Initial temperature
hwall	5[W/(m^2*K)]	5 W/(m²·K)	Heat transfer coefficient
a0	0.127[m]	0.127 m	Piston displacement amplitude
f	0.4	0.4	Frequency
ncycle	16	16	Number of loading cycles
tmax	ncycle/f	40	Total loading time
tstep	0.5/f	1.25	Sampling time interval
Tamb	300[K]	300 K	ambient temperature

FIGURE 7.48 Parameters defined for the damper data.

Object button, located in the Settings window. The geometry of the damper will appear in the Graphics window. Zoom in close to the piston head area to see the narrow region (the orifice) between the piston head and the chamber cylinder. See Figure 7.49.

Now we assign materials to the model. For the fluid we consider Silicon oil with constant density of 950 kg/m³, and heat capacity of 2 kJ/kg.K, but temperature dependent kinematic viscosity, according to $v = v_0 - \alpha \Delta T$, where v_0 is the Silicon kinematic viscosity at 25 °C or 298 K. Using the sketch shown in Figure 7.50, we can define the coefficient α based on a viscosity-temperature coefficient (VTC), a parameter that indicates the change in fluid viscosity due to temperature variation [33] and is defined as

$$1 - \frac{kinematic\ viscosity\ at\ 210^\circ F\left(or \cong 372\,\mathrm{K}\right)}{kinematic\ viscosity\ at\ 100^\circ F\left(or \cong 311\,\mathrm{K}\right)} = 1 - \frac{v_2}{v_1}.$$

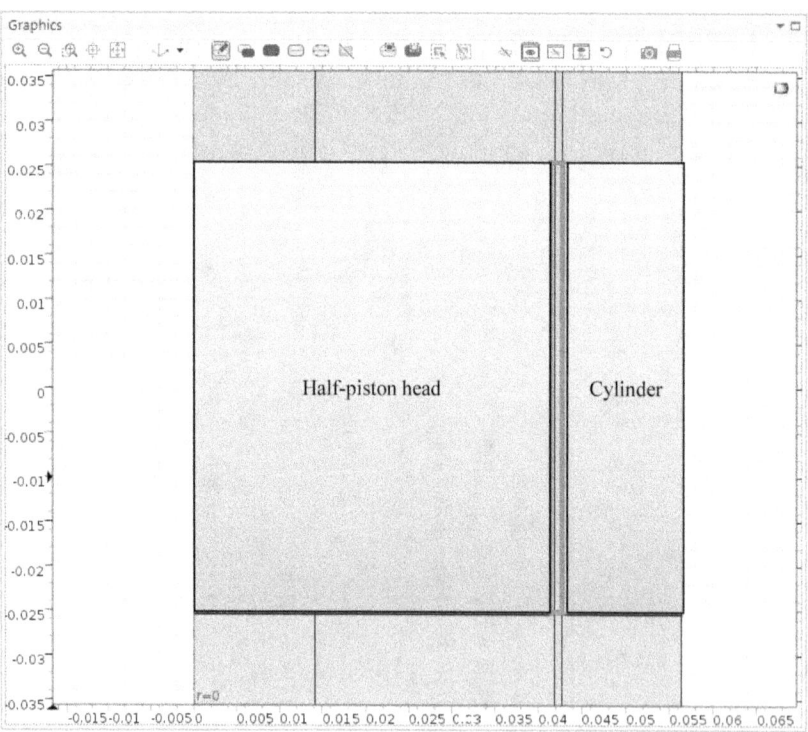

FIGURE 7.49 Detail of the half-piston head and narrow gap/orifice of the damper, highlighted for visualization.

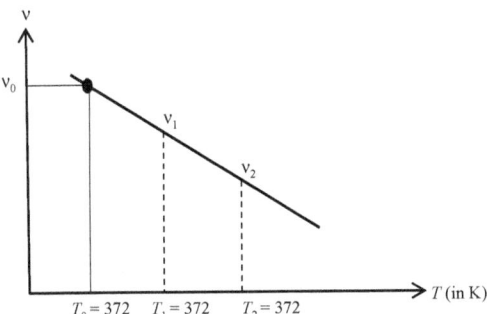

FIGURE 7.50 The sketch for viscosity variation versus temperature for Silicon oil.

The viscosity variation, as a linear function of temperature, can be written as $\dfrac{V_2 - V_1}{372 - 311} = \dfrac{V - V_1}{T - 311}$, or using the definition for VTC, we have $V = V_1 \left[1 - VTC(T - 311)/(372 - 311) \right]$. But this line should passes the data point for , then $V_0 = V_1 \left[1 - VTC(298 - 311)/61 \right] \Rightarrow V_1 = V_0/(1 + 0.21 \times VTC)$. Substituting for V_1, we get:

$$V = V_0 \left[1 - VTC \frac{(T - 311)}{61} \right] \Big/ (1 + 0.21 \times VTC)$$

We use this equation for defining the viscosity of the Silicon oil (with $VTC = 0.6$) in the model.

For the piston and the damper cylinder we use Steel AISI 4340, available in the COMSOL materials library.

6. Click on the Materials toolbar from the toolbar and select Add Material. In the Add Material window, type in Steel AISI 4340 in the Search space. Find and select Steel AISI 4340 from the search result and click on Add to Component. In the corresponding settings window add domains 1–3, 5, and 10–12 (i.e. domains associated with piston's head/rod and cylinder wall) to the Selection list by clicking on them in the Graphics window.

7. To define the fluid Silicon oil properties, click on the Blank Material button from the ribbon toolbar. Rename the node Material 2 (*mat 2*), just created, to Silicon oil. In the corresponding settings window add domains 4 and 6–9 (i.e. domains associated with chamber/container

fluid) to the Selection list. Expand Silicon oil (*mat 2*) node, in the Model Builder window, and click on Basic (*def*) node. In the corresponding settings window locate Output Properties and Model Inputs section and find Model Inputs > Temperature from the list; click the (+) icon to add the selection to the Model inputs list. Now scroll down to locate Local Properties section and type in the data as shown in Figure 7.51.

Now click on Silicon oil (*mat 2*) node and in the corresponding settings window locate the Material Contents section. Enter the data in the table, under the Value column, as shown in Figure 7.52. Note that the material properties nu_25 C and VTC are also shown in the list. For the Dynamic viscosity we use the equation derived above, as: nu_25 C*rho*(1-VTC*(T-311 [K])/(61 [K]))/(1 + VTC*0.2107).

8. To define the physics domains for fluid/Oil flow and heat transfer, click on Laminar Flow (*spf*) node, in the Model Builder window. In the corresponding settings window add only domains 4 and 6–9 to the Selection list. Expand Heat Transfer in Solids (*ht*) node, in the Model Builder window, and click on Heat Transfer in Fluids 1 node. In the corresponding settings window add only domains 4 and 6–9 to the Selections list. Locate Thermodynamics Fluid section and for Ratio of specific heats, select User defined. The default value for γ should read 1.

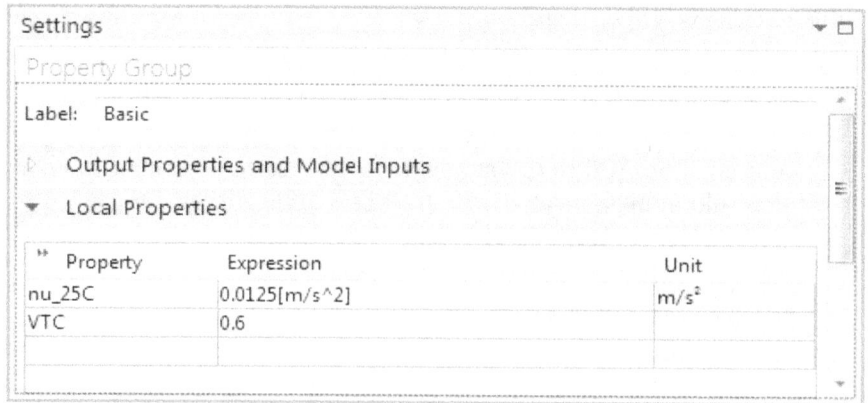

FIGURE 7.51 Kinematic viscosity at 25 °C and VTC setting for Silicon oil.

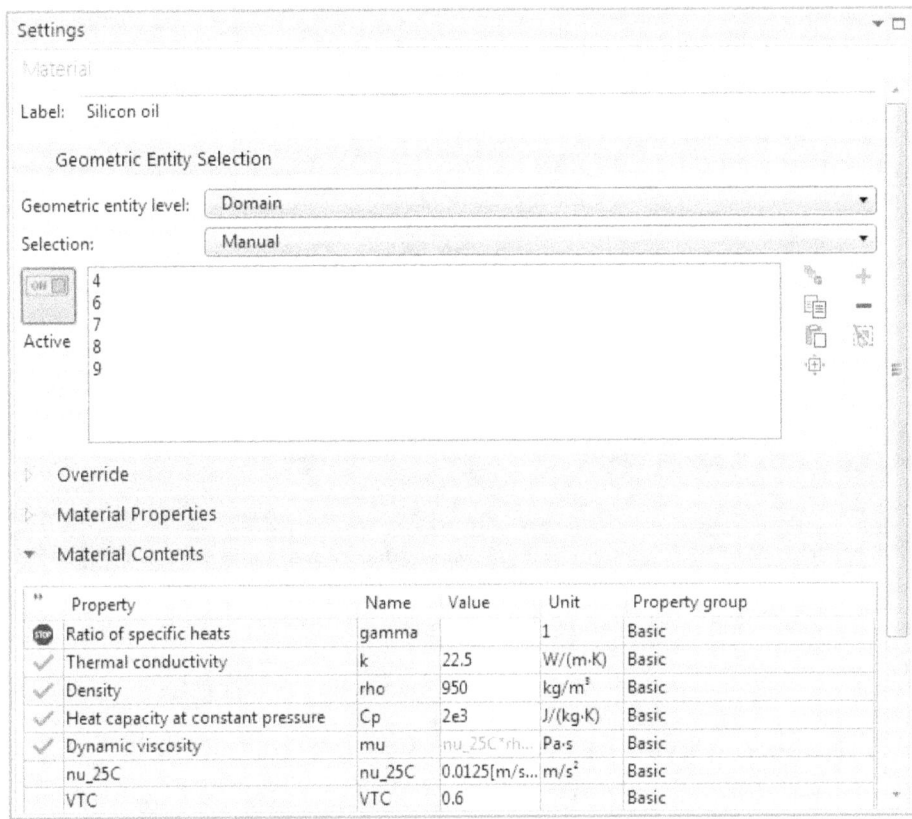

FIGURE 7.52 Material properties settings for Silicon oil.

9. To define vibrational motion of the piston, and correspondingly the moving mesh, we specify them as variables. The piston has a sinusoidal vibrational motion, $Zp = a_0 \sin(2\pi ft)$. Click on the Model tab from the toolbar and select Variables > Local Variables. In the corresponding settings window, locate Variables section and enter the following data, as shown in Figure 7.53. Alternatively, users may upload the file Example 7.7-variables.txt, available from the accompanied disc.

10. Now we define the boundary conditions for fluid flow. Click on Laminar Flow (*spf*) node. From the ribbon toolbar, click on the Boundaries button and select Wall. In the corresponding settings window select and add boundaries 11 and 13 (i.e. boundaries corresponding to the top and bottom of the cylinder head) to the Selection list. Since these boundaries are moving, we define them as moving ones. Locate Boundary

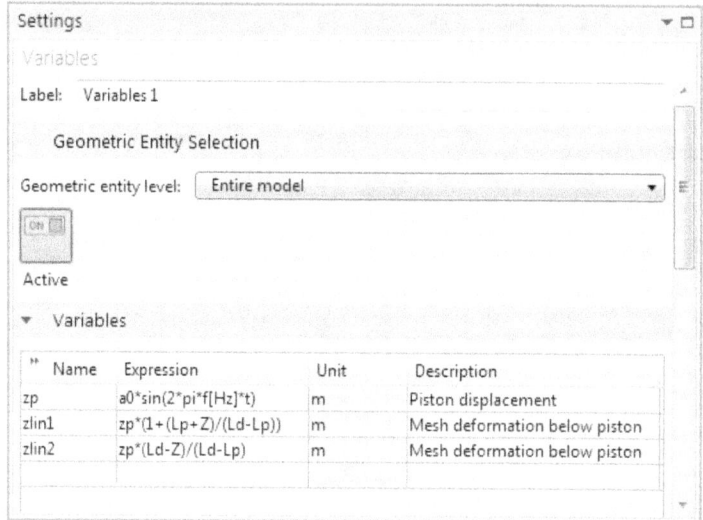

FIGURE 7.53 Variables settings for piston vibration and moving mesh.

Condition section and select Moving wall from the list and enter d (zp, t) for z-component of $\mathbf{u}_w$. This defines the velocity of the piston, obviously in z-direction, as $u_z = \dfrac{d(Zp)}{dt}$. Recall that is already defined as $Zp = a_0 \sin(2\pi f t)$. Similarly, create another Wall.

11. For confined flow, we should define a reference for pressure. Click on the Physics tab from the toolbar and select Points > Pressure Point Constraint. In the corresponding settings window select and add only point 12 to the Selection list. Enter p0 for Pressure, under Pressure Constraint section. We also define the initial pressure. Click on Initial Values 1 node, in the Model Builder window, and in the Settings window enter p0 for Pressure.

12. We now define the heat transfer boundary conditions. Click on Heat Transfer in Solids (*ht*) node, in the Model Builder window. In the corresponding settings window select and add boundaries 8, 12, and 17 (i.e. boundaries representing vertical sides of the piston and rod) to the Selection list. Locate Boundary Condition section and select Sliding wall from the list. Enter d (zp, t) for U_w. Obviously, these boundaries have the same speed as the piston head.

13. We now define boundary conditions and initial values for heat transfer in solids. Expand the Heat Transfer in Solids (*ht*) node, in the Model Builder widow. Click on Initial Values 1 node. In the corresponding settings window locate Initial Values section and enter T0. From the Physics ribbon bar, select Boundaries > Temperature. In the corresponding settings window add boundaries 2, 7, 9, 14, 16, 21, 23, and 28 (i.e. boundaries on the top and bottom of the damper) to the Selection list. Again, locate Temperature section and enter T0 for Temperature. To define the convective cooling boundaries for external sides of the damper, click on Heat Flux 1 node. In the corresponding settings window locate Heat Flux section and click the Convective heat flux button. Enter hwall and Tamb for h and $T_{ext,}$ respectively.

14. For the fluid damper, the energy is damped through viscous dissipation in the Silicon oil. To activate this mechanism for this model, expand Multiphysics node, in the Model Builder window and click on Non-Isothermal Flow 1 (*nitf 1*) node. Locate Flow Heating section and check both boxes for Include work done by pressure changes and Include viscous dissipation.

Since we have moving boundaries, due to motion of the piston head and rod, therefore the fluid and solid domains change with time. This requires that we define a mesh that also moves, consistent with the piston vibrational motion. COMSOL provides tools for moving mesh. This way the mesh domain is also part of the solution and is solved along the main governing equations for the model.

15. Click on Moving Mesh (*ale*) node, in the Model Builder window. In the corresponding settings window locate Frame Settings section and select 1 for Geometry shape order. This defines the order of the polynomials (here linear) used for representing the geometry shape in the spatial frame [34]. To define the boundary conditions for moving mesh domains, click on the Physics tab from the toolbar and select Domains > Prescribed deformation. In the corresponding settings window add domains 2, 5, 8, and 11 only (i.e. domains representing piston head and adjacent domains) to the Selection list. Enter zp for d_z in the Prescribed mesh displacement section.

16. Similarly, create another Prescribed Deformation node and add domains 1, 4, 7, and 10 (i.e. domains representing regions located below the piston head) to the selection list. Enter zlin 1 for d_z in the Prescribed mesh displacement section.

17. Similarly, create another Prescribed Deformation node and add domains 3, 6, 9, and 12 (i.e. domains representing regions located above the piston head) to the selection list. Enter zlin 2 for d_z in the Prescribed mesh displacement section.

The model physics, material properties, and boundary conditions are all set. We now build a mesh using a structured meshing tool, called Mapped mesh in COMSOL.

18. Click on the Mesh tab from the toolbar and select Mapped. Click on Distribution. In the corresponding settings window add boundaries 23, 25, 27, and 28 only to the Selection list. Locate Distribution section and modify the attributes according to Figure 7.54.

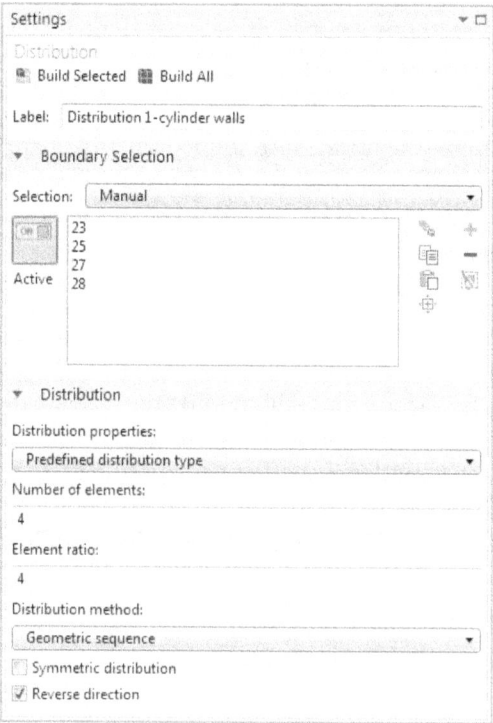

FIGURE 7.54 Structured mesh Distribution settings for cylinder walls.

19. Similarly, create another Distribution by right-clicking on Distribution 1-cylinder walls and selecting Duplicate. In the corresponding settings window add boundaries 1, 5, 8, 15, 19, 22, 26, 29, and 31 only to the Selection list. Locate Distribution section and modify the attributes according to Figure 7.55.

20. Similarly, create another Distribution by right-clicking on Distribution 1-cylinder walls and selecting Duplicate. In the corresponding settings window add boundaries 9, 11, 13, and 14 only to the Selection list. Locate Distribution section and modify the attributes according to Figure 7.56.

21. Similarly, create another Distribution by right-clicking on Distribution 1-cylinder walls and selecting Duplicate. In the corresponding settings window add boundaries 16, 18, 20, and 21 only to the Selection list. Locate Distribution section and modify the attributes according to Figure 7.57.

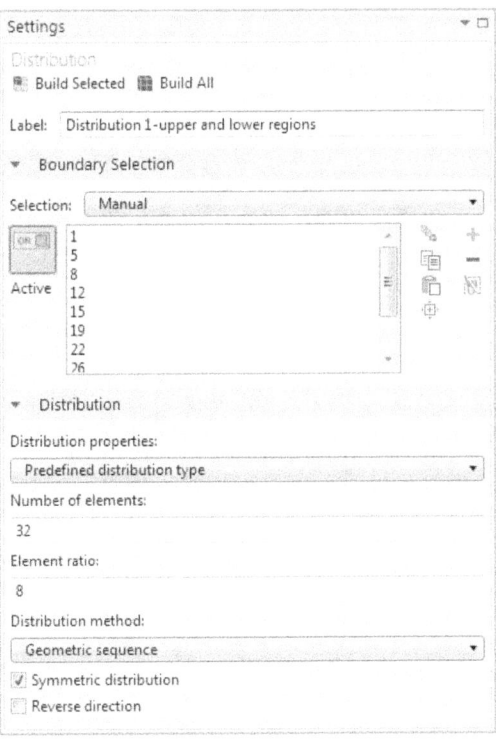

FIGURE 7.55 Structured mesh Distribution settings for upper and lower regions.

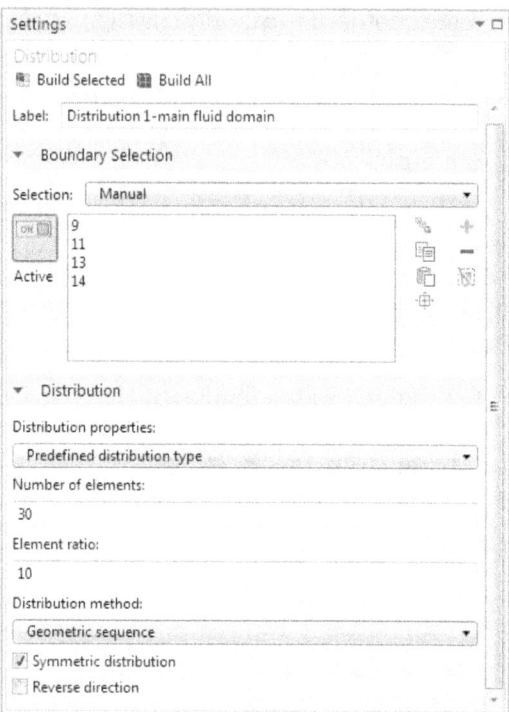

FIGURE 7.56 Structured mesh Distribution settings for main fluid domain.

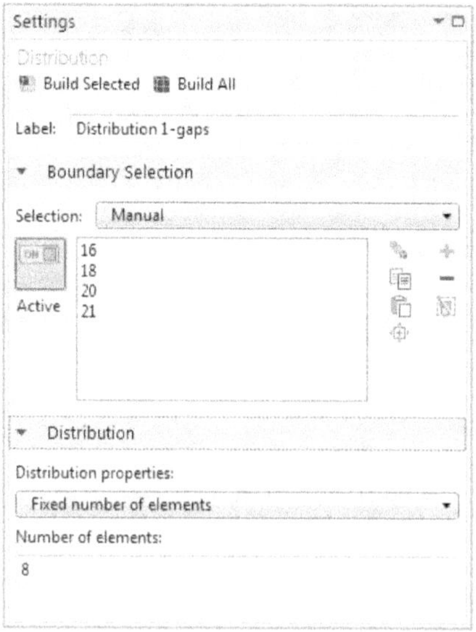

FIGURE 7.57 Structured mesh Distribution settings for orifice fluid domains.

22. Similarly, create another Distribution by right-clicking on Distribution 1-cylinder walls and selecting Duplicate. In the corresponding settings window add boundaries 3, 10, 17, 24, and 30 only to the Selection list. Locate Distribution section and modify the attributes according to Figure 7.58.

23. Similarly, create another Distribution by right-clicking on Distribution 1-cylinder walls and selecting Duplicate. In the corresponding settings window add boundaries 2, 4, 6, and 7 only to the Selection list. Locate Distribution section and modify the attributes according to Figure 7.59.

24. To build all the mesh, click on Build All. Figure 7.60 shows the structured mesh built close to the piston head.

The model is ready for computation. However, we should define a time step that is compatible with the damper motion. For example, half of the frequency or $t\ step = 0.5/f$, as defined as a parameter. Also, we set up some graphs, before running the model, so we can see the results and motion of the piston inside the chamber during the computation.

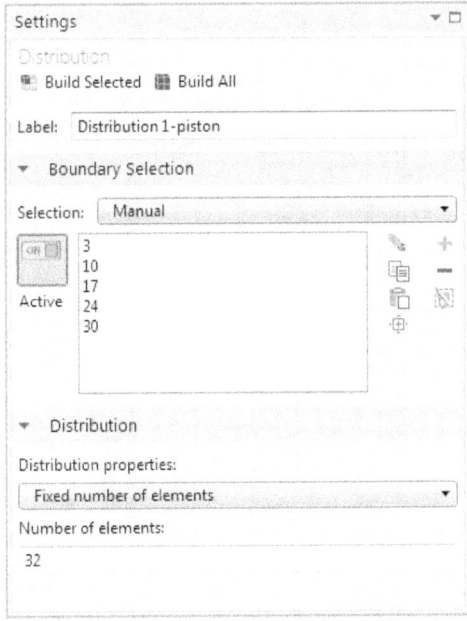

FIGURE 7.58 Structured mesh Distribution settings for piston domains.

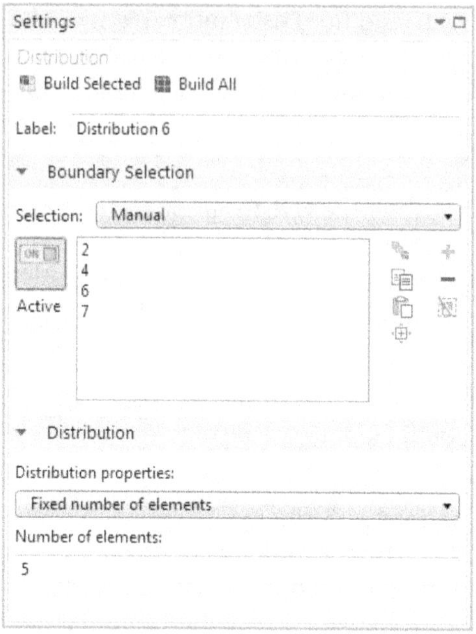

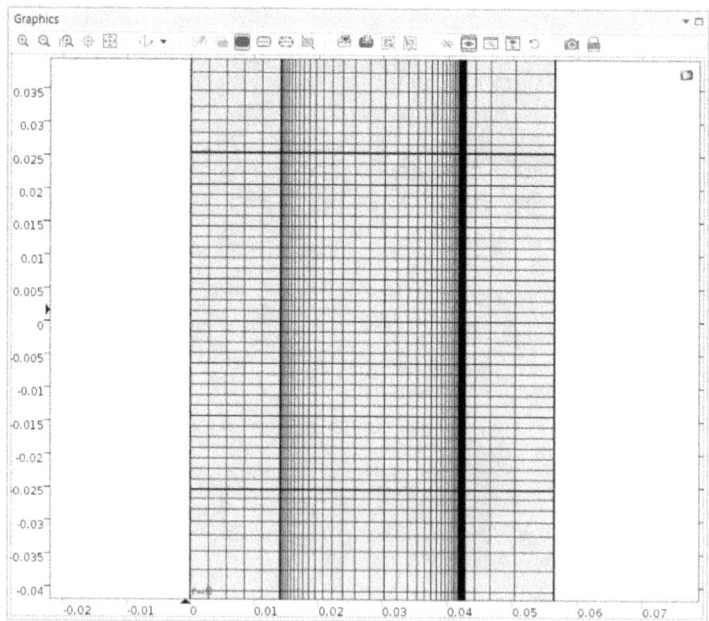

FIGURE 7.59 Structured mesh Distribution settings for rod domains.

FIGURE 7.60 The close-up structured mesh for damper domains (from left to right): rod, piston, orifice, and damper cylinder wall.

25. Expand Study 1 node in the Model Builder window. Click on Step 1: Time Dependent node and in the corresponding settings window locate Study Settings section. Enter (−0.25/f), range (0, t step, t max) for Times. Expand the Results While Solving section and click the Plot check box. Now click on the Study tab from the toolbar and select Show Default Solver by clicking on it. Expand Solution 1 node, in the Model Builder window, and click on Time-Dependent Solver 1 node. In the corresponding settings window locate and expand Time Stepping section. To control the solver marching through the time steps for solving the equations, we would like to manually assign it. Change the settings as shown in Figure 7.61.

26. To create some graphs, in order to visualize the results during running the model, click on the Study tab from the toolbar and click on Get Initial Value button. Several plots are created under Results node, in the Model Builder window. To view, for example, temperature contours during computation, click on Step 1: Time Dependent node and in the corresponding settings window select Isothermal Contours (ht) from the list for Plot group. Click on Compute. Wait until computations are finished. The Graphics window should show the temperature

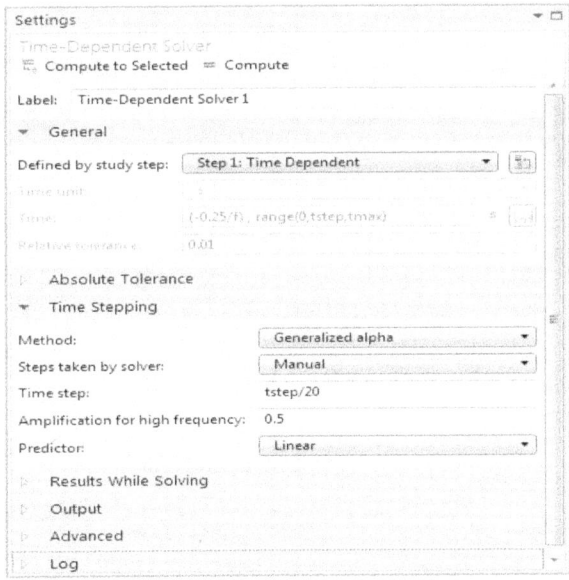

FIGURE 7.61 Settings for time steps of the solver.

contours change during the solution time period. Result for t = 40s is shown in Figure 7.62.

27. We create some graphs to visualize the results further. Click on the Results tab from the toolbar and select Cut Point 2D. In the corresponding settings window locate Point Data section and enter Dd/2-Hw and U0 for r and z, respectively. Click on Velocity (*spf*) node, under Results in the Model Builder window, and rename it to Temperature surface and velocity streamlines, 2D. Click on Surface 1 and in the corresponding settings window change the Expression by typing in T. Right-click on Temperature surface and velocity streamlines, 2D node, and select Streamline from the list. In the corresponding settings window change the Expression by typing in ht.ur and ht.uz for r and z, respectively. Locate Streamline Positioning and, from the

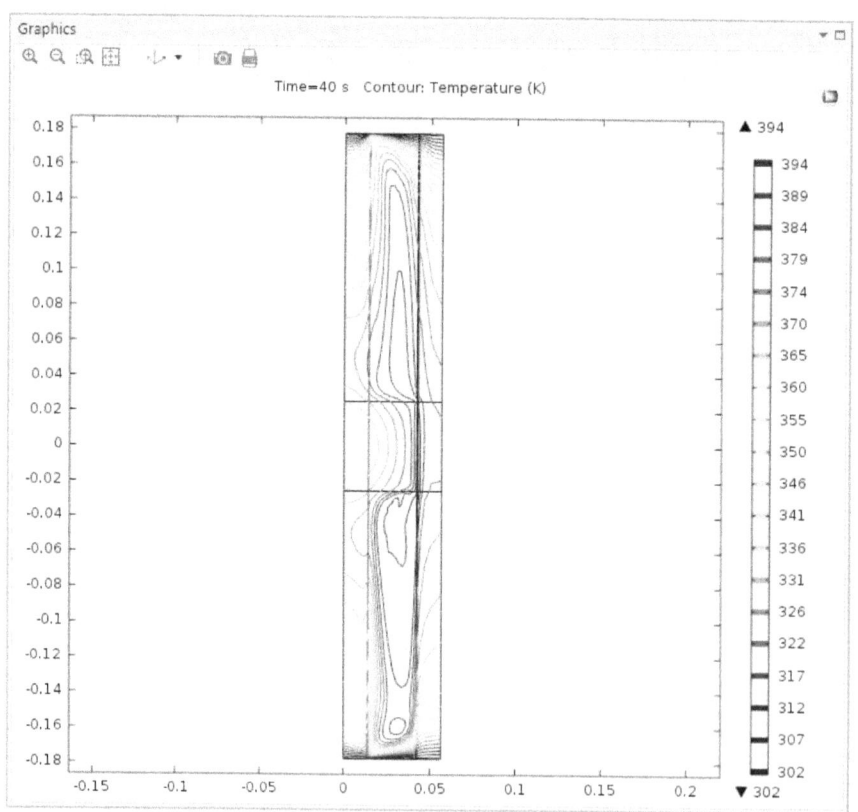

FIGURE 7.62 Temperature contour in the damper at time = 40s.

list for Positioning, select Start point controlled. Click on Results from the toolbar and select 1D Plot Group, and rename it to Temperature along the inner wall. Right-click on this node, and select Line Graph. In the corresponding settings window add boundaries 22, 24, and 26 (i.e. walls of damper chamber). Locate y-axis Data section and type in T for Expression. Similarly, locate x-axis Data section and type in z/U0 for the Expression. Create another 1D Plot Group and rename it to Temperature of inner wall at end-of-stroke position. Right-click on this node and select Point Graph. In the corresponding settings window locate Data section and select Cut Point 2D 1 from the list for Data set. Locate y-Axis Data section and type in T for Expression.

28. Click on Study 1 > Step 1: Time Dependent node. In the corresponding settings window expand Results While Solving section and choose, from the Plot group, the desired plot to be displayed during the run time, for example, Temperature of inner wall at end-of-stroke position. Click on the Study tab from the toolbar and select Compute. The Graphics window should display the temperature increase at the inner wall at the end-of-stroke during the solution time, a temperature increase of about 60° C., as shown in Figure 7.63. To visualize the variation along the wall of the damper, click on Temperature along the inner wall node. In the corresponding settings window locate Data section and select From list under Time selection. From the Times (s) select 5, 10, 20, and 40. Click Plot. The result is shown in Figure 7.64. This result is consistent with those reported by Black and Makris [35].

29. To build an app for this model, launch COMSOL if not already running. From File menu select New. In the New window select Applications Wizard. In the Select Model for Application window click on Browse; locate and open Example 7.7.mph. The New Form window will appear. In this window, select entries Heat transfer coefficient Frequency, and ambient temperature under Parameters, from the Inputs/outputs tab, and move them to Selected window. Click on the Graphics tab; select and move Flow and Stress (*fsi*) to Selected window. Click on the Buttons tab; select and move Compute Study 1 to Selected window. See Figure 7.65. Click on Done. Form Editor desktop window appears. Save the file as App_Example 7.7.

30. Several editing tools are available in the Form Editor window. We use some of these tools in order to lay out the App interface. Click on Grid, from the ribbon bar, to relocate the objects in the Form 1.

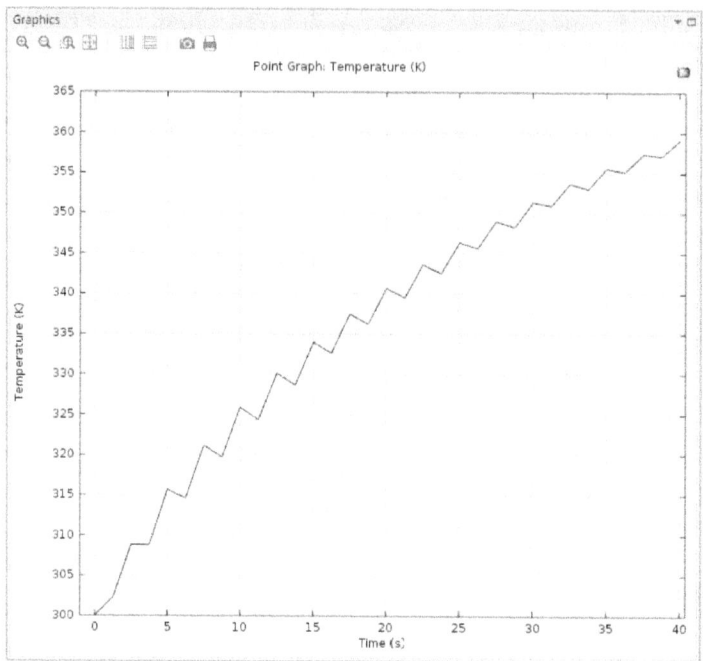

FIGURE 7.63 Temperature at the probe position, inner wall at the end-of-stroke.

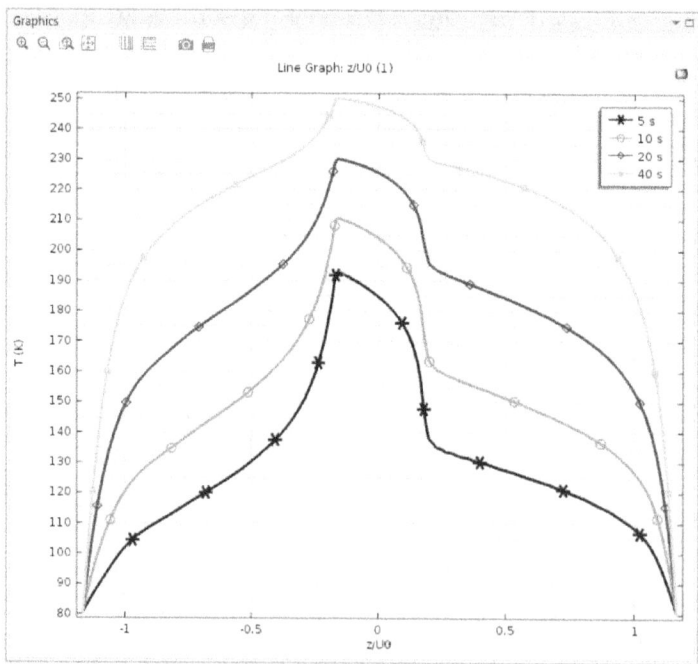

FIGURE 7.64 Temperature of the damper inner wall at the probe position represents z / U0.

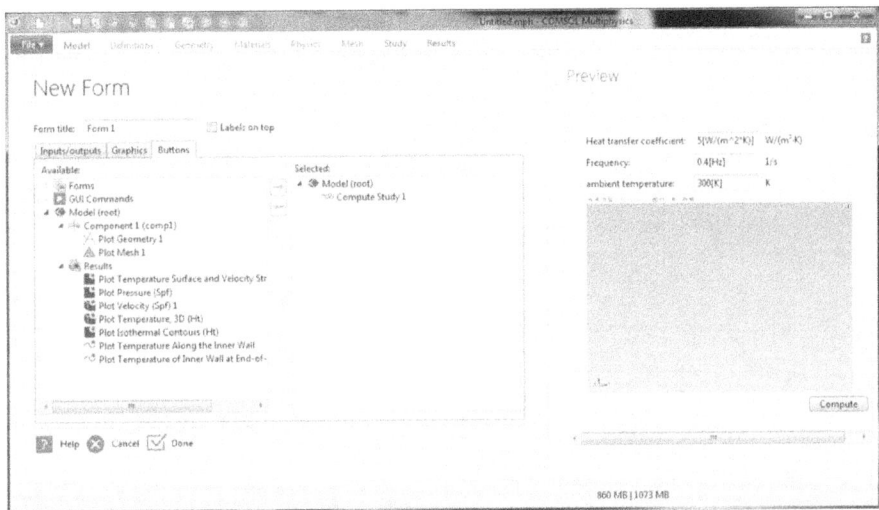

FIGURE 7.65 Application settings for Example 7.7.

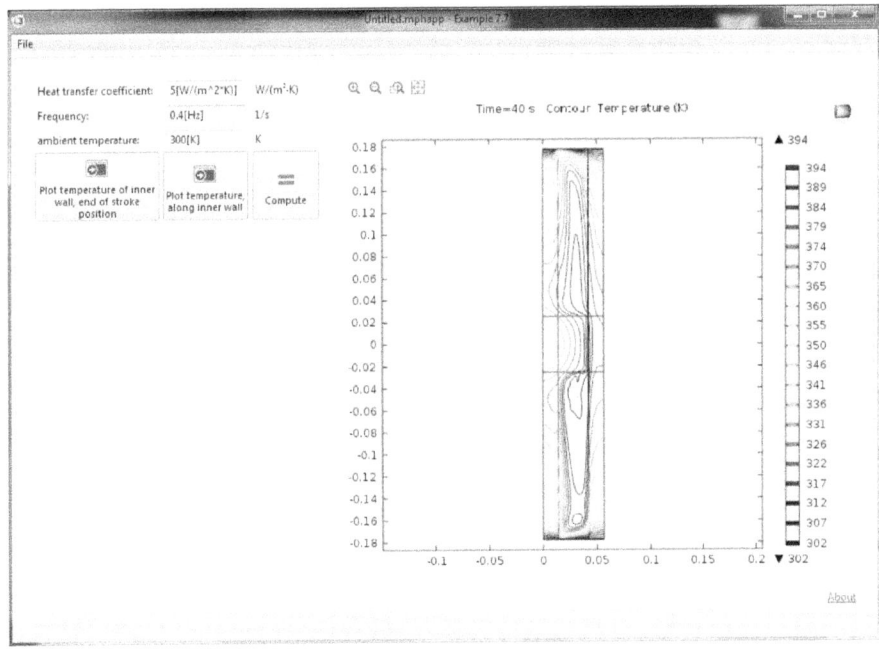

FIGURE 7.66 Application GUI for Example 7.7.

Use the Settings for each object to modify the size, affiliate a picture, and specify the function. Refer to detailed instructions given for Example 2.1, for a guide. Final results for the App interface are shown in Figure 7.66.

EXERCISES

Problem 1: Using the solution to Example 3.1, show the principle stress directions by manipulating the results.

Problem 2: Repeat Example 3.1, using the symmetry of the geometry and loads along the line axis at y = 0.

Problem 3: Repeat Example 3.1 for different boundary conditions, for example, simply supported for vertical sides on the right side of the plate and triangular tension stress applied on the upper edge.

Problem 4: Calculate 10 natural frequencies of the plate given in Example 3.2.

Problem 5: Calculate the plate frequency given in Example 3.2, close to 1500 Hz, and show the corresponding modal shape.

Problem 6: Investigate the effects of different boundary conditions on the natural frequencies for the plate given in Example 3.2.

Problem 7: Repeat Example 3.3 for different sets of applied loads

Problem 8: Repeat Example 3.3 for different ranges of angle theta0.

Problem 9: Check the bracket stress given in Example 3.3 against Tresca yield criterion.

Problem 10: Calculate the moment of inertia of the column cross section given in Example 3.4 and compare those obtained with the Euler's column formula .

Problem 11: Change the applied load given in Example 3.4 to higher values and make a graph of obtained critical load factors versus applied loads.

Problem 12: For Example 3.4, assign different boundary conditions to the column and calculate buckling load. Compare the buckling load with and without lateral loads.

Problem 13: Modify the column given in Example 3.4 to have a variable cross section (tapered). Calculate its critical load factor for 200 Pa compression load (use Scale 0.5 for x_w and y_w in the Extrude). Also in Example 3.4, change the values of point loads and repeat the Eigenfrequency calculations, both with and without loading.

Problem 14: In Example 3.5, design the truss to have the optimum section to satisfy 0.6Fy (Fy is the yield strength equal to 240 MPa) maximum applied stresses. Investigate the buckling for the member with maximum compression load.

Problem 15: In Example 3.5, introduce different damping to the truss vibrations and compare/discuss your results.

Problem 16: In Example 3.6, analyze the truss and compare your results with those in the model.

Problem 17: Repeat Example 3.6 for prescribed displacements, as boundary conditions, equal to 1 cm in x-direction and 2 cm in y-direction at point G.

Problem 18: Calculate the Reynolds number for the flow in Example 4.1.

Problem 19: For Example 4.1, run the model with finer meshes and discuss the results in terms of mesh dependency. Choose a physical point in the flow domain and investigate its numerical values, velocity, and pressure for different mesh sizes.

Problem 20: For a water-jet cutter, pressure of 20,000 psi is typical. Calculate the exit velocity for the nozzle using the model given in Example 4.1. Discuss validity of the assumptions for this model, such as water compressibility, turbulence, and other factors.

Problem 21: For a fly-wheel, the friction between the rotating disk and air is significant. Calculate the total forces applied on the disk due to shear stresses. Consider the dynamic viscosity of the air as a parameter and draw the total friction force versus viscosity. Use the model from Example 4.2.

Problem 22: In Example 4.2, calculate the flow Reynolds number. Discuss the Reynolds number in the context of the laminar flow assumption made for this model.

For Example 4.2, calculate the Reynolds number using total viscosity to be equal to the fluid viscosity and average turbulent viscosity. Define average as $(\mu_{Tmax}+\mu_{Tmin})/2$.

Problem 23: Repeat Example 4.3, calculating the average turbulent dynamic viscosity using the COMSOL Derived Values tool. Also calculate the total shear force exerted on the disk by the fluid.

Problem 24: Rerun Example 4.3 for a wider range of omega (from small to large values) and make a table of the results. Discuss the laminar versus turbulent assumptions with reference to ratio of turbulent maximum viscosity over fluid dynamic viscosity. Also do a literature search for finding the critical Reynolds number for a swirling flow, and compare it against your results. (For reference, see [36]. The transition occurs at about Re~500, according to experimental results in literature.)

Problem 25: Calculate the max Reynolds number for the model in Example 4.4. Also draw streamlines and study the bend effect on the flow. Calculate the "equivalent length" of the bend for having the same pressure drop.

Problem 26: In Example 4.4, extend the last exit part of the pipe (downstream of the bend) and find the length for which the flow becomes fully developed again.

Problem 27: Rebuild the model Example 4.4 for turbulence flow.

Problem 28: Using results given in Example 4.5, draw the contours for the y-component of the velocity vector for Re 50, 100, 400, and 1000.

Problem 29: Using results given in Example 4.6, draw the contours for pressure for Re 50, 100, 400, and 1000.

Problem 30: Run the model given in Example 4.6 and modify boundary conditions to have the two vertical side walls move in opposite directions, instead of the upper and lower walls.

Problem 31: In Example 4.6, calculate the maximum value of Reynolds number and investigate if the laminar flow assumption is justified. Determine the Reynolds number for which the flow becomes turbulent.

Problem 32: In Example 4.6, calculate the pressure change due to water hammer for a series of different boundary conditions for the pipe at the entrance.

Problem 33: In Example 4.6, compare the model results with those obtained by using governing equations given in the example.

Problem 34: In Example 4.6, make an animation of pressure variations along the pipe versus time. Use the Player button in the toolbar to create the animations.

Problem 35: Repeat Example 4.6 with different velocity values for Inlet. Discuss the results based on the flow Reynolds number. Change the pressure boundary condition at the outlet in Example 4.6 and compare the results against the existing obtained results.

Problem 36: In Example 4.7, change the orientation of the baffles inside the mixer and run the model.

Problem 37: In Example 5.1, compare the modeling results with analytical $R_{th} = (r_2-r_1)/(4\pi k\, r_1\, r_2)$, where R_{th} is the thermal resistance for each layer with internal radius r_1 and external radius r_2.

Problem 38: In Example 5.1, fill the void at the center of the sphere and assign a volumetric heat source to this region. Run the model and compare the results with previous boundary conditions.

Problem 39: Repeat Example 5.2 and make use of symmetry for geometry and boundary conditions.

Problem 40: For Example 5.3, do you recommend running this model for more than 60 seconds? Explain your answer.

Problem 41: For Example 5.3, calculate the dimensionless temperature quantity $(T - T_{amb})/(T_{ini} - T_{amb})$ at time equal to 5 sec. Compare the results versus e^{-BiFo}. Let $T_{amb} = (273.15 + 30)$ K and $T_{ini} = (273.15 + 450)$ K. Explain your results.

Problem 42: Using Example 5.3, for this model (with Aluminum wall/fin) the temperature of the fin tip reaches a steady state of about $350°$ C. Modify the dimensions of the fin such that the tip temperature drops to $250°$ C.

Problem 43: For Example 5.3, calculate the heat flux through the side surface of the fin and compare it with that of the tip. Discuss the results.

Problem 44: In Example 5.3, change the wall material to concrete and the fin material to structural steel. Estimate time constant of the problem and build the model.

Problem 45: Using Example 5.4, calculate the equivalent thickness for a layer of air with equivalent resistance to the thin contact resistance layer. Rerun the model using this equivalent air layer and compare the results with original solution results.

Problem 46: For Example 5.4, add another thin layer to the model between concrete and gypsum layers and build a new model.

Problem 47: Compare the results given in Example 6.1 against analytical results. The voltage for an RC circuit is given as . Show that at t = RC = 0.05 sec., the value of voltage across the resistor drops to $1/e$ % (or 36.8%) of its total value of 10 volts. Extract these results for the model and compare them. (Modeling result is 3.651 V.)

Problem 48: In Example 6.1, change the value of resistance to 1000, 2000, 5000, 10000 and calculate/model the circuit responses. Use Parametric sweep and resistance as a parameter.

Problem 49: In Example 6.2, use the analogy between mechanical mass-spring-damper and RCL circuit systems and derive the equivalent mechanical system for the model. Solve the mechanical system by hand and compare your results.

Problem 50: Using Example 6.2, run the model to study the effect of resistance on its behavior for R = 1000, 3000, 5000, 10000 ohm. Use the parametric sweep tool.

Problem 51: Using Example 6.2, run the model to study the effect of resistance on its behavior for L = 10, 100, 1000 mH. Use the parametric sweep tool.

Problem 52: Using Example 6.2, run the model to study the effect of resistance on its behavior for C = 1, 10, 100, 1000 nF. Use the parametric sweep tool.

Problem 53: Using Example 6.3, run the model to study the effect of the winding cross-sectional area on the induced voltage. Also change the material properties of the rotor to study its effects.

Problem 54: Compare the stress and displacements results for the model in Example 7.1 versus results for the model in Example 3.1. Make a table of these results to clearly show the effects of orthotropic materials on these quantities versus isotropic ones.

Problem 55: For Example 7.1, use Voigt for Material data ordering and run the model. What is the difference between Voigt and Standard ordering format? (Refer to the *COMSOL Manual* for definitions.)

Problem 56: In Example 7.1, apply different loading (in plane) scenarios and discuss the results with respect to orthotropic material properties.

Problem 57: In Example 7.3, change the location, size, and material properties of the baffles and compare/discuss the results.

Problem 58: In Example 7.4, change the masses and lengths of the strings of the pendulum and compare/discuss your results.

Problem 59: Build a model for a triple-pendulum, using Example 7.4 and/or Multibody dynamics module available in COMSOL 5.

Problem 60: In Example 7.6, build a perforated plate inside the muffler and study its effects on the acoustic pressure and sound pressure levels.

Problem 61: In Example 7.6, build a second chamber inside the muffler to create a double-wall resonate chamber and study its effects on acoustic pressure and sound power attenuation.

Problem 62: In Example 7.7, perform a parametric analysis using the viscosity of the fluid inside the damper as a parameter and run the model; compare/discuss your results.

SUGGESTED FURTHER READINGS

D. Chapelle and K. J. Bathe, *The Finite Element Analysis of Shells—Fundamentals*, 2nd ed., Springer, 2011.

Ronald W. Lewis, Perumal Nithiarasu, and Kankanhalli N. Seetharamu, *Fundamentals of the Finite Element Method for Heat and Fluid Flow*, Wiley, 2004.

Walter D. Pilkey and Walter Wunderlich, *Mechanics of Structures, Variational and Computational Methods*, 2nd ed., CRC Press, 2002.

Z. P. Bazant and L. Cedolin, *Stability of Structures: Elastic, Inelastic, Fracture and Damage Theories*, World Scientific Publishing Company, 2010.

G. K. Batchelor, *The Theory of Homogeneous Turbulence*, Cambridge: Cambridge University Press, 1982.

TRADEMARK REFERENCES

BIBLIOGRAPHY

[1] M. Tabatabaian, *COMSOL for Engineers*, Dulles, VA: Mercury Learning and Information, 2014.

[2] R. Clough, "Original Formulation of the Finite Element Method," *J. Finite Element Anal. Des.*, vol. 7, no. 2, pp. 89–101, 1990.

[3] J. Robinson, *Early FEM Pioneers*, Dorest, UK: Robinson and Associates, 1985.

[4] Irving H. Shames and Clive L. Dym, *Energy and Finite Element Methods in Structural Mechanics*, Hemisphere Publishing Corp., 1985.

[5] W. Ritz, "Ueber eine neue Methode zur Lösung gewisser Variationsprobleme der mathematischen Physik," *J. Reine Angew. Math.*, vol. 135, pp. 1–61, 1908.

[6] C. Felippa, "An Appreciation of R. Courant's 'Variational Methods for the Solution of Problems of Equilibrium and Vibrations' 1943," *Int. J. Num. Engr.*, vol. 37, pp. 2159–2187, 1994.

[7] R. Courant, "Variational Methods for the Solution of Problems of Equilibrium and Vibrations," *Bulletin of the American Mathematical Society*, vol. 49, pp. 1–23, 1943.

[8] R. D. Cook, *Concepts and Applications of Finite Element Analysis*, 4th ed., John Wiley & Sons, Inc., 2001.

[9] A. Baker, *Finite Elements: Computational Engineering Sciences*, John Wiley & Sons Inc., 2012.

[10] A. Baker, *Finite Element Computational Fluid Mechanics*, McGraw–Hill, 1983.

[11] M. Turner, R. Clough, H. Martin, and L. Topp, "Stiffness and Deflection Analysis of Complex Structures," *J. Aero., Sci.*, vol. 23, no. 9, pp. 805–823, 1956.

[12] R. Clough, "The Finite Element Method in Plane Stress Analysis," in *Proc. 2nd Conf. Electronic Computation, ASCE*, Pittsburg, PA, 1960.

[13] Walter D. Pilkey, and Walter Wunderlinch, *Mechanics of Structures: Variational and Computational Methods*, CRC Press, 1994.

[14] L. Segerlind, *Applied Finite Element Analysis*, 2nd ed., John Wiley & Sons, Inc., 1984.

[15] C. R. Maliska, "On the Physical Significance of Some Dimensionless Numbers Used in Heat Transfer and Fluid Flow," [Online]. Available: http://www.sinmec.ufsc.br/sinmec/site/iframe/pubicacoes/artigos/novos_00s/2000_maliska_encit.pdf.

[16] Ansel C. Ugural and Saul K. Fenster, *Advanced Mechanics of Materials and Applied Elasticity*, 5th ed., Upper Saddle River, NJ: Prentice Hall, 2011.

[17] J. Hartog, *Mechanical Vibrations*, Dover Publications, 1985.

[18] M. Tabatabaian, *CFD Module: Turbulent Flow Modeling*, Dulles, VA: Mercury Learning and Information, 2015.

[19] H. K. Versteeg, and W. Malalasekera, *An Introduction to Computational Fluid Dynamics: The Finite Volume Method*, 2nd ed., Upper Saddle River, NJ: Prentice Hall, 2007.

[20] Y. C. Zhou, B. S. V. Patnail, D. C. Wan, and G. W. Wei, "DSC Solution for Flow in a Staggered Double Lid Driven Cavity," *International Journal for Numerical Methods in Engineering*, vol. 57, pp. 211–234, 2003.

[21] M. Ghidaoui, "A Review of Water Hammer Theory and Practice," in *Applied Mechanics Review, ASME*, 2005.

[22] Yunus Cengel and Afshin Ghajar, *Heat and Mass Transfer: Fundamentals and Applications*, McGraw–Hill, 2014.

[23] William H. Hayt et al., *Engineering Circuit Analysis*, 8th ed., McGraw–Hill, 2011.

[24] J. C. E. Whitaker, *The Electronic Handbook*, 2nd ed., CRC press, 2005.

[25] David E. Keyes et al., "Multiphysics Simulation: Challenges and Opportunities," *The International Journal of High Performance Computing Applications*, vol. 27, no. 1, February 2013.

[26] M. Calkin, *Lagrangian and Hamiltonian Mechanics*, World Scientific Publishing Company, 1996.

[27] Herman F. Mark and Norbert Bikales, *Encyclopedia of Polymer Science and Engineering*, 2nd Edition, John Wiley & Sons, Inc., 1991.

[28] D. E. Rowe, *Thermoelectric Handbook*, CRC Press, 2005.

[29] Elena–Otilia Virjoghe et al., "Numerical Simulation of Thermoelectric System," in *Proceeding ICS'10 Proceedings of the 14th WSEAS International Conference on Systems: Part of the 14th WSEAS CSCC Multiconference—Volume II*, pp. 630–635.

[30] M. Jaegle, "Multiphysics Simulation of Thermoelectric Systems," in *Proceedings of the COMSOL Conference*, Hannover, 2008.

[31] M. Jaegle, "Simulating Thermoelectric Effects with Finite Element Analysis using COMSOL," Fraunhofer–Institute for Physical Measurement Techniques.

[32] [Online]. Available: http://www.bridgeweb.com/MemberPages/article.aspx?id=2547&typeid=3.

[33] Arza Seidel, ed., *Kirk–Othmer Encyclopedia of Chemical Technology*, 5th ed., John Wiley & Sons, Inc., 2007.

[34] COMSOL, "COMSOL 5 Multiphysics Reference Manual," 2014.

[35] C. J. Black and N. Makris, "Viscous Heating of Fluid Dampers Under Small and Large Amplitude Motions: Experimental Studies and Parametric Modeling," *J. Eng. Mech.*, vol. 133, pp. 566–577, 2007.

[36] H. Schlichting, K. Gersten, Boundary–Layer Theory, 8th edition, Springer, 2000.

[37] J. C. e. Whitaker, The Electronic Handbook, second edition, CRC press, 2005.

INDEX